Pitman Research Notes in Mathematics Series

Submission of proposals for consideration
Suggestions for publication, in the form of outlines and representative samples, are invited by the Editorial Board for assessment. Intending authors should approach one of the main editors or another member of the Editorial Board, citing the relevant AMS subject classifications. Alternatively, outlines may be sent directly to the publisher's offices. Refereeing is by members of the board and other mathematical authorities in the topic concerned, throughout the world.

Preparation of accepted manuscripts
On acceptance of a proposal, the publisher will supply full instructions for the preparation of manuscripts in a form suitable for direct photo-lithographic reproduction. Specially printed grid sheets are provided and a contribution is offered by the publisher towards the cost of typing. Word processor output, subject to the publisher's approval, is also acceptable.

Illustrations should be prepared by the authors, ready for direct reproduction without further improvement. The use of hand-drawn symbols should be avoided wherever possible, in order to maintain maximum clarity of the text.

The publisher will be pleased to give any guidance necessary during the preparation of a typescript, and will be happy to answer any queries.

Important note
In order to avoid later retyping, intending authors are strongly urged not to begin final preparation of a typescript before receiving the publisher's guidelines and special paper. In this way it is hoped to preserve the uniform appearance of the series.

Longman Scientific & Technical
Longman House
Burnt Mill
Harlow, Essex, UK
(tel (0279) 26721)

Titles in this series

1 Improperly posed boundary value problems
A Carasso and A P Stone

2 Lie algebras generated by finite dimensional ideals
I N Stewart

3 Bifurcation problems in nonlinear elasticity
R W Dickey

4 Partial differential equations in the complex domain
D L Colton

5 Quasilinear hyperbolic systems and waves
A Jeffrey

6 Solution of boundary value problems by the method of integral operators
D L Colton

7 Taylor expansions and catastrophes
T Poston and I N Stewart

8 Function theoretic methods in differential equations
R P Gilbert and R J Weinacht

9 Differential topology with a view to applications
D R J Chillingworth

10 Characteristic classes of foliations
H V Pittie

11 Stochastic integration and generalized martingales
A U Kussmaul

12 Zeta-functions: An introduction to algebraic geometry
A D Thomas

13 Explicit *a priori* inequalities with applications to boundary value problems
V G Sigillito

14 Nonlinear diffusion
W E Fitzgibbon III and H F Walker

15 Unsolved problems concerning lattice points
J Hammer

16 Edge-colourings of graphs
S Fiorini and R J Wilson

17 Nonlinear analysis and mechanics: Heriot-Watt Symposium Volume I
R J Knops

18 Actions of fine abelian groups
C Kosniowski

19 Closed graph theorems and webbed spaces
M De Wilde

20 Singular perturbation techniques applied to integro-differential equations
H Grabmüller

21 Retarded functional differential equations: A global point of view
S E A Mohammed

22 Multiparameter spectral theory in Hilbert space
B D Sleeman

24 Mathematical modelling techniques
R Aris

25 Singular points of smooth mappings
C G Gibson

26 Nonlinear evolution equations solvable by the spectral transform
F Calogero

27 Nonlinear analysis and mechanics: Heriot-Watt Symposium Volume II
R J Knops

28 Constructive functional analysis
D S Bridges

29 Elongational flows: Aspects of the behaviour of model elasticoviscous fluids
C J S Petrie

30 Nonlinear analysis and mechanics: Heriot-Watt Symposium Volume III
R J Knops

31 Fractional calculus and integral transforms of generalized functions
A C McBride

32 Complex manifold techniques in theoretical physics
D E Lerner and P D Sommers

33 Hilbert's third problem: scissors congruence
C-H Sah

34 Graph theory and combinatorics
R J Wilson

35 The Tricomi equation with applications to the theory of plane transonic flow
A R Manwell

36 Abstract differential equations
S D Zaidman

37 Advances in twistor theory
L P Hughston and R S Ward

38 Operator theory and functional analysis
I Erdelyi

39 Nonlinear analysis and mechanics: Heriot-Watt Symposium Volume IV
R J Knops

40 Singular systems of differential equations
S L Campbell

41 N-dimensional crystallography
R L E Schwarzenberger

42 Nonlinear partial differential equations in physical problems
D Graffi

43 Shifts and periodicity for right invertible operators
D Przeworska-Rolewicz

44 Rings with chain conditions
A W Chatters and C R Hajarnavis

45 Moduli, deformations and classifications of compact complex manifolds
D Sundararaman

46 Nonlinear problems of analysis in geometry and mechanics
M Atteia, D Bancel and I Gumowski

47 Algorithmic methods in optimal control
W A Gruver and E Sachs

48 Abstract Cauchy problems and functional differential equations
F Kappel and W Schappacher

49 Sequence spaces
W H Ruckle

50 Recent contributions to nonlinear partial differential equations
H Berestycki and H Brezis

51 Subnormal operators
J B Conway

52 Wave propagation in viscoelastic media
F Mainardi
53 Nonlinear partial differential equations and their applications: Collège de France Seminar. Volume I
H Brezis and J L Lions
54 Geometry of Coxeter groups
H Hiller
55 Cusps of Gauss mappings
T Banchoff, T Gaffney and C McCrory
56 An approach to algebraic K-theory
A J Berrick
57 Convex analysis and optimization
J-P Aubin and R B Vintner
58 Convex analysis with applications in the differentiation of convex functions
J R Giles
59 Weak and variational methods for moving boundary problems
C M Elliott and J R Ockendon
60 Nonlinear partial differential equations and their applications: Collège de France Seminar. Volume II
H Brezis and J L Lions
61 Singular systems of differential equations II
S L Campbell
62 Rates of convergence in the central limit theorem
Peter Hall
63 Solution of differential equations by means of one-parameter groups
J M Hill
64 Hankel operators on Hilbert space
S C Power
65 Schrödinger-type operators with continuous spectra
M S P Eastham and H Kalf
66 Recent applications of generalized inverses
S L Campbell
67 Riesz and Fredholm theory in Banach algebra
B A Barnes, G J Murphy, M R F Smyth and T T West
68 Evolution equations and their applications
F Kappel and W Schappacher
69 Generalized solutions of Hamilton-Jacobi equations
P L Lions
70 Nonlinear partial differential equations and their applications: Collège de France Seminar. Volume III
H Brezis and J L Lions
71 Spectral theory and wave operators for the Schrödinger equation
A M Berthier
72 Approximation of Hilbert space operators I
D A Herrero
73 Vector valued Nevanlinna Theory
H J W Ziegler
74 Instability, nonexistence and weighted energy methods in fluid dynamics and related theories
B Straughan
75 Local bifurcation and symmetry
A Vanderbauwhede
76 Clifford analysis
F Brackx, R Delanghe and F Sommen
77 Nonlinear equivalence, reduction of PDEs to ODEs and fast convergent numerical methods
E E Rosinger
78 Free boundary problems, theory and applications. Volume I
A Fasano and M Primicerio
79 Free boundary problems, theory and applications. Volume II
A Fasano and M Primicerio
80 Symplectic geometry
A Crumeyrolle and J Grifone
81 An algorithmic analysis of a communication model with retransmission of flawed messages
D M Lucantoni
82 Geometric games and their applications
W H Ruckle
83 Additive groups of rings
S Feigelstock
84 Nonlinear partial differential equations and their applications: Collège de France Seminar. Volume IV
H Brezis and J L Lions
85 Multiplicative functionals on topological algebras
T Husain
86 Hamilton-Jacobi equations in Hilbert spaces
V Barbu and G Da Prato
87 Harmonic maps with symmetry, harmonic morphisms and deformations of metrics
P Baird
88 Similarity solutions of nonlinear partial differential equations
L Dresner
89 Contributions to nonlinear partial differential equations
C Bardos, A Damlamian, J I Díaz and J Hernández
90 Banach and Hilbert spaces of vector-valued functions
J Burbea and P Masani
91 Control and observation of neutral systems
D Salamon
92 Banach bundles, Banach modules and automorphisms of C*-algebras
M J Dupré and R M Gillette
93 Nonlinear partial differential equations and their applications: Collège de France Seminar. Volume V
H Brezis and J L Lions
94 Computer algebra in applied mathematics: an introduction to MACSYMA
R H Rand
95 Advances in nonlinear waves. Volume I
L Debnath
96 FC-groups
M J Tomkinson
97 Topics in relaxation and ellipsoidal methods
M Akgül
98 Analogue of the group algebra for topological semigroups
H Dzinotyiweyi
99 Stochastic functional differential equations
S E A Mohammed

100 Optimal control of variational inequalities
V Barbu
101 Partial differential equations and dynamical systems
W E Fitzgibbon III
102 Approximation of Hilbert space operators. Volume II
C Apostol, L A Fialkow, D A Herrero and D Voiculescu
103 Nondiscrete induction and iterative processes
V Ptak and F-A Potra
104 Analytic functions – growth aspects
O P Juneja and G P Kapoor
105 Theory of Tikhonov regularization for Fredholm equations of the first kind
C W Groetsch
106 Nonlinear partial differential equations and free boundaries. Volume I
J I Díaz
107 Tight and taut immersions of manifolds
T E Cecil and P J Ryan
108 A layering method for viscous, incompressible L_p flows occupying R^n
A Douglis and E B Fabes
109 Nonlinear partial differential equations and their applications: Collège de France Seminar. Volume VI
H Brezis and J L Lions
110 Finite generalized quadrangles
S E Payne and J A Thas
111 Advances in nonlinear waves. Volume II
L Debnath
112 Topics in several complex variables
E Ramírez de Arellano and D Sundararaman
113 Differential equations, flow invariance and applications
N H Pavel
114 Geometrical combinatorics
F C Holroyd and R J Wilson
115 Generators of strongly continuous semigroups
J A van Casteren
116 Growth of algebras and Gelfand–Kirillov dimension
G R Krause and T H Lenagan
117 Theory of bases and cones
P K Kamthan and M Gupta
118 Linear groups and permutations
A R Camina and E A Whelan
119 General Wiener–Hopf factorization methods
F-O Speck
120 Free boundary problems: applications and theory, Volume III
A Bossavit, A Damlamian and M Fremond
121 Free boundary problems: applications and theory, Volume IV
A Bossavit, A Damlamian and M Fremond
122 Nonlinear partial differential equations and their applications: Collège de France Seminar. Volume VII
H Brezis and J L Lions
123 Geometric methods in operator algebras
H Araki and E G Effros
124 Infinite dimensional analysis–stochastic processes
S Albeverio
125 Ennio de Giorgi Colloquium
P Krée
126 Almost-periodic functions in abstract spaces
S Zaidman
127 Nonlinear variational problems
A Marino, L Modica, S Spagnolo and M Degiovanni
128 Second-order systems of partial differential equations in the plane
L K Hua, W Lin and C-Q Wu
129 Asymptotics of high-order ordinary differential equations
R B Paris and A D Wood
130 Stochastic differential equations
R Wu
131 Differential geometry
L A Cordero
132 Nonlinear differential equations
J K Hale and P Martinez-Amores
133 Approximation theory and applications
S P Singh
134 Near-rings and their links with groups
J D P Meldrum
135 Estimating eigenvalues with *a posteriori/a priori* inequalities
J R Kuttler and V G Sigillito
136 Regular semigroups as extensions
F J Pastijn and M Petrich
137 Representations of rank one Lie groups
D H Collingwood
138 Fractional calculus
G F Roach and A C McBride
139 Hamilton's principle in continuum mechanics
A Bedford
140 Numerical analysis
D F Griffiths and G A Watson
141 Semigroups, theory and applications. Volume I
H Brezis, M G Crandall and F Kappel
142 Distribution theorems of L-functions
D Joyner
143 Recent developments in structured continua
D De Kee and P Kaloni
144 Functional analysis and two-point differential operators
J Locker
145 Numerical methods for partial differential equations
S I Hariharan and T H Moulden
146 Completely bounded maps and dilations
V I Paulsen
147 Harmonic analysis on the Heisenberg nilpotent Lie group
W Schempp
148 Contributions to modern calculus of variations
L Cesari
149 Nonlinear parabolic equations: qualitative properties of solutions
L Boccardo and A Tesei
150 From local times to global geometry, control and physics
K D Elworthy

151 A stochastic maximum principle for optimal control of diffusions
U G Haussmann
152 Semigroups, theory and applications. Volume II
H Brezis, M G Crandall and F Kappel
153 A general theory of integration in function spaces
P Muldowney
154 Oakland Conference on partial differential equations and applied mathematics
L R Bragg and J W Dettman
155 Contributions to nonlinear partial differential equations. Volume II
J I Díaz and P L Lions
156 Semigroups of linear operators: an introduction
A C McBride
157 Ordinary and partial differential equations
B D Sleeman and R J Jarvis
158 Hyperbolic equations
F Colombini and M K V Murthy
159 Linear topologies on a ring: an overview
J S Golan
160 Dynamical systems and bifurcation theory
M I Camacho, M J Pacifico and F Takens
161 Branched coverings and algebraic functions
M Namba
162 Perturbation bounds for matrix eigenvalues
R Bhatia
163 Defect minimization in operator equations: theory and applications
R Reemtsen
164 Multidimensional Brownian excursions and potential theory
K Burdzy
165 Viscosity solutions and optimal control
R J Elliott
166 Nonlinear partial differential equations and their applications. Collège de France Seminar. Volume VIII
H Brezis and J L Lions
167 Theory and applications of inverse problems
H Haario
168 Energy stability and convection
G P Galdi and B Straughan
169 Additive groups of rings. Volume II
S Feigelstock
170 Numerical analysis 1987
D F Griffiths and G A Watson
171 Surveys of some recent results in operator theory. Volume I
J B Conway and B B Morrel
172 Amenable Banach algebras
J-P Pier
173 Pseudo-orbits of contact forms
A Bahri
174 Poisson algebras and Poisson manifolds
K H Bhaskara and K Viswanath
175 Maximum principles and eigenvalue problems in partial differential equations
P W Schaefer
176 Mathematical analysis of nonlinear, dynamic processes
K U Grusa
177 Cordes' two-parameter spectral representation theory
D F McGhee and R H Picard
178 Equivariant K-theory for proper actions
N C Phillips
179 Elliptic operators, topology and asymptotic methods
J Roe
180 Nonlinear evolution equations
J K Engelbrecht, V E Fridman and E N Pelinovski
181 Nonlinear partial differential equations and their applications. Collège de France Seminar. Volume IX
H Brezis and J L Lions
182 Critical points at infinity in some variational problems
A Bahri
183 Recent developments in hyperbolic equations
L Cattabriga, F Colombini, M K V Murthy and S Spagnolo
184 Optimization and identification of systems governed by evolution equations on Banach space
N U Ahmed
185 Free boundary problems: theory and applications. Volume I
K H Hoffmann and J Sprekels
186 Free boundary problems: theory and applications. Volume II
K H Hoffmann and J Sprekels
187 An introduction to intersection homology theory
F Kirwan
188 Derivatives, nuclei and dimensions on the frame of torsion theories
J S Golan and H Simmons
189 Theory of reproducing kernels and its applications
S Saitoh
190 Volterra integrodifferential equations in Banach spaces and applications
G Da Prato and M Iannelli
191 Nest algebras
K R Davidson
192 Surveys of some recent results in operator theory. Volume II
J B Conway and B B Morrel
193 Nonlinear variational problems. Volume II
A Marino and M K Murthy
194 Stochastic processes with multidimensional parameter
M E Dozzi
195 Prestressed bodies
D Iesan
196 Hilbert space approach to some classical transforms
R H Picard
197 Stochastic calculus in application
J R Norris
198 Radical theory
B J Gardner
199 The C* – algebras of a class of solvable Lie groups
X Wang

Function spaces, differential operators and nonlinear analysis

L Päivärinta (Editor)

University of Helsinki

Function spaces, differential operators and nonlinear analysis

Longman Scientific & Technical

Copublished in the United States with
John Wiley & Sons, Inc., New York

Longman Scientific & Technical,
Longman Group UK Limited,
Longman House, Burnt Mill, Harlow
Essex CM20 2JE, England
and Associated Companies throughout the world.

Copublished in the United States with
John Wiley & Sons, Inc., 605 Third Avenue, New York, NY 10158

First published in 1989

British Library Cataloguing in Publication Data
Päivärinta, L
Function spaces, differential operators and nonlinear analysis.
1. Mathematics. Nonlinear analysis
I. Päivärinta, L.
515

ISBN 0-582-04568-1

Library of Congress Cataloging-in-Publication Data
Function spaces, differential operators & nonlinear analysis
Päivärinta L (Editor).
p. cm. – (Pitman research notes in mathematics series; 211)
"Proceedings of the Summer School of Function Spaces, Differential Operators, and Nonlinear Analysis . . . held in Sodankylä, Finnish Lapland, in August 1988" – Pref.
ISBN 0470-21351-5
1. Function spaces – Congresses. 2. Differential operators – Congresses. 3. Nonlinear functional analysis – Congresses. I. Päivärinta, L. (Lassi) II. Summer School of Function Spaces, Differential Operators, and Nonlinear Analysis (1988: Sodankylä, Finland) III. Series.
QA323.F86 1989
515.7'3 – dc19 88-36780
CIP

Printed and bound in Great Britain
by Biddles Ltd, Guildford and King's Lynn

CStack SCIMON

Contents

Preface

The present book contains the proceedings of the summer school on Function Spaces, Differential Operators and Nonlinear Analysis which was held in Sodankylä, Finnish Lapland, in August 1988. The material covers three different areas in mathematics intimately related to each other: function spaces, singular boundary value problems and certain nonlinear questions.

The papers under the headline Function Spaces are divided into two different groups. The first section is related to questions of analytic function theory. In the second section topics like atomic decomposition, continuity of pseudodifferential operators, interpolation and certain functional inequalities are examined in many different function spaces. In the third section boundary value problems for non-smooth boundaries are studied for both differential and pseudodifferential operators. Their mathematical theory on singular manifolds is described and several physical examples are discussed. Section 4 deals with nonlinear concepts like monotone operators, degree theoretic approach, equations of Ambrosetti-Prodi type and the use of weighted Sobolev spaces in nonlinear boundary value problems. Also topics from nonlinear potential theory are included as well as applications to carrier transport equations for semiconductors.

Thanks are due to the authors of this volume not only for their thoroughly prepared manuscripts but also for making the summer school possible in the first place. For financial support we are also indebted to the Finnish Mathematical Society, the Finnish Ministry of Education and the Rolf Nevanlinna Institute. Finally it is a pleasure to express my gratitude to Dr Markku Lehtinen, Eero Saksman, Petri Ola and Pirkko Paakkanen for their invaluable help in organizing the summer school.

Helsinki, November 1988

Lassi Päivärinta

List of Contributors

K. Astala
Department of Mathematics
University of Helsinki
Hallituskatu 15
00100 Helsinki
Finland

J. Berkovits
Department of Mathematics
University of Oulu
SF-90570 Oulu
Finland

B. Bojarski
Department of Mathematics
Polish Academy of Sciences
00950 Warsaw
Poland

S. Fisher
Department of Mathematics
Northwestern University
Evanston
Illinois 60208
USA

J. Gajewski
AdW der DDR
Karl-Weierstrass Institute für Mathematik
Mohrenstrasse 39
Berlin 1086
DDR

P. Grisvard
Laboratoire de Mathématiques
University of Nice
I.M.S.P.
Parc Valrose
F-06034 Nice-Cedex
France

P. Gurka
Mathematics Institute Academy of Science
Zitna 25
11567 Prague 1
Czechoslovakia

B. Gustafsson
Matematiska Institutionen
KTH
S-100 44 Stockholm
Sweden

M. Krbec
Mathematics Institute Academy of Science
Zitna 25
11567 Prague 1
Czechoslovakia

A. Kufner
Mathematics Institute Academy of Science
Zitna 25
11567 Prague 1
Czechoslovakia

O. Martio
Department of Mathematics
University of Jyväsklä
Seminaarinkatu 15
SF-40100 Jyväskylä
Finland

V. Mustonen
Department of Mathematics
University of Oulu
SF-90570 Oulu
Finland

L. Nikolova
Department of Mathematics
Sofia University
A. Ivanov 5
1126 Sofia
Bulgaria

B. Opic
Mathematics Institute Academy of Science
Zitna 25
11567 Prague 1
Czechoslovakia

J. Peetre
Department of Mathematics
University of Stockholm
Box 6701
S-113 85 Stockholm
Sweden

L.E. Persson
Department of Mathematics
Luleå University
S-951 87 Luleå
Sweden

L. Päivärinta
Department of Mathematics
University of Helsinki
Hallituskatu 15
00100 Helsinki
Finland

J. Rakosnik | Mathematics Institute Academy of Science
Zitna 25
11567 Prague 1
Czechoslovakia

S. Rempel | AdW der DDR
Karl-Weierstrass Institute of Mathematics
Mohrenstrasse 39
Berlin 1086
DDR

T. Runst | Sektion Mathematik
Universität Jena
Universitätshochhaus
6900 Jena
DDR

H. Triebel | Sektion Mathematik
Universität Jena
Universitätshochhaus
6900 Jena
DDR

1. FUNCTION SPACES I

K. ASTALA

Cauchy-integrals, Lipschitz-classes and Hausdorff-dimension

1. Let Γ be a rectifiable Jordan curve in the complex plane and denote by D the bounded component of $\mathbb{C}\setminus\Gamma$. The *Cauchy integral* of a function f on Γ, integrable with respect to the arc length, is given by

$$C_\Gamma f(z) = \frac{1}{2\pi i}\int_\Gamma \frac{f(\xi)}{\xi - z}\,d\xi, \quad z \in D.$$

When z approaches nontangentially the boundary $\Gamma = \partial D$, $C_\Gamma f(z)$ has a limit at almost every point [2] and thus the Cauchy integral defines a singular integral operator on Γ.

The basic problem in the area of singular integral operators is the question of mapping properties, which function spaces are preserved by a given operator. In the case when Γ is the unit circle it is well known that the Cauchy integral is a bounded operator on many of the classical spaces of functions like L^p $(1 < p < \infty)$, BMO and Λ^α $(0 < \alpha < 1)$. However, for a general curve the situation is often more complicated. For example, we mention the deep result of G. David [Da], who showed that the Cauchy integral is bounded on L^p, $1 < p < \infty$, if and only if Γ is (Ahlfors-) regular, i.e. there is a constant C such that for every disk $B(z_0,R)$ the length of $\Gamma \cap B(z_0,R)$ is at most CR.

In the present paper we shall look for geometric conditions that characterize the boundedness of the Cauchy integral on the Lipschitz classes

$$\Lambda^\omega(\Gamma) = \{f : |f(x) - f(y)| \leqq \omega(|x - y|), \quad x,y \in \Gamma\}.$$

Here $\omega(t)$ is a general modulus of continuity, a positive strictly increasing function of t such that

$$\omega(t)/t \text{ is decreasing in } t \text{ and} \tag{1a}$$

$$\omega(2\delta) \leqq C\omega(\delta), \ \omega(0) = 0. \tag{1b}$$

In addition, we shall assume that the following Dini type conditions are satisfied c.f. [7, p. 127],

$$\int_0^\delta \frac{\omega(t)}{t}\,dt \leq C\omega(\delta), \quad \delta\int_\delta^\infty \frac{\omega(t)}{t^2}\,dt \leq C\omega(\delta). \tag{2}$$

The simplest examples of moduli ω for which (1)-(2) hold are $\omega(t) = t^\alpha$, $0 < \alpha < 1$. In these special cases the boundedness properties of C_Γ have connections with the theory of quasiconformal mappings, see [1].

The spaces $\Lambda^\omega(\Gamma)$ are well defined for every Jordan curve, rectifiable or not. It turns out, see section 2, that with the help of the Stokes' theorem also the Cauchy operator C_Γ can be defined for smooth functions on all Jordan curves. Hence we are faced with the natural question of finding the general Jordan curves Γ on which

$$C_\Gamma : \Lambda^\omega(\Gamma) \to \Lambda^\omega(\Gamma)$$

is a bounded operator.

However, a price must be paid of this generality and thus slight restrictions need to be posed on the class of curves under study: We say that a curve Γ is *biporous*, if there is a constant $\delta < 1$ such that whenever $x_0 \in \Gamma$ and $R < \text{diam}\,(\Gamma)$, then both components $D \cap B(x_0,R)$ and $(\mathbb{C} \setminus D) \cap B(x_0,R)$ contain a disk of radius δR. A basic example of a biporous but highly nonrectifiable Jordan curve is the von Koch snowflake

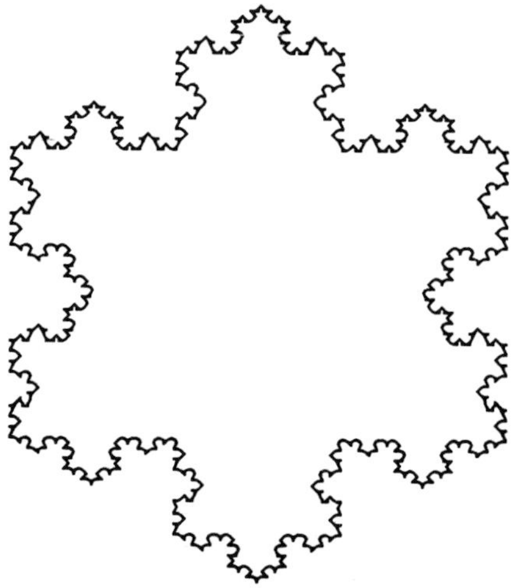

More generally, the so-called quasicircles [8] are all biporous.

As in [1] we describe the geometry of the curves in terms of the Muckenhoupt weights, which were originally introduced to characterize the boundedness of singular integral operators on weighted L^p-spaces [11]. Here we need only the class A_1 consisting of those weight functions w on $\mathbb{C}$ such that

$$\frac{1}{|B|}\int_B w(x)dm(x) \leq Cw(x) \quad \text{a.e. } x \in B \tag{3}$$

holds for all disks $B \subset \mathbb{C}$. Below we shall prove the following

1.1. THEOREM: Let Γ be a biporous Jordan curve in the complex plane and suppose ω satisfies (1), (2). Then the following are equivalent

(a) $C_\Gamma : \Lambda^\omega(\Gamma) \to \Lambda^\omega(\Gamma)$ is a bounded operator

(b) $\omega(d(x,\Gamma))/d(x,\Gamma) \in A_1$.

In (b) $d(x,\Gamma)$ denotes the Euclidean distance from Γ.

For a related but weaker sufficient condition see [5]. It follows from Theorem 1.1 and its proof that if Γ is an Ahlfors-regular Jordan curve and $\omega_\alpha(t) = t^\alpha$, then $C_\Gamma : \Lambda^\alpha \to \Lambda^\alpha$ for each $\alpha \in (0,1)$. In fact, from (b) we see that there are also plenty of non-regular curves with $C_\Gamma : \Lambda^\alpha(\Gamma) \to \Lambda^\alpha(\Gamma)$, $0 < \alpha < 1$.

1.2 COROLLARY: Let Γ and ω be as above and let $\bar{\omega}$ be a modulus of continuity such that $\bar{\omega}(t)/\omega(t)$ is increasing in t. Suppose further that C_Γ is bounded on $\Lambda^\omega(\Gamma)$. Then C_Γ defines a bounded operator

$$C_\Gamma : \Lambda^{\bar{\omega}}(\Gamma) \to \Lambda^{\bar{\omega}}(\Gamma).$$

The condition (b) in Theorem 1.1 is also related to the dimension of the curve Γ. To see this let $\phi: [0,\infty) \to [0,\infty)$ be an increasing function with $\phi(0) = 0$ and write

$$H_\phi(E) = \liminf_{\delta\to 0} \{\sum_{i=1}^{\infty} \phi(r_i) : E \subset \bigcup_{i=1}^{\infty} B(x_i,r_i),\ r_i \leq \delta\}$$

for a set $E \subset \mathbb{C}$. As is well known H_ϕ is a Borel measure; if $\phi(t) = t^s$ then $H_\phi \equiv H_s$ is the s-dimensional Hausdorff measure. Finally, the Hausdorff dimension of E is given by

$$\dim_H(E) = \inf \{s : H_s(E) < \infty\}.$$

1.3. PROPOSITION: Suppose Γ is a biporous Jordan curve and ω a modulus of continuity. If $H_\phi(\Gamma) > 0$ then $C_\Gamma : \Lambda^\omega(\Gamma) \to \Lambda^\omega(\Gamma)$ only if

$$\omega(t) = o(\phi(t)/t).$$

In particular, if $\dim_H(\Gamma) = s < 2$ and C_Γ is bounded on $\Lambda^\omega(\Gamma)$ then $\omega(t) = o(t^{s-1})$.

In the general case the necessary conditions of 1.3 are not sufficient for the boundedness of the Cauchy operator, cf. [1, 1.4]. However, there is a large class of fractal curves Γ, containing the snowflake curve of von Koch, for which the dimension does determine the behaviour of C_Γ on $\Lambda^\omega(\Gamma)$. We have in mind here the *selfsimilar* curves of Hutchinson [10]. First, a Jordan arc Γ_0 is selfsimilar if there are similarities σ_i, $1 \leqq i \leqq n$, such that $\sigma_i(\Gamma_0) \cap \sigma_j(\Gamma_0)$ is either empty or a point, $i \neq j$, and that

$$\Gamma_0 = \bigcup_{i=1}^{n} \sigma_i(\Gamma_0).$$

Then a Jordan curve Γ is called selfsimilar if there exist similarities Σ_j, $1 \leqq j \leqq m$ and a selfsimilar Jordan arc Γ_0 such that $\Gamma = \bigcup_{j=1}^{m} \Sigma_j(\Gamma_0)$.

1.4. PROPOSITION: Let Γ be a selfsimilar Jordan curve with $s = \dim_H(\Gamma)$. Then

$$C_\Gamma : \Lambda^\omega(\Gamma) \to \Lambda^\omega(\Gamma)$$

is bounded if and only if ω satisfies the "Dini condition"

$$\int_0^\delta \omega(t)t^{-s}dt \leqq C\delta^{1-s}\omega(\delta).$$

When $s = 1$ this reduces to the first inequality in (2). Note also that if $\omega(t) = t^\alpha$ then the condition is true if and only if $s - 1 < \alpha < 1$.

The proofs of the above results are given in the next section. For further results in these directions, with $\omega(t) = t^\alpha$, we refer to [1].

2. How can one define the Cauchy integral

$$C_\Gamma f(z) = \frac{1}{2\pi i}\int_\Gamma \frac{f(\xi)}{\xi - z}\, d\xi$$

when the curve Γ is not rectifiable? The idea to overcome the problem of nonrectifiability is due to Zinsmeister [14]: If f is a C^∞-function with compact support in $\mathbb{C}$ and D is the bounded component of $\mathbb{C}\setminus\Gamma$ then by Stokes' theorem

$$C_\Gamma f(z) = -\frac{1}{\pi}\int_{\mathbb{C}\setminus D} \frac{\bar\partial f(\xi)}{\xi - z}\, dm(\xi), \quad z \in D. \tag{4}$$

This expression makes sense for all Jordan curves and hence we have a natural operator defined on smooth functions on Γ. In brief, we say that the Cauchy operator C_Γ is bounded on $\Lambda^\omega(\Gamma)$ if there is a constant $M < \infty$ with

$$\|C_\Gamma f\|_{\Lambda^\omega(\Gamma)} \leqq M\, \|f\|_{\Lambda^\omega(\Gamma)}, \quad f \in C_0^\infty(\mathbb{C}) \tag{5}$$

where $\|f\|_{\Lambda^\omega(\Gamma)} = \inf\{t : |f(x) - f(y)| \leqq t\omega(|x - y|)\ \forall x,y \in \Gamma\}$ and $C_\Gamma f(z)$ as in (4).

To check that all is well defined note that if $f \in C_0^\infty$, then by the generalized Cauchy formula the integral in (4) is continuous in z in the whole complex plane $\mathbb{C}$ and in D it depends only on the boundary values $f|_\Gamma$. Moreover, the norms $\|C_\Gamma f\|_{\Lambda^\omega(\Gamma)}$ and $\|C_\Gamma f\|_{\Lambda^\omega(D)}$ are always equivalent [13]. Note also that if (5) is satisfied, then C_Γ is apriori continuous only in the closure of the C^∞-functions in $\Lambda^\omega(\Gamma)$. However the results below imply that if (5) does hold then $C_\Gamma f(z)$ is defined for all $f \in \Lambda^\omega(\Gamma)$, $z \in \Gamma$ and $\|C_\Gamma f\|_{\Lambda^\omega(\Gamma)} \leqq M\, \|f\|_{\Lambda^\omega(\Gamma)}$.

We turn then to the proof of Theorem 1.1. As a first step we recall some properties of the Whitney extension, c.f. [12, VI.2]; If Γ is any Jordan curve and ω is a modulus of continuity for which (1) holds, then each $f \in \Lambda^\omega(\Gamma)$ has a compactly supported extension $F \in \Lambda^\omega(\mathbb{C})$ with $F|_{\mathbb{C}\setminus\Gamma} \in C^\infty$ and

$$|\operatorname{grad} F(z)| \leqq C_0 \|f\|_{\Lambda^\omega(\Gamma)} \omega(d(z,\Gamma))/d(z,\Gamma).$$

In particular, if $\omega(d(z,\Gamma))/d(z,\Gamma) \in A_1$, $\bar\partial F(z)$ is integrable in $\mathbb{C}$ and thus we can set

$$C_\Gamma f(z) = -\frac{1}{\pi} \int_{\mathbb{C}\setminus D} \frac{\bar\partial F(\xi)}{\xi - z}\, dm(\xi), \quad z \in D.$$

2.1 THEOREM: Let Γ be a Jordan curve and ω a modulus of continuity satisfying (1), (2). If $\omega(d(z,\Gamma))/d(z,\Gamma) \in A_1$, then

$$\|C_\Gamma f\|_{\Lambda^\omega(D)} \leqq M_0 \|f\|_{\Lambda^\omega(\Gamma)}, \quad f \in \Lambda^\omega(\Gamma).$$

PROOF: The argument we apply is based on [4], see also [6], pp. 75-78. To begin, note that if $z \in D$ and $k \in \mathbb{N}$, then

$$\int_{B(z,R)\setminus B(z,r)} \frac{\omega(d(\xi,\Gamma))/d(\xi,\Gamma)}{|\xi - z|^k}\, dm(\xi) \leqq C_k \int_r^R \frac{\omega(t)}{t^k} dt, \quad r \leqq R/2.$$

Indeed, by $\omega(d(\xi,\Gamma))/d(\xi,\Gamma) \in A_1$, this is clear if $2r \leqq R \leqq 4r$ and the general case $2^n r \leqq R \leqq 2^{n+1} r$ follows then by induction.

Now, if $z, z' \in D$

$$|C_\Gamma f(z) - C_\Gamma f(z')| \leqq |z - z'| \int_{\mathbb{C}\setminus D} \frac{|\bar\partial F(\xi)|}{|\xi - z||\xi - z'|}\, dm(\xi) \tag{6}$$

$$\leqq C_0 |z - z'| \, \|f\|_{\Lambda^\omega(\Gamma)} \int_{\mathbb{C}\setminus D} \frac{\omega(d(\xi,\Gamma))/d(\xi,\Gamma)}{|\xi - z|\,|\xi - z'|}\, dm(\xi).$$

Further, if $A_z = \{\xi : |\xi - z| \leqq |z - z'|/2\}$ and $B_z = \{\xi : |\xi - z| \geqq |z - z'|/2,\ |\xi - z| \leqq |\xi - z'|\}$, the union of A_z, $A_{z'}$, B_z and $B_{z'}$ covers the plane and

$$|z - z'| \int_{A_z} \frac{\omega(d(\xi,\Gamma))/d(\xi,\Gamma)}{|\xi - z|\ \ |\xi - z'|}\, dm(\xi) \leqq 2C_1 \int_0^{|z-z'|} \frac{\omega(t)}{t}\, dt, \tag{7a}$$

$$|z - z'| \int_{B_z} \frac{\omega(d(\xi,\Gamma))/d(\xi,\Gamma)}{|\xi - z|\ \ |\xi - z'|}\, dm(\xi) \leqq C_2 |z - z'| \int_{|z-z'|/2}^{\infty} \frac{\omega(t)}{t^2} dt. \tag{7b}$$

Combining (6), (7) with the Dini conditions (2) we obtain $|C_\Gamma f(z) - C_\Gamma f(z')| \leq C_3 \|f\|_{\Lambda^\omega(\Gamma)}\omega(|z - z'|)$. □

2.2. THEOREM: Let Γ be a biporous Jordan curve and let ω be a modulus of continuity for which (1), (2) hold. If the Cauchy operator C_Γ is bounded on $\Lambda^\omega(\Gamma)$, in the sense (5), then

$$\omega(d(z,\Gamma))/d(z,\Gamma) \in A_1.$$

PROOF: We must show that

$$\frac{1}{|B|}\int_B \frac{\omega(d(\xi,\Gamma))}{d(\xi,\Gamma)}dm(\xi) \leq C \inf_{z\in B} \frac{\omega(d(z,\Gamma))}{d(z,\Gamma)}$$

holds for all disks $B \subset \mathbb{C}$. We prove this only for $B = B(z_0,R)$, $z_0 \in \Gamma$ and $R \leq \mathrm{diam}(\Gamma)/5$; the general cases are then easy consequences.

Let $z_0 \in \Gamma$ and $R \leq \mathrm{diam}(\Gamma)/5$. Since Γ is assumed to be biporous we can apply the construction of [1, Lemma 3.4] and obtain for each $n \in \mathbb{N}$ a smooth function u_n satisfying the following three properties. Firstly, the support of u_n is contained in $B(z_0,R) \cap \Gamma + B(2^{-n}R)$, secondly $|u_n(z) - u_n(z')| \leq C_0\min\{1,(2^n/R)|z - z'|\}$ and thirdly, $\bar\partial u_n(z) \geq 0$ in $\mathbb{C}\setminus D$ with

$$\int_{\mathbb{C}\setminus D} \bar\partial u_n dm(z) \geq C_1(2^n/R)|\{z \in B(z_0,R):d(z,\Gamma) < 2^{1-n}R\}|. \tag{8}$$

Set now

$$v_k(z) = \sum_{n=1}^{k} \omega(2^{-n}R)u_n(z), \qquad k \in \mathbb{N}.$$

Clearly $v_k \in C_0^\infty(\mathbb{C})$ and if $z,z' \in \mathbb{C}$ and $2^{-p}R \leq |z - z'| \leq 2^{1-p}R$, $p \in \mathbb{N}$, then

$$\begin{aligned}|v_k(z)-v_k(z')| &\leq C_0 \sum_{n=1}^{p-1} (2^n/R)\omega(2^{-n}R)|z-z'| + C_0 \sum_{n=p}^{k} \omega(2^{-n}R)\\ &\leq C_1|z-z'|\int_{|z-z'|}^{R} \frac{\omega(t)}{t^2}dt + C_0\int_0^{2|z-z'|} \frac{\omega(t)}{t}dt \leq C_2\omega(|z-z'|).\end{aligned}$$

Consequently, $\|v_k\|_{\Lambda^\omega(\mathbb{C})} \leq 3C_2$.

Next, if $z,z' \in D$ are arbitrary,

$$C_\Gamma v_k(z) - C_\Gamma v_k(z') = -\frac{1}{\pi} \cdot \frac{(z' - z)}{(z - z_0)(z' - z_0)} \cdot \int_{\mathbb{C}\setminus D} \frac{\bar\partial v_k(\xi)dm(\xi)}{(1-\frac{\xi-z_0}{z-z_0})(1-\frac{\xi-z_0}{z'-z_0})}$$

and since $v_k \in C_0^\infty$, we have $|C_\Gamma v_k(z) - C_\Gamma v_k(z')| \leqq C\omega(|z-z'|)$ from (5). To obtain an estimate from below we choose a $z_1 \in \Gamma$ with $|z_1 - z_0| = 4R$. As Γ is biporous there exist points $z,z' \in B(z_1,R) \cap D$ with $|z - z'| \geqq C_3R$. Moreover, if ξ is contained in the support of v_k, $\xi \in B(z_0,3R/2)$ and so $|\xi - z_0|/|z - z_0| < 1/2, |\xi - z_0|/|z' - z_0| < 1/2$. Therefore for this special pair z, z' we get

$$\int_{\mathbb{C}\setminus D} \bar\partial v_k(\xi)dm(\xi) \leqq C_4R|C_\Gamma v_k(z) - C_\Gamma v_k(z')| \leqq CR\omega(R).$$

Since that is true for all $k \in \mathbb{N}$,

$$\sum_{n=1}^{\infty} \int_{\mathbb{C}\setminus D} \omega(2^{-n}R)\bar\partial u_n(z)dm(z) \leqq CR\omega(R).$$

Finally, (8) shows that for $B = B(z_0,R)$

$$\int_B \frac{\omega(d(z,\Gamma))}{d(z,\Gamma)}dm(z) \leqq \sum_{n=1}^{\infty} \frac{\omega(2^{-n}R)}{2^{-n}R}|\{z \in B: 2^{-n}R \leqq d(z,\Gamma) < 2^{1-n}R\}|$$

$$\leqq C_5 \sum_{n=1}^{\infty} \omega(2^{-n}R) \int_{\mathbb{C}\setminus D} \bar\partial u_n(z)dm(z) \leqq CR\omega(r). \qquad \square$$

Combining 2.1 and 2.2 we obtain a proof for Theorem 1.1. Furthermore, Corollary 1.2 follows now immediately: Assume that $\bar\omega(t)/\omega(t)$ is increasing in t and that $\omega(d(z,\Gamma))/d(z,\Gamma) \in A_1$. If $d(z_0,\Gamma) = \max\{d(z,\Gamma): z \in B\}$, then by (1a)

$$\frac{1}{|B|}\int_B \frac{\bar\omega(d(z,\Gamma))}{d(z,\Gamma)}dm(z) \leqq \frac{\bar\omega(d(z_0,\Gamma))}{\omega(d(z_0,\Gamma))} \cdot \frac{1}{|B|} \cdot \int_B \frac{\omega(d(z,\Gamma))}{d(z,\Gamma)} dm(z)$$

$$\leqq C\, \frac{\bar\omega(d(z_0,\Gamma))}{d(z_0,\Gamma)}$$

and thus $\bar{\omega}(d(z,\Gamma))/d(z,\Gamma) \in A_1$.

In order to relate the geometric A_1-condition 1.1(b) to the dimension of the curve Γ we need the equality

$$\int_0^\infty \frac{\omega(t)}{t^2}|\{z \in B : d(z,\Gamma) < t\}|dt = \int_B \int_{d(z,\Gamma)}^\infty \frac{\omega(t)}{t^2}dt\ dm(z) \qquad (9)$$

which is a consequence of the Fubini theorem.

<u>PROOF OF PROPOSITION 1.3</u>: Let B be a disk containing a neighborhood of Γ. If $\omega(d(z,\Gamma))/d(z,\Gamma) \in A_1$, we deduce from (9) and (2) that $(\omega(t)/t)|\{z \in B : d(z,\Gamma) < t\}| \leq M < \infty$.

When t is small enough,

$$\{z \in B : d(z,\Gamma) < t\} = \bigcup_{z\in\Gamma} B(z,t).$$

By basic covering theorems, see for instance [9], we can take a subcovering $B(z_i,t)$, $1 \leq i \leq k(t)$ so that at each point $x \in \mathbb{C}$ at most n_0 of these balls intersect. Here n_0 is an absolute constant. As $\pi k(t)t^2 \leq n_0|\{z \in B : d(z,\Gamma) < t\}|$,

$$0 < H_\phi(\Gamma) \leq \liminf_{t\to 0} k(t)\phi(t) \leq (n_0/\pi) \liminf_{t\to 0} \phi(t)|\{z \in B : d(z,\Gamma) < t\}|t^{-2},$$

which yields $\omega(t) = O(\phi(t)/t)$. On the other hand, if $\omega(d(z,\Gamma))/d(z,\Gamma) \in A_1$, then by Muckenhoupts theorem [11, p. 214] also $\omega(d(z,\Gamma))^{1+\varepsilon}/d(z,\Gamma)^{1+\varepsilon} \in A_1$ for some $\varepsilon > 0$. Now $\omega(t)^{1+\varepsilon}/t^\varepsilon = O(\phi(t)/t)$ and thus $\omega(t) = (t/\omega(t))^\varepsilon O(\phi(t)/t) = o(\phi(t)/t)$; according to (2) $\omega(t)/t \to \infty$ as $t \to 0$.

If $\dim_H(\Gamma) = s$, $H_d(\Gamma) > 0$ for all $d < s$. Hence if also $C_\Gamma : \Lambda^\omega(\Gamma) \to \Lambda^\omega(\Gamma)$, there is an $\varepsilon > 0$, independent of $d < s$, such that $\omega(t)^{1+\varepsilon}/t^\varepsilon \leq C_d t^{d-1}$, $0 < t \leq 1$. When $s - \varepsilon(2 - s) < d < s$, $s - 1 < (d + \varepsilon - 1)/(1 + \varepsilon)$ and

$$\omega(t) = o(t^{s-1}).$$

In fact, we have shown that if $C_\Gamma : \Lambda^\omega(\Gamma) \to \Lambda^\omega(\Gamma)$ then

$$\omega(t) = t^{\varepsilon_0}\, o(t^{s-1})$$

for $s = \dim_H(\Gamma)$ and for some $\varepsilon_0 > 0$. □

If Γ is a selfsimilar Jordan curve, then Formula (9) yields a complete characterization of the moduli ω for which $C_\Gamma : \Lambda^\omega \to \Lambda^\omega$. As shown by Hutchinson [10] a selfsimilar Γ with $s = \dim_H(\Gamma)$ supports a canonical probability measure μ such that

$$C_1 R^s \leqq \mu(B(x,R)) \leqq C_2 R^s, \quad x \in \Gamma, \quad R \leqq \operatorname{diam}(\Gamma).$$

Applying this special measure one can estimate the size of the set $\{z \in B : d(z,\Gamma) < t\}$ for $B = B(z_0,R)$, $z_0 \in \Gamma$ and $R \leqq \operatorname{diam}(\Gamma)$. Namely, as in the proof of 1.3 we cover the set by $k(t)$ balls $B(z_i,t)$, $z_i \in \Gamma \cap B$, such that at each point no more than n_0 of them intersect. Then

$$|\{z \in B : d(z,\Gamma) < t\}| \leqq \pi k(t)t^2 \leqq Ct^{2-s} \sum_{i=1}^{k(t)} \mu(B(z_i,t))$$

$$\leqq Ct^{2-s}\mu(B(z_0,2R)) \leqq Ct^{2-s}R^s, \quad t \leqq R.$$

The same reasoning with the roles of the measures reversed gives

$$CR^s t^{2-s} \leqq |\{z \in B : d(z,\Gamma) < t\}|, \quad t \leqq R.$$

<u>PROOF OF PROPOSITION 1.4</u>: Since Γ_0 is invariant under the similarities σ_i we see that Γ is biporous. Therefore, according to Theorem 1.1, $C_\Gamma : \Lambda^\omega(\Gamma) \to \Lambda^\omega(\Gamma)$ if and only if $\omega(d(z,\Gamma))/d(z,\Gamma) \in A_1$. It is necessary to check the Muckenhoupt condition (3) only for $B = B(z_0,R)$ where $z_0 \in \Gamma$ and $R \leqq \operatorname{diam}(\Gamma)$ and then it reads

$$\int_B \frac{\omega(d(z,\Gamma))}{d(z,\Gamma)}\, dm(z) \leqq CR\omega(R).$$

As $\omega(t)/t \leqq \int_t^\infty \omega(s)s^{-2}ds \leqq C\omega(t)/t$ and $\int_R^\infty \omega(s)s^{-2}|\{z \in B : d(z,\Gamma) \leqq s\}|ds$ $\leqq CR\omega(R)$ the boundedness of the Cauchy operator C_Γ is equivalent to

$$\int_0^R \frac{\omega(t)}{t^2}|\{z \in B : d(z,\Gamma) < t\}|dt \leqq CR\omega(R)$$

or, in view of the above estimates, to

$$\int_0^R \omega(t)R^s t^{-s}dt \leq CR\omega(R). \quad \square$$

References

[1] K. Astala, Caldéron's problem for Lipschitz classes and the dimension of quasi-circles. Revista Math. Iber. (1989). To appear.

[2] A.P. Caldéron, Cauchy integrals on Lipschitz curves and related operators. Proc. Nat. Acad. Sci. U.S.A. 74 (1977), 1324-1327.

[3] G. David, Operateurs intégraux singuliers sur certaines courbes du plan complexe. Ann. Sci. de l'E.N.S. 17 (1984), 157-189.

[4] E.P. Dolzhenko, The removability of singularities of analytic functions. Uspeki Math. Nauk. 18 (1963), (112), 135-142 (Russian).

[5] E.M. Dynkin, On the smoothness of Cauchy type integrals. IX. Zap. Nauču. Sem. Leningrad. Otdel. Mat. Inst. Steklov. (LOMI) 92 (1979), 115-133.

[6] J. Garnett, Analytic Capacity and Measure. Springer-Verlag, Berlin and New York, 1972.

[7] J. Garnett, Bounded Analytic Functions. Academic Press, 1981.

[8] F.W. Gehring, Characteristic properties of quasidics. Les Presses de l'Université de Montreal, 1982.

[9] M. de Guzmán, Differentiation of Integrals in R^n. Springer-Verlag, Berlin and New York, 1975.

[10] J.E. Hutchinson, Fractals and Selfsimilarity. Indiana Univ. Math. J. 30 (1981) 713-747.

[11] B. Muckenhoupt, Weighted norm inequalities for the Hardy maximal function. Trans. Amer. Math. Soc. 165 (1972), 207-226.

[12] E.M. Stein, Singular Integrals and Differentiability Properties of Functions. Princeton, New Jersey, 1970.

[13] P.M. Tamrazov, Contour and solid structure properties of holomorphic functions of a complex variable. Russ. Math. Surveys 28 (1973) 141-173.

[14] M. Zinsmeister, Problèmes de Dirichlet, Neumann, Caldéron dans les quasidisques pour les classes holderiennes. Revista Math. Iber 2 (1986), 319-332.

K. Astala
Department of Mathematics
University of Helsinki
Hallituskatu 15
00100 Helsinki,
Finland

B. GUSTAFSSON AND J. PEETRE

Möbius invariant operators on Riemann-surfaces

0. INTRODUCTION AND FIRST EXAMPLES

In the past seven years or so most of my mathematical activities have in one way or other been connected with Hankel operators (or forms). As there have been at least two talks (Arazy, Janson)[2] at this summer school devoted to this subject, I have been forced to pick up things left over ... .

To begin the discussion recall the definition of the classical Schwarz derivative

$$\{F,z\} \equiv -2\sqrt{F'}\,\frac{d^2}{dz^2}\left(\frac{1}{\sqrt{F'}}\right) = \frac{F'''}{F'} - \frac{3}{2}\left(\frac{F''}{F'}\right)^2. \tag{1}$$

This third-order nonlinear DO plays an important role in conformal mapping, in uniformization, in the theory of Kleinian groups, and in the theory of second-order linear DO (all these topics are in fact related!). It has an essentially invariant character, as is manifested in the following identity due to Cayley:

$$\{F,z\} = \{F,\zeta\}\left(\frac{d\zeta}{dz}\right)^2 + \{\zeta,z\}, \tag{2}$$

where $\zeta = \phi(z)$ is a change of coordinate. No workable generalization to higher order of the Schwarz derivative is known. However, if we allow only projective transformations of coordinates the situation changes drastically and we have plenty of other operators which display a similar invariant behavior; in this situation $\{\zeta,z\} = 0$ so (2) simplifies to

$$\{F,z\} = \{F,\zeta\}\left(\frac{d\zeta}{dz}\right)^2. \tag{2'}$$

---

[1] This compilation, prepared for the Sodankylä Summer School, August 1988, was written by J.P. but as many of the results announced were obtained in cooperation with B.G. it has been judged appropriate, following great examples in the past, to include both names here.

[2] As the latter could not come because of illness, his address was actually delivered by Robert Wallstén.

In the general case the exponent 2 has to be replaced by some other integer. To get a global formulation one is led to consider manifolds (either real or complex) equipped with a *projective structure*. Thus in the complex case - to which case we focus our attention in what follows - we are dealing with a *Riemann surface* with such a structure.

In fact, many examples of such *Möbius invariant* DOs can be found on the basis of a general fact, which I have begun to call Bol's lemma or Bol's theorem and which has fascinated me over a period of several years. In a way all that we say here may be viewed as offshoots of this in itself rather elementary observation:

BOL'S LEMMA [7]: If we let the function F transform according to the rule

$$F(z) \mapsto F(\phi(z))e(z)^{\mu-1},$$

where $\phi(z) = (az + b)/(cz + d)$ $(ad - bc = 1)$ is an arbitrary Möbius transformation and where we have put $e(z) = cz + d$, μ being a given integer $\geqq 0$, then its μth derivative $F^{(\mu)}$ transforms according to the rule

$$F^{(\mu)}(z) \mapsto F^{(\mu)}(\phi(z))e(z)^{-\mu-1}.$$

Several proofs of this result are listed in [25] (see also the discussion in section 2 of the present compilation).[3]

Let us consider in detail some examples, interesting in their own right.

EXAMPLE 1: In [41], [42] I suggested the following generalization of the DO in (1):

$$(F^{(\mu)})^{(\lambda-1)/(\mu+1)} \frac{d^\lambda}{dz^\lambda} \left(\frac{1}{(F^{(\mu)})^{(\lambda-1)/(\mu+1)}}\right).$$

---

3 A proof not included in [25] can be based on the idea to prove it first for the function $\phi = 1/z$. See [16], p. 187, exc. 1; the appearance of it there suggests that the result may be quite old.

where λ is another integer ≥ 0. If $\lambda = 2$, $\mu = 1$ we clearly get back the Schwarz derivative (up to a factor). I proposed there (see also [47]) that null solutions of the corresponding DE might be of some interest in analysis, in particular that linear combinations of null solutions might play a similar role in, say, approximation theory as rational functions. It is easy to write down the solutions in question: We find that

$$\frac{1}{(F^{(\mu)})^{(\lambda-1)/(\mu+1)}} = P \quad \text{(a polynomial of degree} < \lambda)$$

or

$$F^{(\mu)} = \frac{1}{P^{(\mu+1)/(\lambda-1)}},$$

whence

$$F(z) = \frac{1}{(\mu-1)!} \int_0^z \frac{(z-\zeta)^{-1} \, d\zeta}{(P(\zeta))^{(\mu+1)/(\lambda-1)}} + Q(z) \quad \text{(a polynomial of degree} < \mu).$$

(Check: P and Q together determine $\lambda + \mu$ independent parameters, which is the order of the equation.) Such functions are, apparently, related to so-called Picard curves (see [29]).

SUBEXAMPLE 1: $\mu = 1$, $\lambda = 2$. Then

$$F'(z) = \frac{1}{(cz+d)^2} \Rightarrow F(z) = \frac{az+b}{cz+d} \quad \text{(a fractional linear function).}$$

EXAMPLE 2: A related construction was suggested by Menahem Schiffer [51]. This depends on a 'polarized' version of Bol's lemma [46]. Let $F(z_1,\ldots,z_{\mu+1})$ denote the μth Newton's divided difference of the function $F(z)$:

$$F(z_1,\ldots,z_{\mu+1}) = \frac{\mu!}{2\pi i} \int \frac{F(\zeta)}{(\zeta - z_1)\ldots(\zeta - z_{\mu+1})} \, d\zeta,$$

where we integrate over a suitable 'contour' encircling the points $z_1,\ldots,z_{\mu+1}$. Apparently

$$F(z,\ldots,z) = F^{(\mu)}(z).$$

More generally (with $D_j = \partial/\partial z_j$)

$$D_1^{a_1} \dots D_{\mu+1}^{a_{\mu+1}} F(z,\dots,z) = \frac{\mu! a_1! \dots a_{\mu+1}!}{(\mu + \Sigma a_j)!} F^{(\mu+\Sigma a_j)}(z),$$

thus in particular (in slightly abusive notation)

$$D_1 F = \frac{\mu!}{(\mu + 1)!} F^{(\mu+1)}, \; D_1 D_2 F = \frac{\mu!}{(\mu + 2)!} F^{(\mu+2)}, \dots . \tag{4}$$

We have the transformation rule (F transforms as in the lemma):

$$F(z_1,\dots,z_{\mu+1}) \mapsto F(\phi(z_1),\dots,\phi(z_\mu))e(z_1)^{-1}\dots e(z_{\mu+1})^{-1}.$$

Thus if all points z_j are equal to z

$$F^{(\mu)}(z) \mapsto F^{(\mu)}(\phi(z))e(z)^{-(\mu+1)},$$

which is one of the proofs of Bol's lemma (theorem) given in [25].

Consider now

$$D_1 \dots D_\lambda \log F$$

which expression, apparently, if $\lambda > 1$ is changed, under the influence of ϕ, into

$$D_1 \dots D_\lambda \log F\circ\phi = (D_1 \dots D_\lambda \log F)\circ\phi\cdot\phi'(z_1)\dots\phi'(z_\lambda).$$

Denoting by M (M for Menahem) the restriction to the diagonal, the DO is obtained is transformed as

$$M \mapsto M\circ\phi\cdot(\phi')^\lambda,$$

thus as a form of degree λ. M is thus a nonlinear homogeneous Möbius invariant DO of order $\lambda + \mu$.

SUBEXAMPLE 2: $\mu = 1$, $\lambda = 2$. Then

$$D_1 \log F = \frac{D_1 F}{F}$$

$$D_1 D_2 \log F = \frac{D_1 D_2 F}{F} - \frac{D_1 F \cdot D_2 F}{F^2}$$

gives (see (4))

$$M = \frac{1!}{3!}\frac{F'''}{F'} - \left(\frac{1!}{2!}\right)^2\left(\frac{F''}{F'}\right)^2 = \frac{1}{6}\frac{F'''}{F'} - \frac{1}{4}\left(\frac{F''}{F'}\right)^2,$$

which can be rewritten as

$$-\frac{1}{3}\sqrt{(F')}D^2\left(\frac{1}{\sqrt{F'}}\right),$$

as

$$D\left(\frac{1}{\sqrt{F'}}\right) = -\frac{1}{2}\frac{F''}{(\sqrt{F'})^3},$$

$$D^2\left(\frac{1}{\sqrt{F'}}\right) = -\frac{1}{2}\frac{F'''}{(\sqrt{F'})^3} + \frac{1}{2}\cdot\frac{3}{2}\frac{(F'')^3}{(\sqrt{F'})^5}.$$

Therefore we get essentially back the Schwarzian.

SUBEXAMPLE 3: $\mu = 2$, $\lambda = 3$. Now

$$D_1 D_2 D_3 \log F = \frac{D_1 D_2 D_3 F}{F} - \frac{D_1 D_2 F \cdot D_3 F}{F^2}$$

$$- \frac{D_1 D_3 F \cdot D_2 F}{F^2} - \frac{D_2 D_3 F \cdot D_1 F}{F^2} + \frac{2 D_1 F \cdot D_2 F \cdot D_3 F}{F^3}$$

gives (see again (4))

$$M = \frac{(2/120)F^{V}}{F''} - 3\frac{(2/6)\cdot(2/24)F^{IV}F'''}{(F'')^2} + \frac{2((2/6)F''')^3}{(F'')^3}$$

$$= \frac{(1/60)\,(F'')^2F^{V} - (1/12)\,F''F'''F^{IV} + (2/27)\,(F''')^3}{(F'')^3}.$$

This expression, again, can be transformed into

$$-\frac{1}{40}(F'')^{2/3}D^3[(F'')^{-2/3}],$$

which is readily found by carrying out the last derivation.

The above suggests that the operators in examples 1 and 2 always coincide (up to a factor). However, continuing the calculation in subexamples 2 and 3 shows that this is not the case. In section 3 we will write down the general form of a (homogeneous) Möbius invariant DO. Actually, this is already at least implicit in Morikawa [39] in an invariant theoretic context. We believe however that our presentation (we propose in fact two slightly different avenues) is more transparent. We give also an abundance of concrete examples.

All the DOs encountered previously have been homogeneous (in F). Now we mention an interesting instance of a nonhomogeneous Möbius invariant DO.

EXAMPLE 3: The following DE was encountered by Jacobi [32] in the theory of theta functions

$$C^2D(\log C^3C'') = \surd(16C^3C'' + 1) \qquad (5)$$

which, apparently, has a Möbius invariant character provided we let C transform as a form of degree -(1/2) (the case $\mu = 2$ of Bol's lemma). In [32] Jacobi shows that (5) is satisfied with $C = (2/\pi)\theta_{e\epsilon}^{-2}(0,t)$, where $\theta_{e\epsilon}(0,t)$, $e\epsilon = 0$, is any '*Thetanullwerth*' (theta constant), and that the general solution is obtained by application of a Möbius transformation. This example will be analysed in detail in Appendix 2. So far, however, Jacobi's equation (5) stands out as an isolated special case.

The rest of the present compilation is organized as follows. However, generally speaking, the paper has no 'plan' so that its various subdivisions (including the appendices), even parts of them, may to some extent be read independently of each other.

In section 1 we briefly recall some salient facts about projective structures on manifolds and, in particular, their connection to uniformization and so second-order linear DO.

Section 2 reviews the main contents of the paper [25], centering around

Bol's lemma and related issues, for instance, the notion of transvectant.

In section 3, as we already told above, a description of 'all' Möbius invariant operators is obtained.

In section 4 we show how the transvectant can be exploited in connection with Hankel theory. We also point out the parallel between Hankel theory and operator calculi ('quantization'), the latter subject being briefly reviewed in section 5.

In section 6 we discuss reproducing and 'coreproducing' kernels in Hilbert spaces of analytic functions.

Finally, the appendices, six in number, contain auxiliary material more or less loosely related to the main body of our paper.

1. COMPLEX MANIFOLDS WITH A PROJECTIVE STRUCTURE

We collect here some basic facts about complex manifolds equipped with a projective structure. It is however only out of convenience that we have restricted attention to the complex case only. Projective structures are also of interest in the real case, at least in dimension 1, for instance in the oscillatory theory of second-order linear DE, for which we refer to the book [9]. (A brief mention of projective structure can further be found in the excellent book [3], pp. 42-56, where also (chiefly) nonlinear equations are considered. Compare further [15]).

So let Ω be a complex manifold of (complex) dimension n. We say that we have a *projective structure* on Ω if there is given a covering of Ω with coordinate neighborhoods $\{U\}$ and corresponding local coordinates $\{z = (z^1,\ldots,z^n)\}$ such that the change of coordinates is mediated by projective (fractional linear) maps: if $U \cap U' \neq \emptyset$ then z and z' are connected by a relation of the form

$$z'^j = \frac{a_{j0} + a_{j1}z^1 + \ldots + a_{jn}z^n}{a_{00} + a_{01}z^1 \cdot \ldots + a_{0n}z^n} \quad (j = 1,\ldots,n); \tag{1}$$

we can always require that $\det(a_{jk}) = 1$.

In the same way one defines for instance *affine structure*. For example, a complex torus has a canonical affine structure.

More generally, in the book [24] one considers 'structures' associated with any Lie pseudogroup of differentiable (smooth) transformations of $\mathbb{C}^n$.

Let us return to the projective situation and fix attention to the case $n = 1$. So we have a Riemann surface with a projective structure. Formula (1) reduces to

$$z' = \frac{az + b}{cz + d} \quad (ad - bc = 1). \tag{1'}$$

In particular, let us make clear the relation to uniformization (for more details see [22], [23]). Let us start with some particular projective coordinate z defined in the coordinate neighborhood U. Then if $U \cap U' \neq \emptyset$ the function z can using (1') be continued analytically to $U \cup U'$ and, in general, along any path issuing from U. In this way one gets a map $\tilde{\Omega} \mapsto \hat{\mathbb{C}}$, where $\tilde{\Omega}$ is the universal cover of Ω and $\hat{\mathbb{C}}$ the extended complex plane (Riemann sphere, conformally equivalent to the projective line $\mathbf{P}^1$), with the property that germs lying over the same point of Ω are related by projective transformations; it is called the geometric realization of Ω by Gunning [22]. Conversely, given any such map we can define a projective structure on Ω by (locally) pulling back to Ω the identity function on $\hat{\mathbb{C}}$. (See also Tyurin's lectures [53] which came to our attention at a rather late stage while compiling this report.)

If Ω is a multiply connected planar domain bounded by finitely many smooth or even analytic arcs, a 'regular' domain in the sense of [2], then there are several natural projective structures on Ω which compete with each other. First, we have the one which comes from the uniformization theorem (we map $\tilde{\Omega}$ onto the unit disk D). Second, we take for the geometric realization simply a 'circular' model for Ω; by a classic theorem (see e.g. [27], pp. 481-488 for a proof) every such domain is conformally equivalent to one bounded by finitely many (generalized) circles. It is clear that the circular model is unique up to an arbitrary Möbius transformation (an element of the group $PL(2,\mathbb{C})$).

Projective structures arise further classically in connection with linear DE.

First, consider the case of second order. Then as a projective coordinate one can (locally) take the quotient of two independent solutions (the denominator is required not to vanish). Thus by a local change of independent variable and multiplying the dependent variable with a suitable factor any second-order linear DE can be reduced to the norm form

$$\frac{d^2F}{dz^2} = 0.$$

More computationally: assuming that the equation is already in the form

$$\frac{d^2F}{dz^2} + q(z)F = 0$$

(this first reduction is easily achieved by introducing a suitable multiplier), the final reduction is obtained by solving the third-order equation

$$\frac{1}{2}\{\zeta,z\} = q(z),$$

where { , } , as before, stands for the Schwarz derivative.

HISTORICAL REMARK: This was already known to Kummer [37] and, at least in a special case, it can be found in Jacobi [31]. (A portion of Kummer's paper is reproduced in German translation in [9], pp. 102-103.) The Schwarz derivative appears also, before Schwarz, in Riemann's long unpublished lectures as well as in the work of the young Poincare. More about the history in [20] and, more briefly, in the marvellous book [28], chapter 10.

As for higher (μth) order linear DE one easily proves that they can be brought on the canonical form

$$F^{(\mu)} + a_{\mu-3}(z)F^{(\mu-3)} + \ldots + a_0(z)F = 0; \qquad (2)$$

the coordinate systems in which the equation has this form obviously determine a projective structure on our manifold Ω. More about this in section 2.

2. THE BOL OPERATOR AND GREEN'S FORMULA

We now review part of the contents of the paper [25].

Consider a Riemann surface Ω equipped with a projective structure. Let κ be the canonical sheaf on Ω, i.e. (local) sections of κ are of the form $s = f(z)dz$ where z is a local coordinate and f an analytic function (in the overlap of two coordinate neighborhoods U and V with local coordinates z and ζ the corresponding coefficients f and g are related by an equation

$g(\zeta) = f(\phi(\zeta))\phi'(\zeta)$ if $z = \phi(\zeta)$). More generally, sections of powers κ^n (forms of integer degree n) are of the form $f(z)(dz)^n$, with an analogous transition rule. If we select a square root of κ, i.e. an invertible sheaf λ such that $\lambda^2 = \kappa$, then one can also talk of half-integer forms.

It follows now from Bol's lemma (see Introduction) that for each $\mu \geqq 0$ one can define a linear operator L from $\lambda^{1-\mu}$ into $\lambda^{1+\mu}$: if z is a projective coordinate on Ω then the form $F(z)(dz)^{(1-\mu)/2}$ is mapped onto the form $F^{(\mu)}(z)(dz)^{(1+\mu)/2}$. (Notice that formally each successive derivation accounts for another factor dz!)

<u>REMARK</u> (on Eichler cohomology): We have a short exact sequence

$$0 \to \Pi_{\mu-1} \to \lambda^{1-\mu} \to \lambda^{1+\mu} \to 0$$

so we can consider the corresponding exact sequence of cohomology groups. It turns out that the only nontrivial cohomology group is $H^1(\Omega,\Pi_{\mu-1})$ (see e.g. [22,23]). Eichler cohomology, introduced by Eichler in [17], plays a great role, e.g. in Kleinian groups (see e.g. [36]). Eichler himself viewed his theory as a sort of amplification of the classical theory of Abelian integrals and Abelian differentials. This is also why we here distinguish between capital letters, such as F, B,..., for 'integrals' and small ones, such as f,b,..., for 'differentials'.

In [25] it is investigated how L_μ looks in a general coordinate z (not necessarily a projective one). First of all, it is clear that $L_0 \equiv \text{id}$ (identity) and further that $L_1 = d$ (differential). They are independent of the projective structure. On the other hand, if $F = F(z)(dz)^{-\frac{1}{2}}$ then $L_2F = (F''(z) + q(z)F(z))(dz)^{3/2}$, and the functions q(z), a different function for each coordinate neighborhood, determine the projective structure. In [25] it is shown that if F(z) is the coefficient of a form F of degree $(1-\mu)/2$ then the coefficient of $L_\mu F$, a form of degree $(1+\mu)/2$, is of the form

$$F^{(\mu)}(z) + A_2F^{(\mu-2)}(z) + \ldots + A_\mu F(z),$$

where the coefficients $A_i = A_i^{(\mu)}$ $(i = 2,\ldots,\mu)$ are certain universal polynomials in the derivatives $q(z), q'(z),\ldots,q^{(i)}(z)$ of q(z) with respect to z.

For example, we have

$$L_3F = (F''' + 4qF' + 2q'F)(dz)^{5/2},$$

$$L_4F = (F^{IV} + 10qF'' + 10q'F' + (9q^2 + 3q'')F)(dz)^{7/2}$$

and so forth.

REMARK: This result can also be formulated as follows. Let F_1, F_2 be a basis for the solution of the DE $F'' + qF = 0$. Then the functions $F = P(F_1, F_2)$, where P runs through all polynomials homogeneous of degree $\mu - 1$, satisfy a linear μth-order DE, whose coefficients depend only on q. We were led to this formulation while reading the review [50] of the book [48] (the case $\mu = 3$).

One can further invoke a certain bilinear 'covariant' introduced by Gordan [19] in classical invariant theory, known as the *transvectant* (German: *Überschiebung*). This depends on the following fact:

GORDAN'S LEMMA: Let f_k $(k = 1,2)$ transform under Möbius transforms according to the rule

$$f_k(z) \mapsto f_k(\phi(z))e(z)^{-\nu_k} \quad \text{where} \quad \nu_k \in \mathbb{Z}.$$

Then

$$J_s(z) \overset{\text{def}}{\equiv} \sum_{i=0}^{s} (-1)^{s-i}\binom{s}{i} \frac{f_1^{(i)}\, f_2^{(s-i)}}{(\nu_1)_i(\nu_2)_{s-i}},$$

where $s \geq 0$ is any integer such that $\nu_k \neq 0, -1, \ldots, -(s-1)$ and, generally speaking, $(x)_i = x(x+1)\ldots(x + i - 1)$, transforms according to the rule

$$J_s(z) \mapsto J_s(\phi(z))e(z)^{-\nu},$$

where $\nu = \nu_1 + \nu_2 + 2s$.

For the proof see [25] or give your own. There this is used in the following way in the global situation of a Riemann surface Ω endowed with a projective structure. Taking $\nu_1 = -(\mu - 1)$, $\nu_2 = 2(\mu - s)$ where now $0 \leqq s \leqq \mu - 1$, one obtains for each form Θ of degree k a linear map

$$M^k_\Theta : \lambda^{1-\mu} \to \lambda^{1+\mu}.$$

Then one can define, given forms $\Theta_1,\ldots,\Theta_\mu$ of degree $1,\ldots,\mu$ respectively, a μth-order linear DO $L : \lambda^{1-\mu} \to \lambda^{1+\mu}$ given by

$$L = L_\mu + M^1_{\Theta_1} + \ldots + M^\mu_{\Theta_\mu}.$$

The point is that this process can be reversed. That is, essentially all μth-order linear DOs arise in this way. In fact, thereby one recaptures the classical *Laguerre-Forsyth invariants* (see the remarkable book [60], long fallen into oblivion).

In [7] a different approach to invariants of higher-order linear DOs is indicated. Suppose the DO is already in the normal form (2) of section 1. Subtract from it the $(\mu - 2)$th-order linear DO

$$\sqrt{(a_{\mu-2}(z))}\left(\frac{d}{dz}\right)^{\mu-2} \sqrt{(a_{\mu-2}(z))}$$

and continue by induction. From Bol's lemma (see Introduction) it is clear that this is an invariant (coordinate-independent) procedure. The drawback is of course the ambiguity in the definition of the square roots involved. Even worse, near points where a coefficient vanishes, a branch point is introduced. Nevertheless, it might be worth while to make a closer comparison of the invariants arising in this way with the Laguerre-Forsyth invariants.

A further noteworthy point in [25] is an integral formula for the Bol operator. As before, let Ω be a Riemann surface with a projective structure and let $\mathcal{O}$ be an open set on Ω bounded by an analytic curve $\partial\mathcal{O}$. Assume that we have on $\mathcal{O}$ a complete Hermitian metric of constant curvature, say, (in local coordinates) $ds = |dz|/\omega(z)$. It is easy to see that

$$q(z) = -\frac{\partial^2\omega(z)/\partial z^2}{\omega(z)}$$

transforms as the coefficient connected with a projective structure. We assume that this projective structure on $\mathcal{O}$ agrees with the one induced from the given projective structure on Ω. Then one can show that (if $\mu \geq 1$)

$$\int_{\mathcal{O}} L_\mu F\bar{g}\omega^{\mu-1}\, dz d\bar{z} = \text{const} \cdot \int_{\partial\mathcal{O}} F\bar{g}(dz)^{(1-\mu)/2}(d\bar{z})^{(1+\mu)/2}.$$

(The constant depends on μ only.) Obviously, this reduces to the ordinary Green's formula if $\mu = 1$. In [26] it is used to prove the theorem that the reproducing kernel in weighted Bergman space $A^{\alpha,2}(\Omega)$ (α integer ≥ 0) over a multiply connected plane domain $\mathcal{O}$ in $\mathbb{C}$ admits a meromorphic continuation to the Schottky double Ω of $\mathcal{O}$, and this result again is used to study Hankel forms over the said space (see the discussion in section 4).

3. DETERMINATION OF ALL MÖBIUS INVARIANT OPERATORS

We make now an assault to find all Möbius invariant operators of the type appearing in Examples 1 and 2 in the Introduction. First we apply the Bol operator so the 'integral' $F(z)(dz)^{-(\mu-1)/2}$ gets replaced by a 'differential' $f(z)(dz)^{(\mu+1)/2}$, where $f(z) = F^{(\mu)}(z)$. The problem is therefore to find all 'covariant' operators of the type

$$f(dz)^{(\mu+1)/2} \mapsto \sum_{\substack{k_1 \geq 0, \ldots, k_\lambda \geq 0 \\ k_1 + \ldots + k_\lambda = \lambda}} a_{k_1 \ldots k_\lambda} \frac{f^{(k_1)} \cdots f^{(k_\lambda)}}{f^\lambda} (dz)^\lambda.$$

These operators form a finite dimensional vector space M_λ. One can also multiply two such operators so that one has an operator $M_\lambda \otimes M_{\lambda'} \to M_{\lambda+\lambda'}$. In other words, $\sum^{\oplus}_{\lambda \geq 0} M_\lambda$ is a commutative graded ring. In what follows we will uncover its structure.

It turns out that the problem is essentially independent of μ. If f transforms according to the rule

$$f \mapsto (f \circ \phi) e^{-\mu-1},$$

then the kth derivative transforms as

$$f^{(k)} \mapsto \sum_{j=0}^{k} \binom{k}{j} \frac{(\mu + k)!}{(\mu + j)!} \epsilon^{k-j} (f^{(j)} \circ \phi) e^{-\mu-1-2k},$$

where $\epsilon = -e \cdot c = (1/2)(\phi''/\phi'^2)$ and, as before, $e = cz + d = 1/\sqrt{\phi'}$ (generalization of Bol's lemma - the case $k = \mu$; see [25] or [39] or [52]). If we set

$$\hat{D}^k\hat{f} = f^{(k)} = \frac{f^{(k)}}{(\mu + k)!},$$

this can be written as

$$\hat{D}^k f \mapsto (\hat{D} + \epsilon)^k f\circ\phi\cdot e^{-\mu-1-2k},$$

with $\hat{D}$ and ϵ treated as commuting operators. More generally,

$$P(\hat{D},\ldots,\hat{D})f \mapsto P(\hat{D} + \epsilon,\ldots,\hat{D} + \epsilon)f\circ\phi\cdot e^{-(\mu+1)\lambda-2\lambda}$$

if

$$P(x_1,\ldots,x_\lambda) = \sum \hat{a}_{k_1,\ldots,k_\lambda} x_1^{k_1} \ldots x_\lambda^{k_\lambda}$$

with

$$\hat{a}_{k_1,\ldots,k_\lambda} = a_{k_1,\ldots,k_\lambda} \cdot (k_1 + \mu)!\ldots(k_\lambda + \mu)!.$$

REMARK: This may be viewed as a generalization of the transformation rule for the bracket $\{\ ,\ \}_1$ (cf. [23], [25]), viz.

$$\frac{f'}{f} \mapsto \left(\frac{f'}{f} \circ \phi + \frac{1}{2}(\mu + 1)\frac{\phi''}{(\phi')^2}\right)\phi'.$$

Thus the condition for covariance comes in the form

$$\boxed{P(x_1 + 1,\ldots,x_\lambda + 1) = P(x_1,\ldots,x_\lambda)}$$

or, equivalently, as

$$\boxed{\sum_{i=1}^{\lambda} \frac{\partial P}{\partial x_i} = 0}$$

or again, in terms of coefficients, as

$$\sum \hat{a}_{k_1\ldots k_\lambda} \binom{k_1}{j_1} \cdots \binom{k_\lambda}{j_\lambda} = \begin{cases} \hat{a}_{j_1\ldots j_\lambda} & \text{if } j_1 + \ldots + j_\lambda = \lambda \\ 0 & \text{if } j_1 + \ldots + j_\lambda < \lambda \end{cases} .$$

In particular, we find

$$\dim M_\lambda = p(\lambda) - p(\lambda - 1),$$

where $p(\lambda)$ is the number of partitions of λ. This gives the table

λ	$p(\lambda)$	$\dim M_\lambda$
1	1	0
2	2	1
3	3	1
4	5	2
5	7	2
6	11	4
7	15	4
8	22	7
9	30	8
10	42	12

which, in particular, explains why the operators in the Introduction (examples 1 and 2) coincide if $\mu = 3, 4$ but not in general.

The calculations are facilitated by the remark that the polynomial P can be taken to be symmetric. Notice also that $P(1,1,\ldots,1) = 0$. This can be checked at the hand of the examples below.

EXAMPLE 1: $\lambda = 2$. Write

$$P(x,y) = A(x^2 + y^2) + 2Bxy.$$

Then

$$\frac{\partial P}{\partial x} = 2Ax + 2By, \quad \frac{\partial P}{\partial y} = 2Ay + 2Bx,$$

so that

$$\frac{\partial P}{\partial x} + \frac{\partial P}{\partial y} = 2(A + B)(x + y)$$

This gives $A = -B$ and $P = A(x - y)^2$. We thus obtain the DO

$$\frac{\hat{f}''}{\hat{f}} - \frac{\hat{f}'^2}{\hat{f}^2}$$

or

$$\frac{\mu!}{(\mu + 2)!} \frac{f''}{f} - \frac{\mu!^2}{(\mu + 1)!^2} \frac{f'^2}{f^2} .$$

$\mu = 1$ corresponds to the Schwarzian:

$$\frac{f''}{f} - \frac{3}{2} (\frac{f'}{f})^2.$$

<u>EXAMPLE 2</u>: $\lambda = 3$. Now

$$P(x,y,z) = 2A(x^3+y^3+z^3) + B(x^2y+x^2z+y^2x+y^2z+z^2x+z^2y) + 6Cxyz$$

and

$$\frac{\partial P}{\partial x} = 6Ax^2 + 2Bx(y+z) + B(y^2 + z^2) + 6Cyz \text{ etc.}$$

Thus

$$\frac{\partial P}{\partial x} + \frac{\partial P}{\partial y} + \frac{\partial P}{\partial z} = (6A+2B)(x^2+y^2+z^2) + 2(2B+3C)(xy + xz + yz),$$

yielding $B = -3A$, $C = -(2/3)B = 2A$. The corresponding DO is

$$\frac{\hat{f}'''}{\hat{f}} - 3\frac{\hat{f}''\hat{f}'}{\hat{f}^2} + 2\frac{(\hat{f}')^3}{\hat{f}^3}$$

or

$$\frac{\mu!}{(\mu+3)!}\,\frac{f'''}{f} - 3\,\frac{(\mu)!^2}{(\mu+2)!(\mu+1)!}\,\frac{f''f'}{f^2} + 2\,\frac{(\mu)!^3}{((\mu+1)!)^3}\,\frac{f'^3}{f^3}\,.$$

$\mu = 2$ gives the expression we found earlier (see Subexample 3, Introduction):

$$\frac{1}{3\cdot 4\cdot 5}\,\frac{f'''}{f} - 3\,\frac{1}{3\cdot 4\cdot 5}\qquad \frac{f''f'}{f^2} + 2\,\frac{1}{3\cdot 3\cdot 3}\,\frac{f'^3}{f^3}\,.$$

In these two cases, therefore, these operators must coincide with Schiffer's 'logarithmic' operator M (see Introduction).

<u>EXAMPLE 3</u>: $\lambda = 4$. The direct eliminations become so complicated that we limit ourselves to give the end result. One finds two independent covariant operators:

$$\delta_1 = \frac{\hat{f}^{IV}}{\hat{f}} - 4\,\frac{\hat{f}'''\hat{f}'}{\hat{f}^2} + 3\,\frac{\hat{f}''^2}{\hat{f}^2}$$

and

$$\delta_2 = \frac{\hat{f}''^2}{\hat{f}^2} - 2\,\frac{\hat{f}''\hat{f}'^2}{\hat{f}^3} + \frac{\hat{f}'^4}{\hat{f}^4}$$

$$= \left(\frac{\hat{f}''}{\hat{f}} - \frac{\hat{f}'^2}{\hat{f}^2}\right)^2 = \text{the square of the operator in the case } \lambda = 2.$$

In this case the operator M is essentially $\delta_1 - 6\delta_2$.

Before continuing this series of examples let us write down the general formula for computing the coefficients. If $P = \sum A_k x^k = \sum A_{k_1,\ldots,k_\lambda} x_1^{k_1}\ldots x_\lambda^{k_\lambda}$ is the polynomial corresponding to a covariant DO holds

$$\sum_i (\ell_i + 1)A_{\ell+e_i} = 0,$$

where $e_i = (0,\ldots,\underset{\text{position } i}{1},\ldots,0)$.

EXAMPLE 4: $\lambda = 5$. The possible partitions of the number 4 correspond to the vectors ℓ = (40000), (31000), (22000), (21100), (11110), yielding the system of equations

$$\begin{aligned} 5A_5 + 4A_{41} &= 0, \\ 4A_{41} + 2A_{32} + 3A_{311} &= 0, \\ 2 \cdot 3A_{32} + 3A_{2111} &= 0, \\ 3A_{311} + 2 \cdot 2A_{221} + 2A_{2111} &= 0, \\ 4 \cdot 2A_{2111} + A_{11111} &= 0, \end{aligned}$$

'superfluous' zeros being omitted ($A_5 = A_{50000}$ etc.). There are five equations and seven unknowns corresponding to two independent solutions, as predicted. Set $A_5 = 24A$, $A_{41} = 6B$, $A_{32} = 6C$, $A_{311} = 4D$, $A_{211} = 4E$, $A_{2111} = 6F$, $A_{11111} = 120G$. Then we can write our system as

$$\begin{aligned} 5A + B &= 0, \\ 2B + C + D &= 0, \\ 3C + E &= 0, \\ 3D + 4E + 3F &= 0, \\ 2F + 5G &= 0, \end{aligned}$$

which gives at once (first, third and fifth equations) $B = -5A$, $E = -3C$, $G = -(2/5)F$. If $A = B = 0$ we can express the remaining coefficients in terms of C: $D = -C$, $E = -3C$, $F = 5C$, $G = -2C$. Thus one gets the covariant DO

$$f''' f'' - f''' f'^2 - 3f''^2 f' + 5f''(f')^3 - 2(f')^5 \equiv (f''-f'^2)(f''' -3f''f'+2f'^3);$$

here and in the next formula we *omit* the ^ in the notation for the derivative. If $F = G = 0$ one expresses instead the coefficients in terms of A: $B = -5A$, $C = 2A$, $D = 8A$, $E = -6A$. The operator now reads:

$$f^V f^4 - 5f^{IV} f' f^3 + 2f''' f'' f^3 + 8f''' f'^2 f^2 - 6f''^2 f' f^2.$$

In the same way as above one can also treat multilinear expressions of the type

$$\sum a_{k_1 \ldots k_\lambda} f_1^{(k_\lambda)} \ldots f_\lambda^{(k_\lambda)}$$

yielding an analogous result (P need not any longer be symmetric).

EXAMPLE 5: $\lambda = 3$. Write

$$P = 2A_1x^3 + 2A_2y^3 + 2A_3z^3 + B_{12}x^2y + B_{13}x^2z + B_{21}y^2x + B_{23}y^2z$$
$$+ B_{31}z^2x + B_{32}z^2y + 6Cxyz.$$

The conditions for covariance are:

$$6A_1 + B_{12} + B_{13} = 0,$$
$$6A_2 + B_{21} + B_{23} = 0,$$
$$6A_3 + B_{31} + B_{32} = 0,$$
$$2B_{12} + 2B_{21} + 6C = 0,$$
$$2B_{31} + 2B_{13} + 6C = 0,$$
$$2B_{23} + 2B_{32} + 6C = 0.$$

If $C = 0$ we get $B_{21} = -B_{12}$, etc., whence $A_1 = -(1/6)(B_{12} - B_{21})$, etc. Thus

$$f = B_{21}[(1/3)(x^3 - y^3) - xy(x - y)] + \ldots = (1/3)B_{21}(x - y)^3 + \ldots,$$

which gives the differential expression

$$B_{21}[(1/3)(f'''gh - fg'''h) - (f''g'h - f'g''h)] + B_{31}[(1/3)(f'''hg - fgh''')$$
$$- (f''gh' - f'gh'')] + B_{32}[(1/3)(fg'''h - fgh''') - (fg''h' - fg'h'')].$$

Let us return to the 'logarithmic' operator M. Let F(z) be the $(\mu + 1)$th integral of f(z) $(F^{(\mu)} = f)$ and let $F \equiv F(z_1,\ldots,z_\mu)$ be the μth divided difference of F(z) (see Introduction). We may write

$$D_1 \ldots D_\lambda \log F = \sum C_{\alpha_1,\ldots,\alpha_r} \frac{D_{\alpha_1}F \ldots D_{\alpha_r}F}{F^r},$$

where the summation is carried over all families $\alpha = \{\alpha_1,\ldots,\alpha_r\}$ of disjoint nonempty subsets α_i of the set $\{1,\ldots,\lambda\}$ with $\alpha_1 \cup \ldots \cup \alpha_r = \{1,\ldots,\lambda\}$ and D_{α_i} stands for the partial derivative with respect to the indices in α_i.
Taking $z_1 = \ldots = z_{\mu+1} = z$ we obtain the covariant DO

$$\sum C_{\alpha_1\ldots\alpha_r} \frac{\hat{f}^{(|\alpha_1|)} \ldots \hat{f}^{(|\alpha_r|)}}{\hat{f}^r}$$

where $|\alpha_i|$ is the number of elements in α_i. The coefficients C can be found recursively as follows:

<u>Case 1</u>. If β is given by $\{1,\ldots,\lambda + 1\} = \beta_0 \cup \ldots \cup \beta_r \equiv \{\lambda\} \cup \alpha_1 \cup \ldots \cup \alpha_r$ then

$$C_\beta = -rC_\alpha.$$

<u>Case 2</u>. If β is given by $\{1,\ldots,\lambda + 1\} = \beta_1 \cup \ldots \cup \beta_r \equiv \alpha_1 \cup \ldots \cup (\alpha_i \cup \{\lambda+1\}) \cup \ldots \cup \alpha_r$ (for some index i) then

$$C_\beta = C_\alpha.$$

The corresponding polynomial P is obtained recursively according to the scheme

$$P(x_1,\ldots,x_\lambda) \mapsto \sum_{i=1}^{\lambda+1} (x_1+\ldots+x_{\lambda+1} - (\lambda+1)x_i)P(x_1,\ldots,\hat{x}_i,\ldots,x_{\lambda+1}).$$

<u>EXAMPLE 6</u>: $P(x,y) = (x - y)^2 \mapsto (x + y - 2z)^2(x - y)^2 + (x + z - 2y)^2(x - z)^2 + (y + z - 2x)(y - z)^2$. Writing $(x + y - 2z)(x - y)^2 = (x^2 - y^2)(x - y) - 2z(x^2 - 2xy + y^2) = x^3 - x^2y - xy^2 + y^3 - 2x^2z + 4xyz - 2zy^2$, etc., form

the sum. One obtains the polynomial

$$2(x^3 + y^3 + z^3) - 3(x^2y + \dots (6 \text{ terms})) + 12xyz,$$

corresponding to the covariant DO

$$\hat{f}'''\hat{f}^2 - 3\hat{f}''\hat{f}'\hat{f} + 2\hat{f}^3$$

found earlier.

It seems more difficult to incorporate the operator of the primitive example 1 of the Introduction, viz.

$$D^\lambda(f^{-(\lambda-1)/(\mu+1)}),$$

in the general picture, as the corresponding polynomials in general are μ-*dependent.*

The derivative of f^α is given by a formula of the type

$$D^\lambda f^\alpha = \sum_{k=0}^{\lambda} [\alpha]_k f^{\alpha-k} Q_k^\lambda f,$$

where $[\alpha]_j = \alpha(\alpha - 1)\dots(\alpha - (j - 1))$ and Q_k^λ is the DO given by the recursion

$$Q_k^\lambda f = f' Q_{k-1}^{\lambda-1} f + (Q_k^{\lambda-1} f)'.$$

<u>EXAMPLE 7</u>: $\lambda = 4$. Taking $\alpha = -2/(\mu+1)$, one finds, after some calculations (and omitting a constant factor),

$$(\mu+3)(\hat{f}^{IV}\hat{f}^3 - 4\hat{f}'''\hat{f}'\hat{f}^2 + 3\hat{f}''^2\hat{f}^2) - 3(5 + 2\mu)(\hat{f}'^4 - 2\hat{f}''\hat{f}'^2\hat{f} + \hat{f}''^2\hat{f}^2),$$

where the second term also may be written as a square $(\hat{f}'^2 - \hat{f}''\hat{f})^2$.

<u>QUESTION</u>: Is there a general formula?

Now we proceed to give an entirely different approach to the problem of finding all Möbius invariant operators of the type considered, which is akin to the procedure in [39].

If f is the coefficient of a (1/2)-form ($\mu = 0$ is sufficient!) write

(near $z = 0$)

$$f(z) = \sum_{i=0}^{\infty} s_i z^i,$$

so that $s_i = \hat{f}^{(i)}(0)$, and introduce

$$V = s_0 \frac{\partial}{\partial s_1} + 2s_1 \frac{\partial}{\partial s_2} + 3s_2 \frac{\partial}{\partial s_3} + \dots ,$$

one of the *Cayley-Aronhold operators* (see [39]), so that $V_{s_0} = 0$, $V_{s_1} = s_0$, $V_{s_2} = 2s_1$, etc. The operators considered by us are automatically invariant for translation and dilation. Thus it suffices to consider transformations of the type

$$z \mapsto \frac{z}{1 + \epsilon z}$$

only. These form, apparently, a group whose infinitesimal generator is up to sign V. Whence the condition

$$\boxed{VP = 0}\ ; \qquad (*)$$

the polynomial P is now viewed as a function of $s_0, s_1, \dots$.

<u>EXAMPLE 8</u>: $\lambda = 2$, $P = as_2s_0 + bs_1^2$. Then $VP = 2a_2s_1s_0 + 2bs_1s_0 = 0$ yielding $a = b$. We obtain the DO $f''f - f'^2$, which of course corresponds to the Schwarzian.

<u>EXAMPLE 9</u>: $\lambda = 3$. $P = as_3s_0^2 + bs_2s_1s_0 + cs_1^3$. We find $VP = 3as_2s_0^2 + 2bs_1^2s_0 + bs_2s_0^2 + 3cs_1^2s_0 = 0$ yielding $3a + b = 0$, $2b + 3c = 0$ or $b = -3a$, $c = -(2/3)b = 2a$, corresponding to the DO $f'''f^2 - 3f''f'f + 2f'^3$.

How does the 'log' operator enter into the new picture? Set $\sigma_k \stackrel{\text{def}}{=} s_ks_1 - s_{k+1}s_0$, so that $V\sigma_k = k\sigma_{k-1}$. Consider the DOs L_k given by

$$L_k = (\text{ad}V)^{k-1}L_1, \quad L_1 = \frac{\partial}{\partial s_1}\left(1 + \frac{1}{1!}\frac{\partial}{\partial s_1} + \frac{1}{2!}\frac{\partial^2}{\partial s_2^2} + \dots\right).$$

Then follows that if P satisfies (*) then also $Q = \sum_{k\geqq 1}(-1)^k(\sigma_k/k!)L_kP$ satisfies (*).

PROOF: We obtain

$$VQ = \sum(-1)^k \left(\frac{V\sigma_k}{k!} L_kP + \frac{\sigma_k}{k!} L_kVP - \frac{\sigma_k}{k!}[L_k, V]P\right)$$

$$= \sum(-1)^k \frac{\sigma_{k-1}}{(k-1)!}L_kP - \sum(-1)^{k+1} \frac{\sigma_k}{k!} L_{k+1}P = 0. \qquad \square$$

Further, we consider briefly the transition between the two models. If $k = (k_1,\ldots,k_\lambda)$ is a partition of the number λ, set

$$m_k = \sum x_{i_1}^{k_1} \ldots x_{i_\lambda}^{k_\lambda} ,$$

where the summation extends over all *different* permutations of $x_1^{k_1} \ldots x_\lambda^{k_\lambda}$; let the number of such permutations be N_k. If

$$U = \sum \frac{\partial}{\partial x_i} ,$$

then apparently

$$Um_k = k_1 \frac{N_k}{N_{k-e_1}} m_{k-e_1} + \ldots,$$

which formula has to be juxtaposed to the relation

$$Vs_k = k_1 s_{k-e_1} + \ldots,$$

where $s_k = s_{k_1} s_{k_2} \ldots s_{k_\lambda}$. If we consider the map

$$T : P = \sum \frac{1}{N_k} a_k m_k \mapsto Q = \sum a_k s_k,$$

we therefore have

$$\boxed{TU = VT} .$$

This readily yields the sought relationship.

Let us also say a few words about the (graded) ring N^r of invariant DOs of given order r. We claim that it is a polynomial ring in s_0, s_0^{-1} and r - 1 more 'unknowns'. For instance, N^2 is generated (apart from the identity operator) by the Schwarz operator $G_2 = G_2(f) = f''f - f'^2$, N^3 by G_2 and $G_3 = G_3(f) = f'''f^2 - 3f''f'f + 2f'^3$, and so forth.

<u>PROOF</u>: Choose, quite generally, $G_2, G_3, G_4, \ldots$ in such a way that

$$G_\nu(f) = f^{(\nu)}f^{\nu-1} + \text{operators of lower order};$$

such operators do exist if $\nu \geqq 2$; see below. Let g_ν be the corresponding polynomial in the variables $s_0, s_1, s_2, \ldots$. A little thinking reveals that one can as well make the substitution $s_0 = 1$; this somewhat facilitates the following computations. If

$$P = \sum_{k=(k_1,\ldots,k_r)} a_k s_1^{k_1} \ldots s_r^{k_r}$$

contains a term with $k_r > 0$ write $s_r = g_r$ + polynomial in $s_1, \ldots, s_{r-1}$. Then P equals an expression which is a polynomial in $g_r, s_1 \ldots, s_{r-1}$. Continue by induction. We see that P can be written in the form

$$P = s_1^m Q_1 + s_1^{m-1} Q_2 + \ldots + Q_m,$$

where $Q_1, Q_2, \ldots, Q_m$ are polynomials in $g_2, \ldots, g_r$ only. Hence

$$0 = VP = ms_1^{m-1} Q_1 + (m - 1)s_1^{m-2} Q_2 + \ldots + 0.$$

It follows that all Q_i vanish except Q_m. Thus P also is a polynomial in $g_2, \ldots, g_r$. It is clear that there are no relations between the latter (and s_0). □

Here is an *explicit* system of generators. Make the 'Ansatz'

$$g_n = s_n + a_1 s_{n-1} s_1 + a_2 s_{n-2} s_1^2 + \ldots + a_{n-2} s_2 s_1^{n-2} + a_{n-1} s_1^n.$$

Then

$$Vg_n = ns_{n-1} + a_1((n-1)s_{n-2}s_1 + 1\cdot s_{n-1}) + a_2((n-2)s_{n-3}s_1^2 + 2s_{n-2}s_1) + \dots$$

$$+ a_{n-2}(2s_1^{n-1} + (n-2)\cdot 2s_2 s_1^{n-3}) + a_{n-1}ns_1^{n-1},$$

which yields the recursions

$$\begin{aligned}
n + a_1 &= 0\\
(n-1)a_1 + 2a_2 &= 0\\
(n-2)a_2 + 3a_3 &= 0\\
(n-3)a_3 + 4a_4 &= 0\\
&\dots\\
3a_{n-3} + (n-2)a_{n-2} &= 0\\
2a_{n-2} + na_{n-1} &= 0
\end{aligned}$$

whence

$$a_k = (-1)^k \binom{n}{k} \ (k < n-1), \quad a_{n-1} = (-1)^{n-1}\left(\binom{n}{n-1} - 1\right).$$

(Notice that this is in agreement with that the sum of the coefficients has to be 0, as $\sum_{k=0}^{n} (-1)^k \binom{n}{k} = 0$.) The corresponding DO is thus

$$G_n = \sum_{k=0}^{n-1} (-1)^k \binom{n}{k} f^{(n-k)}(f')^k f^{n-k-1} + (-1)^n (f')^n.$$

A general invariant DO thus comes as a polynomial

$$\sum_{\substack{k=(k_0,k_2,\dots,k_r)\\ k_0+k_2+\dots+k_r=\text{const}}} C_k f^{k_0} G_2^{k_2} G_3^{k_3} \dots G_p^{k_p}.$$

4. THE TRANSVECTANT AND (GENERALIZED) HANKEL OPERATORS

The classical theory of Hankel operators (or forms), see e.g. [40], appendix 4, is usually formulated for operators (or forms) living on the Hardy space $H^2(\mathbb{T})$ (an analytic function f on the unit disk D is in $H^2(\mathbb{T})$ iff its trace on $\mathbb{T} = \partial D$ is in $L^2(\mathbb{T})$). More precisely, if B is an analytic function in D and if P_- denotes the orthogonal projection in $L^2(\mathbb{T})$ onto the complement $H^2_-(\mathbb{T})$ of $H^2(\mathbb{T})$ in that space, the Hankel operator H_B of symbol B is defined by the formula

$$H_B f = P_-(\bar{B}f). \tag{1}$$

Of vital importance for further developments of the theory and its ramifications is the following covariance property

$$U_\phi H_B U_\phi^{-1} = H_{B\circ\phi} \quad (\phi \in SU(1,1)). \tag{2}$$

Here U stands for the natural (unitary) action of the Möbius group SU(1,1) on $H^2(\mathbb{T})$ and $H^2_-(\mathbb{T})$,

$$U_\phi f(z) = f(\phi z)e(z)^{-1}.$$

The general character of formula (1) suggests many generalizations. In the first place, what comes to one's mind is replacing $H^2(\mathbb{T})$ by a weighted Bergman (or Dzhrbashyan) space $A^{\alpha,2}(D)$ $(\alpha > -1)$: an analytic function f on D is in $A^{\alpha,2}(D)$ iff it is square integrable with respect to the probability measure

$$d\mu_\alpha(z) = (\alpha + 1)(1 - |z|^2)^\alpha dA(z),$$

where $dA(z) = dxdy/\pi$ is the normalized area measure. The action of SU(1,1) is then given by

$$U_\phi f(z) = f(\phi z)e(z)^{-(\alpha+2)},$$

but, due to the usual ambiguity in defining powers of complex numbers, we have only a projective representation (a genuine representation of a suitable covering group). To be on the safe side let us fix our attention on the case α = integer = 0,1,... . Now the symbol too transforms with a weight:

$$B \mapsto (B\circ\phi)e^{\alpha+1},$$

and is unique up to a polynomial of degree $\alpha + 1$. Bol's lemma suggests that we may alternatively, instead of B, take $b = B^{(\alpha+2)}$ as 'symbol'. This also has several advantages. For instance, b rather than B is used in the general theory of Hankel forms (on arbitrary domains, even in higher dimensions) developed in [35].

One can furthermore consider generalizations of higher weight. This was first suggested in [33]. It is convenient to express things in terms of bilinear forms, rather than linear operators. Consider as in section 2 the transvectant J_s, now with $\nu_1 = \nu_2 = \alpha + 2$. Then the (generalized) Hankel form of weight s and symbol b is defined by the integral

$$\Gamma_B(f_1,f_2) = \int_{\mathbb{T}} \bar{B}J_s z^{2\alpha+2}dz/2\pi i.$$

(Equivalently, one could have studied what in the literature are called 'little' Hankel operators.) Notice that the word 'weight' here is used in the sense of Cartan's theory. From the point of view of group representations the Hankel forms of higher weight are of importance, because they provide the decomposition of the 'regular' representation of the group SU(1,1) on the space of Hilbert-Schmidt operators on $A^{\alpha,2}(D)$.

One can also define [8], mimicking the primitive definition (1), so-called 'big' Hankel operators, even in the case of higher weight, operators which map the Hilbert space $A^{\alpha,2}(D)$ into its orthogonal complement $(A^{\alpha,2}(D))^{\perp}$ in $L^2(D,\mu_\alpha)$. More precisely, one considers operators of the form

$$T_B^r f(z) = \int_D \overline{K(z,\zeta)\Delta^r B(z,\zeta)} f(\zeta)d\mu_\alpha(\zeta),$$

where $K(z,\zeta) = (1 - z\bar{\zeta})^{-(\alpha+2)}$ is the reproducing kernel in $A^{\alpha,2}(D)$ and where we have put

$$\Delta^{(r)}B(z,\zeta) = \sum_{j+k=r-1} \binom{-r}{j} \frac{1}{k!}[B^{(k)}(z) + (-1)^{k+1}B^{(k)}(\zeta)](z - \zeta)^{-j}. \quad (3)$$

In particular,

$$\Delta^{(1)}B(z,\zeta) = B(z) - B(\zeta),$$

$$\Delta^{(2)}B(z,\zeta) = B'(z) + B'(\zeta) - 2\frac{B(z) - B(\zeta)}{z - \zeta}$$

$$\cdots$$

For certain reasons (see Appendix 1) we call $\Delta^{(r)}$ the *differential-difference operators of Lagrange*. They are related to Newton's divided differences (see section 0) in a simple way:

$$\Delta^{(r)}B(z,\zeta) = (z - \zeta)^r B(\underbrace{z,\ldots,z}_{r}, \underbrace{\zeta,\ldots,\zeta}_{r}).$$

The properties (boundedness, compactness, membership in Schatten classes, etc.) of the operators T_B^r were studied in [2] for $r = 1$ and in [8] for $r > 1$. If $r = 1$ we have $T_B^1 = B - P_\alpha B = [B,P_\alpha]$, denoting by P_α the orthogonal projection onto $A^{\alpha,2}(D)$ in $L^2(D,\mu_\alpha)$. In this case one can also allow non-analytic symbols [2]. The decomposition of $L^2(D,\mu_\alpha)$ into irreducible subspaces is however not yet fully understood. (The operators T_B^r do not do the whole job, as in the case of 'little' Hankel operators, because they correspond to discrete summands in the decomposition and there must be some continuous ones too.)

REMARK: In [43] a generalization of the above definition of the operators T_B^r in the case of the unit ball in $\mathbb{C}^d$ is proposed. This involves also a corresponding generalization of the transvectant.

Until now we have confined our attention to the case when the underlying space is the unit disk D in the complex plane (except for the above brief allusion to the ball). Now a few words about 'regular' multiply connected planar domains. Hankel forms (or equivalently, little Hankel operators) [26] and big Hankel operators [1] in the case of lowest weight are defined as before. However, if we wish to define the corresponding objects of higher weight, we must first select a projective structure. In the case of big Hankel operators, however, only the projective structure associated with the 'circular model' (section 1) seems to work, due to the 'global' definition of the operators Δ^r in formula (3).

As for higher dimensions, besides the ball, one can probably use similar considerations with any symmetric domain, not only with one of rank 1. As for 'curved' situations, what comes to one's mind are the first place strictly pseudoconvex domains, in some sense 'modeled' on the ball. About the only information known to us in that case are some observations due to Ligocka [38].

5. GENERAL OPERATOR CALCULI AND QUANTIZATION

The general character of the relation (2) in the previous section not only leads, as we have seen, to various generalizations of the classical notion of Hankel operator, but also puts Hankel operator theory with its various offshoots on equal footing with operator calculi (the theory of ΨDO). Let us therefore say a few words about this, in particular, about Unterberger's program of quantization of symmetric spaces (see e.g. [54], [55], [56].

Let us begin by recalling some salient facts about the Weyl calculus, which has origin in Weyl's ideas about quantum mechanics [59].

<u>REMARK</u>: A different approach to 'quantization', also of interest in Hankel (and Toeplitz) theory, was advocated by the late Berezin in a number of publications (see e.g. the survey [4] and further the book [57], lecture 10).

As everybody knows, quantization is something which has to do with the interplay between complex-valued functions on a 'phase space' (the symbols) and operators in a Hilbert space. In Weyl's version the phase space is 'flat', a symplectic vector space, and the Hilbert space is the one where the 'complex wave representation' of the CCR (canonical commutation relations) acts, in other words, Fock space. In the case of two 'degrees of freedom' it is the space $F^{a,2}(\mathbb{C})$ of entire functions on $\mathbb{C}$ which are square integrable with respect to the Gaussian measure

$$d\gamma_a(z) = ae^{-a|z|^2}dA(z),$$

where a is a positive real number whose inverse (*sic*!) has the interpretation of 'Planck's constant'. If g is any given symbol, then its Weyl transform is the operator S_b defined by the formula

$$S_g = \int_{\mathbb{C}} g(\zeta) S_\zeta \, dA(\zeta),$$

where again the operators S_ζ are defined by the formula

$$S_\zeta f(z) = f(2\zeta - z) e^{2az\bar{\zeta}} e^{-2a|\zeta|^2} \quad (f \in F^{a,2}(\mathbb{C})). \tag{1}$$

The symplectic group Sp(2) (in this case isomorphic to SL(2,$\mathbb{R}$)) or rather a double cover of it, the metaplectic group Mp(2), has a natural action on $F^{a,2}(\mathbb{C})$ via unitary operators which we denote by V_ψ ($\psi \in$ Mp(2)), say, and then, in analogy with formula (2) in section 4,

$$V_\psi S_g V_\psi^{-1} = S_{g\circ\psi} \quad (\psi \in \mathrm{Mp}(2)). \tag{2}$$

Unterberger's basic observation is now that the operators S_ζ are associated with the spacial symmetries $z \mapsto 2\zeta - z$ of the underlying manifold $\mathbb{C}$ (reflexions about the point ζ). Therefore exactly the same game can be played with any symmetric space, in particular, with the classical symmetric domains of Cartan.

EXAMPLE: In the case of the unit disk D the spatial symmetries are given by

$$s_\zeta z = \frac{\eta - z}{1 - z\bar{\eta}},$$

where ζ has to coincide with the hyperbolic midpoint of the line segment with endpoints 0 and η. Therefore the analog of formula (1) is

$$S_\zeta f(z) = f(s_\zeta(z))(s_\zeta'(z))^{(\alpha+2)/2}. \tag{1'}$$

The Hilbert space is now of course our friend $A^{\alpha,2}(D)$. (Parenthetically, we remark that in Berezin's interpretation [4] it is the quantity $1/(\alpha+2)$ that plays the role of Planck's constant!)

The point we wish to make here is now that the analogy between formula (2) above and (2) of section 4 forces upon us the view that Hankel operators (or forms) and operator calculi ought to be looked upon from a unified point of view. A difference is of course that in the case of calculi one considers linear maps from the Hilbert space into itself, while in the Hankel case one has maps from one Hilbert space into another, in the case of small Hankel operators a space which can be identified with the *dual* (not the anti-dual) of the given Hilbert space. This is about the same as the distinction between collineations and correlations in classical geometry. Indeed, in quantum theory states may be viewed as points of the associated projective space and observables (usually realized by linear operators) map projective points into projective points, and thus correspond to collineations. So one may ask the question what correlations do for quantum theory. Another difference is that in Hankel theory one deals with irreducible families of operators (under the corresponding group action), not so in the case of calculi. This explains why in the case of calculi one in general expects only 'one-sided' results (S_p criteria, etc.). At any rate, time is still not mature to say if the analogy established really has any deeper implications or not.

6. REPRODUCING AND COREPRODUCING KERNELS

Let Ω be any plane domain and μ a suitable positive measure on it. We denote by $A^2(\Omega,\mu)$ the Hilbert space of all analytic functions in Ω which are square integrable with respect to μ. The orthogonal complement $(A^2(\Omega,\mu))^\perp$ (in $L^2(\Omega,\mu)$) consists of all functions of the form $\bar{D}^*g$ with g 'vanishing' on the boundary. (We write $\bar{D}$ or $\bar{\partial}$ for the Cauchy-Riemann operator $\partial/\partial\bar{z}$, and D or ∂ for $\partial/\partial z$.)

DETERMINATION OF $\bar{D}^*$: Partial integration yields (g is a test function)

$$\int_\Omega \frac{\partial f}{\partial \bar{z}}\, \bar{g}\lambda dA = - \int_\Omega f\, \frac{\partial \bar{g}\lambda}{\partial \bar{z}}\, dA.$$

Now

$$\frac{\partial \bar{g}\lambda}{\partial \bar{z}} = \overline{\frac{\partial}{\partial z}\, g\lambda} = \overline{\frac{\partial g}{\partial z}\lambda + g\,\frac{\partial \lambda}{\partial z}}\ .$$

Hence

$$\boxed{\bar{D}^* = -D - \frac{\partial}{\partial z}\,(\log \lambda)}\ .$$

(Here $\lambda = d\mu/dA$ is the Radon-Nikodym derivative and $dA = dxdy/\pi$.)

EXAMPLE 1: If $\mu = \mu_\alpha$, $\Omega = D$ (weighted Bergman case; see section 4) then

$$\bar{D}^* = -D + \frac{\alpha\bar{z}}{1 - |z|^2}.$$

REMARK: This operator appears also in [25] as a covariant derivative taking $(\alpha/2)$-forms into $((\alpha+2)/2)$-forms.

We write

$$\delta = K + \bar{D}^* J, \tag{1}$$

where δ is *point evaluation* (*not* delta function). This gives

$$f(\zeta) = \int_\Omega \overline{K(z,\bar{\zeta})} f(z) d\mu(z) + \int_\Omega \overline{J(z,\bar{\zeta})} \frac{\partial f(z)}{\partial \bar{z}} d\mu(z),$$

which incidentally solves the $\bar{\partial}$ problem.

REMARK: K is usually called the reproducing kernel. Accordingly, J might be termed the 'coreproducing' kernel. It is of interest also in Hankel theory (see [1]).

DETERMINATION OF J (weighted Bergman case):

$$\left(\frac{\partial}{\partial z} - \alpha\frac{\bar{z}}{1 - z\bar{z}}\right)J = K = (1 - z\bar{\zeta})^{-(\alpha+2)}.$$

'Ansatz':

$$J = \frac{1}{\bar{z}} f(z\bar{z}) \text{ if } \zeta = 0.$$

This gives

$$\frac{1}{\bar{z}} f'(z\bar{z})\bar{z} - \alpha\frac{\bar{z}}{1 - z\bar{z}} \cdot \frac{1}{\bar{z}} f(z\bar{z}) = 1.$$

Thus f = f(t) satisfies the DE

$$f' - \frac{\alpha}{1 - t} f = 1$$

or

$$((1 - t)^{\alpha} f)' = (1 - t)^{\alpha}.$$

Integration:

$$(1 - t)^{\alpha} f = C - \frac{1}{\alpha + 1} (1 - t)^{\alpha+1},$$

$$f(1) = 0 \Rightarrow C = 0,$$

$$\boxed{f = - \frac{1}{\alpha + 1} (1 - t)} \quad .$$

From the transformation properties of J (see below) follows

$$\boxed{J = - \frac{1}{\alpha + 1} \cdot \frac{1 - z\bar{z}}{(1 - z\bar{\zeta})^{\alpha+1} (\bar{z} - \bar{\zeta})}} \quad .$$

<u>REMARK</u>: As $\partial/\partial z(1/\bar{z}) = \delta$ (in our normalization of A!) we see that we have the right polar strength. Moreover, if

$$G = 2 \log \left| \frac{1 - z\bar{\zeta}}{z - \zeta} \right| \quad \text{('rationalized' Green's function)}$$

then

$$\frac{\partial G}{\partial \bar{\zeta}} = - \frac{z}{1 - z\bar{\zeta}} + \frac{1}{\bar{z} - \bar{\zeta}} = \frac{1 - z\bar{z}}{(1 - z\bar{\zeta})(\bar{z} - \bar{\zeta})} .$$

Hence the formula can also be written

$$\boxed{K_{\alpha} = \delta + \frac{1}{\alpha + 1} \bar{\partial}^{*}_{\alpha z} (K_{\alpha-2} \bar{\partial}_{\zeta} G)} .$$

For $\alpha = 0$ this is well known [5]. One reason for carrying out all these calculations has been precisely to detect possible generalizations of this formula.

The elements of $A^{\alpha,2}(D)$ behave as forms of degree ν, where $\nu = (\alpha+2)/2$. Thus, if f is in $A^{\alpha}(D)$ then $\bar{D}f$ transforms as a form of bidegree $(\nu,1)$ and has to be integrated against forms g of degree $\nu - 1$. (In general, the $A^{\alpha,2}$-pairing extends to a pairing $\kappa^{(p,q)} \times \kappa^{(p',q')} \to \mathbb{C}$ where $p+q' = q+p' = \nu$; here $\kappa^{(\ ,\ )}$ stands for the (C^{∞}) sheaf of the bidegree indicated. In our case $p = \nu$, $q = 1$, $p' = \nu - 1$, $q' = 0$.) It follows that J transforms likewise in the variable z and as a form of degree $(0,\nu)$ in ζ:

$$J(z,\zeta) \mapsto J(\phi z,\phi\zeta)e(z)^{-\alpha}\bar{\epsilon}(\zeta)^{-(\alpha+2)}$$

where

$$\phi = \begin{pmatrix} a & b \\ c & d \end{pmatrix}, \quad e(z) = cz + d, \quad \epsilon(\zeta) = c\zeta + d.$$

We check the transformation properties of $\bar{D}^* = -D + \alpha z/(1-z\bar{z})$. Let

$$g(z) \mapsto g(\phi(z))e(z)^{-\alpha}.$$

On the one hand:

$$D(g(\phi(z))e(z)^{-\alpha}) = Dg(\phi(z))e(z)^{-\alpha-2} - \alpha c g(\phi(z))e(z)^{-\alpha-1},$$

and on the other

$$\alpha \frac{\bar{z}}{1 - |z|^2} g(\phi(z))e(z)^{-\alpha}.$$

The difference involves a factor

$$c + \frac{\bar{z}e(z)}{1 - |z|^2} = \frac{c(1-z\bar{z})+\bar{z}(cz+d)}{(1-|\phi z|^2)e(z)\overline{e(z)}} = \frac{c-cz\bar{z}+cz\bar{z}+d\bar{z}}{(1-|\phi(z)|^2)e(z)\overline{e(z)}}$$

$$= \frac{\overline{\phi(z)}}{1 - |\phi(z)|^2}\frac{1}{e(z)},$$

as

$$\phi(z) = \frac{az + b}{e(z)}, \quad \overline{\phi(z)} = \frac{c + d\bar{z}}{\overline{e(z)}} .$$

Thus we find

$$\bar{D}^* g(z) \mapsto -[Dg(\phi(z)) - \alpha \frac{\overline{\phi(z)}}{1 - |\phi z|^2} g(\phi(z))]e(z)^{-\alpha-2}.$$

LIMITING CASE: $\alpha \to \infty$, $R \to \infty$. For a disk D_R of radius R:

$$J = -\frac{R^2}{\alpha + 1} \frac{1 - z\bar{z}/R^2}{(1 - z\bar{\zeta}/R^2)^{\alpha+1}(\bar{z} - \bar{\zeta})} \to -\frac{e^{z\bar{\zeta}}}{\bar{z} - \bar{\zeta}},$$

$$\bar{D}^* = -D + \frac{R^2/\alpha\bar{z}}{1 - |z|^2/R^2} \to -D + \bar{z}.$$

We can check directly:

$$(D - \bar{z}) \frac{e^{z\bar{\zeta}}}{\bar{z} - \bar{\zeta}} = \frac{\bar{\zeta}e^{z\bar{\zeta}}}{\bar{z} - \bar{\zeta}} - \frac{\bar{z}e^{z\bar{\zeta}}}{\bar{z} - \bar{\zeta}} = -e^{z\bar{\zeta}}.$$

THE CASE OF AN ANNULUS $\Omega = \Omega_R = \{z : 1 < |z| < R\}$: It is better to rewrite equation (1)

$$\partial u/\partial z = -\lambda K + \text{usual delta}$$

with $u = \lambda J$; λ is assumed to be radial, $\lambda = \lambda(r^2)$, $r^2 = |z|^2 = z\bar{z}$. Set

$$M_n(r) = \int_1^r r_1^{2n}\lambda(r_1^2)dr_1^2.$$

Then

$$K = \sum_{n=-\infty}^{\infty} \frac{(z\bar{w})^n}{M_n(R)} .$$

We first solve for each n the equation

$$\frac{\partial u_n}{\partial z} = - \frac{(\bar{w}z)^{n-1}}{M_{n-1}(R)} \lambda(r^2).$$

Put

$$u_n = f_n(z\bar{z})z^n$$

so that

$$\frac{\partial u_n}{\partial z} = [f_n'(z\bar{z})z\bar{z} + nf_n(z\bar{z})]z^{n-1}.$$

Hence (writing $t = r^2$)

$$tf_n' + nf_n(t) = - \frac{\bar{w}^{n-1}}{M_{n-1}(R)} \lambda(t)$$

yielding the particular solution ($C = 0$)

$$f_n(t) = -t^{-n} \int_1^t t_1^{n-1} \lambda(t_1) dt_1 \cdot \frac{\bar{w}^{n-1}}{M_{n-1}(R)}$$

or

$$u_n = - \frac{M_{n-1}(r)}{M_{n-1}(R)} \cdot \frac{z^n \bar{w}^{n-1}}{r^{2n}} .$$

On the other hand

$$\partial v/\partial z = \text{delta}$$

is solved by

$$v = \frac{1}{\bar{z} - \bar{w}} = \begin{cases} -\Sigma_{n=0}^{\infty} \dfrac{\bar{z}^n}{\bar{w}^{n+1}} & (1 < |z| < |w|) \\ \Sigma_{n=-\infty}^{-1} \dfrac{\bar{z}^n}{\bar{w}^{n+1}} & (|w| < |z| < R) \end{cases} \qquad (2)$$

and the homogeneous equation

$$\partial h/\partial z = 0$$

is solved by any anti-analytic function

$$h = \sum_{-\infty}^{\infty} a_n \bar{z}^n. \tag{3}$$

Summing ((1) + (2) + (3)) we get

$$u = \sum_{n=-\infty}^{\infty} u_n + v + h$$

where we have to adjust the coefficients a_n so as to meet the boundary condition. Thus, finally, we get

$$u = \begin{cases} -\sum_{-\infty}^{\infty} \dfrac{M_{n-1}(r)}{M_{n-1}(R)} \cdot \dfrac{z^n \bar{w}^{n-1}}{r^{2n}} & (1 < |z| < |w|) \\ \sum_{n=-\infty}^{\infty} \left(1 - \dfrac{M_{n-1}(r)}{M_{n-1}(R)}\right) \cdot \dfrac{z^n \bar{w}^{n-1}}{r^{2n}} & (|w| < |z| < R) \end{cases}.$$

REMARK: This technique of finding a fundamental solution via a series development has a general character and can be applied in many other situations.

EXAMPLE 2: $\lambda \equiv 1$. Then

$$J = \begin{cases} -\sum_{n=-\infty}^{\infty} \dfrac{1 - r^{-2n}}{R^{2n} - 1} z^n \bar{w}^{n-1} & (|z| < |w|) \\ \sum_{n=-\infty}^{\infty} \dfrac{(R/r)^{2n} - 1}{R^{2n} - 1} z^n \bar{w}^{n-1} & (|z| > |w|) \end{cases}.$$

It is easy to express these series in terms of theta functions; see the formula for Green's function given in [14], S. 335-357.

SEVERAL VARIABLES: Take $\Omega = \mathcal{B}$ = unit ball in $\mathbb{C}^n$ (equipped with its usual Hermitian metric $\|\cdot\|$, the corresponding inner product being $(\cdot,\cdot)$) and $\lambda = (\alpha + 1)(1 - \|z\|^2)^\alpha$. We have to solve the equation

$$\delta = K + \sum_{k=1}^{n} \left(-\frac{\partial J^k}{\partial z^k} + \alpha \frac{\bar{z}^k}{1 - \|z\|^2} J^k\right)$$

corresponding to the integral formula

$$f(\zeta) = \int_{\mathcal{B}} \overline{K(z,\bar{\zeta})} f(z) d\mu(z) + \int_{\mathcal{B}} \sum_{k=1}^{n} \overline{J^k(z,\bar{\zeta})} \frac{\partial f(z)}{\partial z^k} dV(z);$$

dV is Euclidean volume measure conveniently normalized.

'Ansatz':

$J^k = z^k f(\|z\|^2)$ for $\zeta = 0$.

Then

$$\frac{\partial J^k}{\partial z^k} = f(\|z\|^2) + z^k \bar{z}^k f'(\|z\|^2),$$

$$\sum \frac{\partial J^k}{\partial z^k} = nf(t) + tf'(t) \quad (t = \|z\|^2),$$

$$\sum \alpha \frac{\bar{z}^k}{1 - \|z\|^2} J^k = \alpha \frac{tf(t)}{1 - t},$$

which gives the DE

$$tf'(t) + (n - \frac{\alpha t}{1-t}) f(t) = 1$$

with the integrating factor $t^n(1 - t)^\alpha$. Thus we find

$$f(t) = t^{-n}(1 - t)^{-\alpha} \int_0^t t_1^{n-1}(1 - t_1)^\alpha dt_1. \tag{4}$$

In particular

$$f(t) = \frac{1}{n}(1 - t^{-n}) \text{ if } \alpha = 0.$$

Invariance properties of K and J:

$$K(z,\zeta) = K(\phi(z),\overline{\phi(\zeta)})(\det\phi'(z))^{1+\alpha/(n+1)}(\overline{\det\phi'(\zeta)})^{1+\alpha/(n+1)},$$

$$\sum_{k=1}^{n} J^k(z,\zeta)\frac{\partial \phi^\ell}{\partial z^k} = J^\ell(\phi(z),\overline{\phi(\zeta)})(\det\phi'(z))^{1+\alpha/(n+1)}(\overline{\det\phi'(\zeta)})^{1+\alpha/(n+1)}.$$

Here ϕ is any element of the group of biholomorphic automorphisms of $\mathcal{B}$ (known to be isomorphic to the group PSU(n,1)).

We apply this to the fundamental symmetry ϕ interchanging 0 and ζ, i.e. $\phi(0) = \zeta$, $\phi(\zeta) = 0$, $\phi^2 = \text{id}$. We omit the calculations, which are very similar to the ones in [43], and write down only the end result:

$$\boxed{J(z,\zeta) = f(\rho(z,\zeta))\frac{z - \zeta}{(1 - \|\zeta\|^2)(1 - (z,\zeta))^{\alpha+n}}}\ .$$

Here the function f is as in (4) and the invariant distance between the points z and ζ is given by

$$\rho(z,\zeta) = 1 - \frac{(1 - \|z\|^2)(1 - \|\zeta\|^2)}{|1 - (z,\zeta)|^2}\ .$$

APPENDIX 1. LAGRANGE'S PROOF OF THE ADDITION THEOREM FOR ELLIPTIC FUNCTIONS

In the appendices we take up various issues more or less loosely connected with what we have discussed in the main body of the paper. We begin with Lagrange's beautiful proof of the addition theorem for elliptic functions (due to Euler). It was subsequently superseded by other, more powerful proofs, for instance the one based on Abel's theorem (see e.g. [58], S. 27-32), so but for a brief mention in Houzel's masterly survey of the classical theory of elliptic and Abelian functions ([30], p. 9) it seems to be completely forgotten nowadays (Houzel writes: 'ce qui provoqua l'admiration d'Euler').

Consider an elliptic curve given by the equation

$$\xi^2 = P(x) \equiv Ex^4 + Dx^3 + Cx^2 + Bx + A.$$

(Such an equation is invariant under the transformations

$$x' = \frac{ax + b}{cx + d}\ , \quad \xi' = \frac{\xi}{(cx + d)^2}\ .$$

Therefore it is natural to consider the curve to lie on the projective completion of the bundle κ^{-1}, where κ is the canonical sheaf over $\mathbf{P}^1$, which is known to be a rational ruled surface (see [21], pp. 514-520).)

Consider the differentials

$$\frac{dx}{X} = dt,\ \frac{dy}{Y} = ds,\ \frac{dz}{Z} = dr,$$

where

$$t + s + r = 0 \quad (\Rightarrow dt + ds + dr = 0)$$

and where we have put

$$X = \sqrt{P(x)},\quad Y = \sqrt{P(y)},\quad Z = \sqrt{P(z)}.$$

Then

$$dX = \frac{P'(x)}{2\sqrt{P(x)}} \cdot \sqrt{P(x)}\ dt = \frac{1}{2} P'(x)dt \quad \text{etc.}$$

and we can write

$$d\left(\frac{X - Y}{x - y}\right)^2 = 2\frac{X - Y}{x - y} \cdot \left(\frac{dX - dY}{x - y} - \frac{(X - Y)(dx - dy)}{(x - y)^2}\right) \tag{1}$$

$$= \frac{(X - Y)(P'(x)dt - P'(y)ds)}{(x - y)^2} - 2\frac{(X - Y)^2(Xdt - Yds)}{(x - y)^3}.$$

On the other hand, as P is a polynomial of degree at most 4, we have

$$\frac{P'(x) + P'(y) - 2[P(x)-P(y)]/(x-y)}{(x - y)^2} = 2E(x + y) + D.$$

We can write this as

$$d(E(x+y)^2 + D(x+y)) = \frac{P'(x)+P'(y)-2[(X-Y)(X+Y)]/(x-y)}{(x - y)^2}(Xdt + Yds). \tag{2}$$

Subtracting (2) from (1) we find

$$d\left[\left(\frac{X - Y}{x - y}\right)^2 - E(x + y)^2 - D(x + y)\right]$$

$$= -\frac{YP'(x) + XP'(y)}{(x - y)^2} - 4\frac{(X - Y)XY}{(x - y)^3}\,(dt + ds) \qquad (3)$$

$$= 2\,\frac{dX/dx + dY/dy - 2(X-Y)/(x-y)}{(x - y)^2}\,XYdr.$$

Thus, the Lagrange differential-difference operator appears in two different ways. Formula (3) is a way of expressing the addition theorem for elliptic functions.

In particular, taking r = const it follows that

$$\frac{dx}{X} + \frac{dy}{Y} = 0$$

has a solution y which is an algebraic function of x. Indeed, in integrated form (3) gives

$$\left(\frac{X - Y}{x - y}\right)^2 = E(x + y)^2 + D(x + y) + G. \qquad (3')$$

This is how the addition theorem for elliptic functions was formulated by Euler (see [30]).

Consider the special case when there is no x^4 term (E = 0), that is, one of the four roots of the polynomial P sits at infinity. Then we have virtually the elliptic curve in Weierstrass's normal form. In this case one further takes D = 4, C = 0, thus the curve has the equation

$$\xi^3 = 4x^3 + Bx + A.$$

Analysing the behavior of the curve at ∞ one readily sees that G = -4z. Thus (3') takes the symmetric form

$$x + y + z = k^2,$$

where k denotes the slope of the three collinear points (x,ξ) etc. This is how the addition theorem is usually stated, along with its geometric interpretation (see again [58], loc. cit. or [21], p. 227).

APPENDIX 2. JACOBI'S DE

In this appendix we reproduce the essentials of the proof of Jacobi's theorem [32] in Example 3 of the Introduction to the effect that the '*Thetanullwerthe*' (theta constants) satisfy a third-order *algebraic* DE. This is because it is something which, apparently, is little known nowadays (in this context it is perhaps amusing to have a peek at Rubel's paper [49]) and, on the other hand, definitely belongs to our subject. After some thinking it is not as formidable as it looks at first sight - we should bear in mind that Jacobi was a master of DEs, both ODEs and PDEs, at a time when the theory of DEs still was finding explicit solutions.

The proof goes via the theory of complete elliptic integrals and their representation in terms of the '*Thetanullwerthe*'. Recall that the complete elliptic integral of the first kind is defined as

$$K(k) = \int_0^{2\pi} \frac{d\phi}{\sqrt{(1 - k^2 \sin^2\phi)}} ;$$

the number k is known as the modulus. In terms of theta values one has (see e.g. [10], th. 2.1, p. 35)

$$K(k) = \frac{2}{\pi}\,\theta_{00}^2(0,t) \quad \text{where} \quad k' = \theta_{01}^2(0,t)/\theta_{10}^2(0,t).$$

Here and in the sequel $k' = \sqrt{(1 - k^2)}$ is the complementary modulus. The basis of the method is now the fact that K satisfies a second-order linear DE (Legendre's or the Fuchs-Picard equation: see e.g. [12], pp. 58-62) which may be written as

$$\frac{d(k^2k'^2\, dK/dk^2)}{dk^2} = \frac{1}{4}K.$$

or with

$$\frac{dk^2}{k^2k'^2} = d\log\frac{k^2}{k'^2} = dz, \qquad (1)$$

again as

$$\frac{d^2K}{dz^2} = \frac{1}{4} k^2k'^2K. \tag{2}$$

Because of the symmetry it follows that a second solution is $K'(k) = K(k')$.

The first step of the proof is quite general, so let us for a while consider the general equation

$$\frac{d^2F}{dz^2} + q(z)F = 0. \tag{3}$$

We know that a local 'uniformizing' parameter can be obtained by putting

$$t = \frac{F_1}{F}$$

where F and F_1 are any two linearly independent solutions of (3). Also, in view of the special form of equation (3) (no 'middle' term with the first derivative!), the Wronskian is constant:

$$[F,F_1] = FF_1' - F'F_1 = \alpha. \tag{4}$$

(In Jacobi's case, viz. equation (2), one takes $F = (2/\pi)K$, $F_1 = -2K'$, with z and k related by (1), and, considering the expansion of these functions for small values of k, one finds $\alpha = 1$; this was done by Euler [18].) Write then (4) as

$$\left(\frac{F_1}{F}\right)' = \frac{\alpha}{F^2},$$

that is,

$$dt = \frac{\alpha}{F^2} dz \tag{5}$$

or, with

$$C = \frac{1}{F},$$

again as

$$dt = \alpha C^2 dz. \tag{5'}$$

It follows that

$$\frac{dF}{dt} = \frac{1}{\alpha} F^2 \frac{dF}{dz}$$

or

$$\frac{dC}{dt} = -\frac{1}{\alpha}\frac{dF}{dz} .$$

Continuing the differentiation gives

$$\frac{d^2C}{dt^2} = -\frac{1}{\alpha^2} F^2 \frac{d^2F}{dz^2} = \frac{1}{\alpha^2} F^3 q = \frac{1}{\alpha^2} C^{-3} q$$

or

$$\boxed{C^3 C'' = \alpha^{-2} q} . \tag{6}$$

In the case of equation (2) (in which case $q = -(1/4)k^2k'^2$, $\alpha = 1$) this gives

$$C^3 C'' = -\frac{1}{4} k^2 k^2. \tag{7}$$

To proceed further we must invoke the inverse function, say, Q to q. Differentiating the relation $z = Q(q)$ and using (5') and (6) one finally finds

$$\boxed{\alpha C^2 \frac{dQ(C^3C''/\alpha^2)}{dt} = 1} . \tag{8}$$

It remains to determine Q and Q' in the case of equation (2). As $k^4 - k^2 - 4q = 0$ we find

$$z = \log \frac{k^2}{k'^2} = \log \frac{1 + \sqrt{(1 + 16q)}}{1 - \sqrt{(1 + 16q)}}$$

yielding

$$dz = \left(\frac{1}{(1 + \sqrt{(1 + 16q)}} + \frac{1}{1 - \sqrt{(1 + 16q)}}\right) \frac{16dq}{2\sqrt{(1 + 16q)}} = \frac{d \log q}{\sqrt{(1 + 16q)}} .$$

Thus (6) (or (7)) gives

$$C^2 \frac{d}{dt} \log(C^3C'') = \sqrt{(1 + 16C^3C'')}$$

as in example 1 of the Introduction. (This for the principal theta $\theta = \theta_{00}$; the calculations for the remaining thetas are similar.) If we, following Jacobi, put $C = y^{-2}$ one gets after some 'simplifications'

$$(y^2y''' - 15yy'y'' + 30y'^3)^2 + 32(yy'' - 3y'^2)^3 = y^{10}(yy'' - 3y'^2)^2,$$

which does not seem to be very illuminating. This DE is thus satisfied by theta series. The rest of Jacobi's proof is devoted to exhibiting the general solution, but we have not examined this part in detail. Perhaps there is a general principle saying that the solutions of a Möbius third-order DE can be generated from a single particular solution? At least a naive count of parameters supports such a belief.

We conjecture that theta series in two variables can be treated in a similar fashion. This is also suggested by the parallel between the arithmetic-geometric means of Gauss and Borchardt (see [44]).

APPENDIX 3. A TRANSFORMATION THEORY FOR THE HEAT EQUATION

We know that the second-order linear DO has an interesting transformation theory, connected with names such as Jacobi, Kummer, Riemann, Schwarz, etc. (cf. section 1). It is perhaps less well known that the heat equation is susceptible to a similar treatment, to which we now turn.

In the equation

$$\frac{\partial u}{\partial a} = \frac{1}{2} \frac{\partial^2 u}{\partial c^2} + qu \quad (q = q(a,c))$$

let us make the substitution

$$v = u(\phi,\psi)m \quad (\text{where } \phi = \phi(a),\ \psi = \psi(a,c),\ m = m(a,c)).$$

(The perhaps strange looking choice of the letters a and c for the independent variables is in accordance with [45].) When does it go over into an equation of the same type? Derivation yields

$$\frac{\partial v}{\partial a} - \frac{1}{2}\frac{\partial^2 v}{\partial c^2} - \tilde{q}v$$

$$= \frac{\partial u}{\partial a}\phi' m - \frac{1}{2}\frac{\partial^2 u}{\partial c^2}\left(\frac{\partial \psi}{\partial c}\right)^2 m + \frac{\partial u}{\partial c}\left(\frac{\partial \psi}{\partial a}m - \frac{1}{2}\frac{\partial^2 \psi}{\partial c^2}m - \frac{\partial \psi}{\partial c}\frac{\partial m}{\partial c}\right) + u\left(\frac{\partial m}{\partial a} - \frac{1}{2}\frac{\partial^2 m}{\partial c^2} - \tilde{q}m\right).$$

Thus one gets the following conditions:

$$\phi' = \left(\frac{\partial \psi}{\partial c}\right)^2, \tag{1}$$

yielding

$$\boxed{\psi = \sqrt{(\phi')}c + r} \qquad (r = r(a)).$$

$$\frac{\partial \psi}{\partial a}m - \frac{1}{2}\frac{\partial^2 \psi}{\partial c^2}m - \frac{\partial \psi}{\partial c}\frac{\partial m}{\partial c} = 0, \tag{2}$$

which, if we take into account that

$$\frac{\partial \psi}{\partial a} = \frac{1}{2}\frac{\phi''}{\sqrt{\phi'}}c + r', \quad \frac{\partial^2 \psi}{\partial c^2} = 0, \quad \frac{\partial \psi}{\partial c} = \sqrt{\phi'},$$

gives

$$\boxed{m = N\exp\left(\frac{1}{4}\frac{\phi''}{\phi'}c^2 + \frac{r'}{\sqrt{\phi'}}c\right)} \qquad (N = N(a).$$

Our transformation is thus determined by the data ϕ, r, N. Moreover, one obtains the following relation between q and $\tilde{q}$:

$$\frac{\partial m}{\partial a} - \frac{1}{2}\frac{\partial^2 m}{\partial c^2} - \tilde{q}m = -q\phi' m. \tag{3}$$

Thus holds:

$$\frac{\partial v}{\partial a} - \frac{1}{2}\frac{\partial^2 v}{\partial c^2} - \tilde{q}v \equiv \left(\frac{\partial u}{\partial a} - \frac{1}{2}\frac{\partial^2 u}{\partial c^2} - qu\right)\phi' m.$$

Inserting the two boxed formulae into (3) gives

$$\frac{1}{4}\left(\frac{\phi''}{\phi'}\right)' c^2 N + \left(\frac{r'}{\sqrt{\phi'}}\right)' cN + N'$$

$$= \frac{1}{2}\left(\frac{1}{2}\frac{\phi''}{\phi'}c + \frac{r'}{\sqrt{\phi'}}\right)^2 N + \frac{1}{2}\cdot\frac{1}{2}\frac{\phi''}{\phi'} N + \tilde{q}N - q\phi' N.$$

Write:

$$\tilde{q} - q\phi' = \frac{1}{2}Pc^2 + Qc + R \quad (P = P(a) \text{ etc.}).$$

For the c^2 coefficient:

or

$$\frac{1}{4}\left(\frac{\phi''}{\phi'}\right)' - \frac{1}{8}\left(\frac{\phi''}{\phi'}\right)^2 = \frac{1}{2}P$$

$$\boxed{\frac{1}{2}\{\phi,a\} = P}\ .$$

For the c-coefficient:

$$\left(\frac{r''}{\sqrt{\phi'}}\right)' - \frac{1}{2}\frac{\phi''}{\phi'}\frac{r'}{\sqrt{\phi'}} = Q$$

or

$$\boxed{\frac{r''}{\sqrt{\phi'}} - \frac{r'\phi''}{(\sqrt{\phi'})^3} = Q}\ .$$

It will be convenient to set

$$r = s\sqrt{\phi'},\quad r' = s\frac{\phi''}{2\sqrt{\phi'}} + s'\sqrt{\phi'},$$

$$r'' = s\left(\frac{\phi'''}{2\sqrt{\phi'}} - \frac{(\phi'')^2}{4(\sqrt{\phi'})^3}\right) + 2s'\frac{\phi''}{2\sqrt{\phi'}} + s''\sqrt{\phi'},$$

yielding

$$s\left(\frac{\phi'''}{2\phi'} - \frac{(\phi'')^2}{4(\phi')^2}\right) + s'\frac{\phi''}{\phi'} + s'' - s\frac{(\phi'')^2}{2(\phi')^2} - s'\frac{\phi''}{\phi'} = Q$$

or

$$s \cdot \frac{1}{2}\{\phi,a\} + s'' = Q$$

or again, taking account of the expression for P,

$$\boxed{s'' + Ps = Q}\ .$$

For the c^0 coefficient:

$$N' = \left(\frac{1}{2}\frac{(r')^2}{\phi'} + \frac{1}{4}\frac{\phi''}{\phi'} + R\right)N$$

or in terms of s

$$\boxed{N' = \left[\frac{1}{8}s^2\left(\frac{\phi''}{\phi'}\right)^2 + \frac{1}{2}\frac{\phi''}{\phi'}ss' + \frac{1}{2}(s')^2 + \frac{1}{4}\frac{\phi''}{\phi'} + R\right]N}\ .$$

From the last boxed formulae it follows that P, Q, R essentially determine ϕ, r, N. But the formula for N can be put in a more clearly visible? (convincing?) form. Write

$$\psi = \sqrt{(\phi')}(c + s)$$

with as before $r = \sqrt{(\phi')}s$. We thus may view our transformation as a composition, a 'translation' followed by a 'dilation'. In an analogous way we can reform the expression for m:

$$m = N\exp\left(\frac{1}{4}\frac{\phi''}{\phi'}(c+s)^2 + s'c - \frac{1}{4}\frac{\phi''}{\phi'}s^2\right).$$

It is natural to absorb the c-independent term $-(1/4)(\phi''/\phi')s^2$ in the factor N. Thus let us write

$$N = N^*\exp\left(\frac{1}{4}\frac{\phi''}{\phi'}s^2\right)$$

so that

$$\boxed{m = N^*\exp\left(\frac{1}{4}\frac{\phi''}{\phi'}(c+s)^2 + s'c\right)}\quad .$$

The DE for N* now becomes

$$\frac{N^{*\prime}}{N^*} + \frac{1}{4}\left(\frac{\phi''}{\phi'}\right)' s^2 + \frac{1}{2}\frac{\phi''}{\phi'}ss' = \frac{1}{8}\left(\frac{\phi''}{\phi'}\right)^2 s^2 + \frac{1}{2}\frac{\phi''}{\phi'}ss' + \frac{1}{2}(s')^2 + \frac{1}{4}\frac{\phi''}{\phi'} + R$$

or

$$\frac{N^{*\prime}}{N^*} = -\frac{1}{4}\{\phi,a\}s^2 + \frac{1}{2}s'^2 + \frac{1}{4}\frac{\phi''}{\phi'} + R = \frac{1}{4}\frac{\phi''}{\phi'} + L + R,$$

where $L = (1/2)[(s')^2 - Ps^2]$ may be viewed as a *Lagrangian*. Integration gives

$$\boxed{N^* = (\phi')^{1/4}\exp\left(\int L\,da + \int R\,da\right)}\quad ,$$

where $\int L\,da$ again may be interpreted as an *action*.

In the examples below we take $q = 0$, $Q = R = 0$.

<u>EXAMPLE 1</u>: $P = 0$. $\partial v/\partial a = (1/2)\partial^2 v/\partial c^2$ (heat equation). Take

$$\phi(a) = \frac{1}{a}\quad \left(\text{implying } \phi' = -\frac{1}{a^2},\ \phi'' = \frac{2}{a^3},\ \frac{\phi''}{\phi'} = -\frac{2}{a}\right)$$

and

$$s(a) = -d = \text{constant} \quad (\text{implying } s' = 0,\ L = 0).$$

Taking u = 1 gives then (the fundamental solution of the heat equation)

$$\boxed{v = \frac{1}{a^{\frac{1}{2}}} \exp\left(-\frac{(c-d)^2}{2a}\right)} \quad .$$

EXAMPLE 2 (generalization of example 1): $P = -1$. $\partial v/\partial a = (1/2)\partial^2 v/\partial c^2 - v$ (Gibb's equation for the harmonic oscillator). Take

$$\phi(a) = \coth a \quad (\text{implying } \phi' = -\frac{1}{\sinh^2 a},\ \phi'' = \frac{2\cosh a}{\sinh^3 a},\ \frac{\phi''}{\phi'} = -2\coth a)$$

and

$$s(a) = -d \cosh a$$

implying

$$s' = -d \sinh a,\quad L = \tfrac{1}{2}d^2((\cosh a)^2 + (\sinh a)^2) = \frac{d^2}{4}(e^{2a} + e^{-2a}),$$

$$\int L da = \frac{d^2}{8}(e^{2a} - e^{-2a}) = \frac{d^2}{4}\sinh 2a = \frac{1}{2} d^2 \cosh a \sinh a \ .$$

We find the fundamental solution (Mehler's formula)

$$\boxed{v = \frac{1}{(\sinh a)^{\frac{1}{2}}} \exp\left(-\frac{\cosh a(c^2 + d^2) - 2cd}{2\sinh a}\right)} \quad .$$

QUESTION. Are there any other interesting examples?

APPENDIX 4. WAVE PACKETS VERSUS GAUSS-WEIERSTRASS FUNCTIONS[4]

In this appendix we will compare wave packets in the sense of Cordoba and Fefferman [13] with the Gauss-Weierstrass functions (see [45], [34]). For the sake of simplicity we shall confine our attention to the one-dimensional case. Alternative names for the same objects are: Gabor wavelet, (canonical) coherent state, Gaussian density, etc.

We consider the Fock space $F^{1,2}(\mathbb{C})$ of entire functions in $\mathbb{C}$ with the metric

$$\|f\|^2 = \frac{1}{\pi}\iint_{\mathbb{C}} |f(z)|^2 e^{-|z|^2}\,dxdy.$$

Thus, compared to section 5, 'Planck's constant' $1/\alpha$ is (out of convenience) taken to be 1. Make the substitution

$$f(z) = f_1(z)e^{z^2/2}.$$

Using the identity $z^2 = x^2 - y^2 + 2ixy$ the metric then takes the form

$$\|f_1\|^2 = \frac{1}{\pi}\iint_{\mathbb{C}} |f_1(z)|^2 e^{-2y^2}\,dxdy,$$

which in view of Parseval's theorem and Fubini can be written as

$$\frac{1}{\pi}\int_{\mathbb{R}} [\frac{1}{2\pi}\int_{\mathbb{R}} |\hat{f}_1(\xi)e^{-y\xi}|^2 d\xi]e^{-2y^2}dy$$

$$= \frac{1}{2\pi^2}\int_{\mathbb{R}} |\hat{f}_1(\xi)|^2(\int_{\mathbb{R}} e^{-2y\xi-2y^2}dy)d\xi = \frac{1}{2\pi^2}\sqrt{(\tfrac{\pi}{2})}\int_{\mathbb{R}} |\hat{f}_1(\xi)|^2 e^{\xi^2/2}d\xi.$$

So finally setting

$$\hat{f}_1(\xi) = f_2(\xi)e^{-\xi^2/4}$$

---

4 The following is the outcome of a discussion that the writer had with A. Cordoba, to whom he expresses his sincere thanks for his patience.

we get the metric

$$\|f_2\|^2 = \int_{\mathbf{R}} |f_2(\xi)|^2 d\xi.$$

Summing up, we have deduced (I hope, in a comprehensive way) the well-known *Bargmann transform*. It connects the Bargmann-Segal and Schrödinger representations of the Heisenberg group.

Next, take (*Gauss-Weierstrass function* [45])

$$f = e_{ac} = e^{az^2/2+cz} \qquad (|a| < 1).$$

Then

$$f_1 = e^{az^2/2+cz}e^{-z^2/2} = e^{(a-1)z^2/2+cz};$$

$$\hat{f}_1(\xi) = \int_{\mathbf{R}} \exp(-ix\xi + \tfrac{1}{2}(a-1)x^2 + cx)dx$$

$$= \int_{\mathbf{R}} \exp[-\tfrac{1}{2}(1-a)(x + \tfrac{i\xi-c}{1-a})^2 + \tfrac{1}{2}\tfrac{(i\xi-c)^2}{1-a}]dx = \sqrt{(\tfrac{2\pi}{1-a})}\exp(\tfrac{1}{2}\tfrac{(i\xi-c)^2}{1-a});$$

$$f_2(\xi) = \frac{1}{\sqrt{(1-a)}}\exp(\tfrac{1}{2}\tfrac{(i\xi-c)^2}{1-a} + \tfrac{\xi^2}{4}) = \frac{1}{\sqrt{(1-a)}}\exp(-\tfrac{1}{4}\tfrac{1+a}{1-a}\xi^2 - \tfrac{i\xi c}{1-a} + \tfrac{1}{2}\tfrac{c^2}{1-a}).$$

This has to be compared with the *wave packet*

$$\phi_{(x_0,\xi_0,g)}(\xi) = \exp[ix_0(\xi-\xi_0) + \tfrac{i}{2}g(\xi-\xi_0)^2],$$

where x_0, ξ_0 are *real* quantities and Im $g > 0$ (see [13]; notice that compared to that paper here the Latin and Greek letters happen to have the opposite meaning.) This suggests putting

$$g = \frac{i}{2}\frac{1+a}{1-a}; \quad g\xi_0 - x_0 = \frac{c}{1-a}.$$

The first relation is the Cayley transform, while the second relation expresses, in classical parlance, that ξ_0, $-x_0$ are the *characteristics* of the complex number $c/(1-a)$. Look now at the constant exponentials. On the one hand one has

$$\frac{1}{2}\frac{c^2}{1 - a},$$

and on the other

$$\frac{1}{2}ig\xi_0^2 - ix_0\xi_0.$$

How do we explain this discrepancy?

It will be expedient to pass via the transformation theory of the heat equation (see Appendix 3). Consider quite generally the PDE

$$\frac{\partial F}{\partial a} = \frac{1}{2}\frac{\partial^2 F}{\partial c^2}.$$

The substitution

$$G(g,z) = F(a,c) \cdot \exp\left(-\frac{c^2}{2(1 - a)}\right) \surd(1 - a) \text{ where } g = \frac{i}{2}\frac{1+a}{1-a}, \; z = \frac{c}{1-a}$$

gives the equation

$$\frac{\partial G}{\partial g} = -\frac{i}{2}\frac{\partial^2 G}{\partial z^2};$$

this is the 'Cayley transformed' heat equation. Next set

$$\phi = G(g, g\xi - x)\exp(\tfrac{i}{2}g\xi^2 - ix\xi).$$

Then

$$\frac{\partial\phi}{\partial g} = \left(\frac{\partial G}{\partial g} + \xi\frac{\partial G}{\partial z} + \frac{i}{2}\xi^2 G\right) \cdot \exp(-\tfrac{i}{2}g\xi^2 + ix\xi),$$

$$\frac{\partial^2\phi}{\partial x^2} = \left(\frac{\partial^2 G}{\partial z^2} + 2i\xi\frac{\partial G}{\partial z} - \xi^2 G\right) \cdot \exp\left(-\frac{i}{g}g\xi^2 + ix\xi\right).$$

Thus, taking the difference, we end up with the 'Schrödinger equation'

$$\boxed{\frac{\partial\phi}{\partial g} = -\frac{i}{2}\frac{\partial^2\phi}{\partial x^2}}\ ,$$

which occurs in [13]. The exponential factors are the same as above. We have thus adequately established the essential identity of the two concepts, wave packets and Gauss-Weierstrass functions.

APPENDIX 5. THE CROSS RATIO OF FOUR NEARBY POINTS ON A LINE

Let $x = x(t)$ be the coordinate of a moving point on a projective line (or on a Riemann surface equipped with a projective structure).

The cross ratio of any four of the points is

$$D(x(t_1),\dots,x(t_4)) = \frac{x(t_1)-x(t_3)}{x(t_1)-x(t_4)} : \frac{x(t_2)-x(t_3)}{x(t_2)-x(t_4)} \prod_{\substack{i=1,2\\k=3,4}} (x(t_i)-x(t_k))^{(-1)^{i+k}} .$$

Near $t = 0$ we have the expansion

$$x(t_i) - x(t_k) = (t_i - t_k)x'(0) + \frac{t_i^2 - t_k^2}{2} x''(0) + \dots$$

$$= (t_i-t_k)x'(0)[1 + \frac{1}{2}\frac{t_i^2-t_k^2}{t_i-t_k}\frac{x''(0)}{x'(0)} + \frac{1}{6}\frac{t_i^3-t_k^3}{t_i-t_k}\frac{x'''(0)}{x'(0)} + \frac{1}{24}\frac{t_i^4-t_k^4}{t_i-t_k}\frac{x^{IV}(0)}{x'(0)} + \dots]$$

$$= (t_i-t_k)x'(0)[1 + a_1^{ik} + a_2^{ik} + a_3^{ik} + \dots],$$

where we have put

$$a_r^{ik} \overset{\text{def}}{=} \xi_r \frac{t_i^{r+1} - t_k^{r+1}}{t_i - t_k}$$

with

$$\xi_r \overset{\text{def}}{=} \frac{1}{(r+1)!}\frac{x^{(r+1)}(0)}{x'(0)} .$$

We therefore find

$$\log D(x(t_1),\ldots,x(t_4)) : D(t_1,\ldots,t_4)$$

$$= \sum_{\substack{i=1,2\\k=3,4}} (-1)^{i+k}[(a_1^{ik} + a_2^{ik} + \ldots) - \tfrac{1}{2}(a_1^{ik} + a_2^{ik} + \ldots)^2 + \tfrac{1}{3}(a_1^{ik} + a_2^{ik} + \ldots)^3 - \ldots]. \quad (1)$$

Let us further introduce the notation

$$T^{\alpha\beta} \stackrel{def}{=} (t_1^\alpha - t_2^\beta)(t_3^\beta - t_4^\beta) = \sum_{\substack{i=1,2\\k=3,4}} (-1)^{i+k} t_i^\alpha t_k^\beta.$$

Notice that

$$T^{\alpha 0} = 0, T^{0\beta} = 0.$$

We find

$$\sum(-1)^{i+k} a_r^{ik} = \sum(-1)^{i+k}(t_i^r + t_i^{r-1} t_k + \ldots t_k^r)\xi_r (T^{r-1,1} + T^{r-2,2} + \ldots + T^{1,r-1})\xi_r;$$

$$\sum(-1)^{i+k} a_r^{ik} a_s^{ik} = \sum(-1)^{i+k}(t_i^r + t_i^{r-1} t_k + \ldots + t_k^r)(t_i^s + t_i^{s-1} t_k + \ldots + t_k^s)\xi_r\xi_s$$

$$= (0 + T^{r+s-1,1} + T^{r+s-2,2} + \ldots + T^{r,s}$$

$$+ T^{r+s-1,1} + T^{r+s-2,2} + T^{r+s-3,3} + \ldots T^{r-1,s+1} + \ldots$$

$$+ T^{s,r} + T^{s-1,r+1} + T^{s-2,r+2} + \ldots 0)\xi_r\xi_s$$

$$= (2T^{r+s-1,1} + 3T^{r+s-2,2} + \ldots + 3T^{2,r+s-2} + 2T^{1,r+s-1})\xi_r\xi_s$$

$$+ [\sum_{n=1}^{r+s-1} (1 + \min(n,r,s,r+s-n))T^{r+s-n,n}]\xi_r\xi_s.$$

Keeping terms up to order 2 in (1) this gives

$$(-1)^{i+k}(a_2^{ik} - \tfrac{1}{2}(a_1^{ik})^2)$$

$$= T^{1,1}(\xi_2 - \tfrac{1}{2}2\xi_1^2) = T^{1,1}(\xi_2 - \xi_1^2) = T^{1,1}\left(\frac{1}{6}\frac{x'''(0)}{x'(0)} - \left(\frac{1}{2}\frac{x''(0)}{x'(0)}\right)^2\right)$$

$$= \frac{1}{6}T^{1,1} \left(\frac{x'''(0)}{x'(0)} - \frac{3}{2} \left(\frac{x''(0)}{x'(0)}\right)^2 \right) = \frac{1}{6} T^{1,1} Sx(0),$$

where S stands for the Schwarzian.

Let y be a function of x. Thus, kinematically speaking, we consider a relative motion. Then

$$\log \frac{D(x(t_1),\ldots,x(t_4))}{D(t_1,\ldots,t_4)} = \frac{1}{6}(t_1 - t_2)(t_3 - t_4)Sx(0) + \ldots \text{ (near } t = 0),$$

$$\log \frac{D(y(x_1),\ldots,y(x_4))}{D(x_1,\ldots,x_4)} = \frac{1}{6}(x_1-x_2)(x_3-x_4)Sy(x(0)) + \ldots \text{ (near } x = x(t)).$$

Thus we find

$$\log \frac{D(y(x(t_1)),\ldots,y(x(t_4)))}{D(t_1,\ldots,t_4)} = \frac{1}{6}(x(t_1)-x(t_2))(x(t_3)-x(t_4))Sy(x(0))$$

$$+ \frac{1}{6}(t_1 - t_2)(t_3 - t_4)Sx(0) + \ldots$$

Passing to the limit we thus get as an application Cayley's formula mentioned in the Introduction (rewritten in the present notation)

$$\boxed{(S(y\circ x))(t)(dt)^2 = (Sy)(x(t))(dx(t))^2 + (Sx)(t)(dt)^2} \quad .$$

REMARK: This connection between the cross ratio and the Schwarzian is of course classical (see e.g. [11]). The point is that we are also interested in the higher-order terms.

The sums of higher-order products of factors a_r^{ik} can be treated in an analogous way as above in the case of just one or two factors, and it is easy to write down a recursion for the coefficients involved. It suffices to consider the case of three factors. Thus consider the expression

$$\sum(-1)^{i+k} a_r^{ik} a_s^{ik} a_v^{ik} = \sum_{n=1}^{r+s+v-1} D_{n,r,s,v} T^{r+s+v-n,n} \xi_r \xi_s \xi_v$$

where

$$D_{n,r,s,v} = \sum_{\substack{a+b+c=n \\ 0\leq a\leq r, 0\leq b\leq s, 0\leq c\leq v}} 1.$$

To find a closed expression for $D_{n,r,s,v}$ we consider the identity

$$(1 + x +\ldots+ x^r)(1 + z +\ldots+ x^s)(1 + x +\ldots+ x^v) = \frac{(1-x^{r+1})(1-x^{s+1})(1-x^{v+1})}{(1-x)^3}.$$

Now

$$(1 - x)^{-3} = \sum_{n=0}^{\infty} D_n x^n \quad \text{where} \quad D_n = \frac{(n + 1)(n + 2)}{2},$$

so the r.h.s. can be written

$$\sum_{n=0}^{\infty} D_n x^n \cdot (1-x^{r+1} - x^{s+1} - x^{v+1} + x^{r+s+2} + x^{r+v+2} + x^{s+v+2} - x^{r+s+v+3}).$$

It follows that ($D_n = 0$ if $n < 0$)

$$D_{n,r,s,v} = D_n - D_{n-r-1} - D_{n-s-1} - D_{n-v-1} + D_{n-r-s-2} + D_{n-r-v-2} + D_{n-s-v-2} - D_{n-r-s-v-3}.$$

Consider now the third-order terms in (1), viz.

$$\sum(-1)^{i+k}(a_3^{ik} - a_1^{ik}a_2^{ik} + \frac{1}{3}(a_1^{ik})^2)$$

$$= (T^{2,1} + T^{1,2})\xi_3 - (2T^{1,2} + 2T^{2,1})\xi_1\xi_2 + \frac{1}{3}(3T^{2,1} + 3T^{1,2})\xi_1^3$$

$$= (T^{2,1}+T^{1,2})(\xi_3-2\xi_1\xi_2+\xi_1^3) = (T^{2,1}+T^{1,2})(\frac{1}{24}\frac{x^{IV}}{x'} -2\cdot\frac{1}{2}\frac{x''}{x'}\cdot\frac{1}{6}\frac{x'''}{x'} + (\frac{1}{2}\frac{x''}{x'})^3)$$

$$= (T^{2,1}+T^{1,2})\cdot\frac{1}{24}(\frac{x^{IV}}{x'} - 4\frac{x''x'''}{(x')^2} + 3(\frac{x''}{x'})^3) = (T^{2,1} + T^{1,2})\cdot\frac{1}{24}(Sx)',$$

as

$$S = \frac{x'''}{x'} - \frac{3}{2}(\frac{x''}{x'})^2,$$

$$S' = \frac{x^{IV}}{x'} - \frac{x''' x''}{x'^2} - \frac{3}{2}\cdot 2\frac{x''}{x'}(\frac{x'''}{x'} - \frac{x''\cdot x''}{x'^2}) = \frac{x^{IV}}{x'} - 4\frac{x''' x''}{x'^3} + 3(\frac{x''}{x'})^3.$$

Thus

$$\log \frac{D}{D} = \frac{1}{6}T^{1,1}S + \frac{1}{24}(T^{2,1} + T^{1,2})S' + \dots .$$

We have also considered fourth-order terms and by similar calculations to those just done we have found that now there appears the additional term

$$\frac{1}{120}(T^{3,1} + T^{2,2} + T^{1,3})(S'' + \frac{1}{3}S^2) - \frac{1}{72}T^{2,2}S^2$$

in the corresponding formula.

We conjecture that in general the coefficient of $T^{\alpha\beta}$ is a polynomial in $S,S',S'',\dots,S^{(\nu)}$ where $\nu = \alpha + \beta$, but this we have not proved. Another question: How does Cayley's formula generalize? Apparently, results of this kind can be obtained using the method we just used (see above) in the case of that formula.

APPENDIX 6. A BOL'S LEMMA FOR THE 'POLY-HEAT' EQUATION[5]

Bojarski has in [6] obtained an interesting result about the transformation of the poly-harmonic equation under the Möbius (conformal) group in any number of dimensions. In the case of two (real) variables it essentially reduces to our Bol's lemma (see Introduction). We now state a counterpart of this result for the iterated heat operator, which seems to be new. Introduce the notation

$$H = \frac{\partial}{\partial t} - \frac{1}{2}\frac{\partial^2}{\partial x^2} .$$

Then one has the formula

$$H^m \left[(ct + d)^{m-3/2} \exp(-\frac{1}{2}\frac{cx^2}{ct+d}) \; u(\frac{at+b}{ct+d}, \frac{x}{ct+d}) \right]$$

$$= (ct + d)^{-m-3/2} \exp(-\frac{1}{2}\frac{cx^2}{ct+d}) H^m \; u(\frac{at+b}{ct+d}, \frac{x}{ct+d}).$$

---

5 Outcome of a discussion with Bogdan Bojarski on the occasion of a hike to the Pyhätunturi mountain.

Here $\begin{pmatrix} a & b \\ c & d \end{pmatrix}$ is any unimodular matrix ($ad - bc = 1$) and $m = 1,2,\ldots$ The case $m = 1$ is well known and, of course, implicit in our Appendix 3, where the variables were denoted a,c. The general case follows easily from it by induction. Notice that this result agrees with the fundamental solution of the 'poly-heat' equation which is easy to write down and which generalizes the classical fundamental solution $(1/\pi)\exp(-x^2/2t)$ in the case of the heat equation. (Each iteration produces a new factor t.)

References

[1] J. Arazy, S. Fisher, S. Janson and J. Peetre, An identity for reproducing kernels in a planar domain and Hilbert-Schmidt Hankel operators, Technical report, Uppsala (1988) (submitted).

[2] J. Arazy, S. Fisher and J. Peetre, Hankel operators on planar domains, Amer. Math. J. (to appear).

[3] V.I. Arnold, Geometrical Methods in the Theory of Ordinary Differential Equations (Grundlehren 250), Springer (1983).

[4] F.A. Berezin, General concept of quantization, Comm. Math. Phys., 40 (1975), 153-174.

[5] S. Bergman, The Kernel Function and Conformal Mapping, Amer. Math. Soc. (1950).

[6] B. Bojarski, Remarks on polyharmonic operators and conformal maps in space, In Trudyi Vsesoyuznogo Simpoziuma v Tbilici, 21-23 aprelya 1982, pp. 49-56 (Russian).

[7] G. Bol, Invarianten linearer Differentialgleichungen. Abh. Math. Sem. Univ. Hamburger Univ., 16 (1949), 1-28.

[8] J. Boman, S. Janson and J. Peetre, Big Hankel operators of higher weight, Technical report, Stockholm (1987) (submitted).

[9] O. Borůvka, Lineare Differentialtransformationen 2. Ordnung. VEB Deutscher Verlag (1967); English translation: English Universities Press (1971).

[10] J. Borwein and P. Borwein, Pi and the AGM, Wiley (1987).

[11] É. Cartan, Leçons sur la Théorie des Espaces à Connexions Projectifs, Gauthier-Villars (1937).

[12] H. Clemens, A Scrap Book of Complex Curve Theory, Plenum (1980).

[13] A. Cordoba and C. Fefferman, Wave packets and Fourier integral operators, Comm. Partial Differential Equations, 3 (1978), 979-1005.

[14] R. Courant and D. Hilbert, Methoden der mathematischen Physik I, Springer (1924).

[15] P. Deligne, Équations Différentielles à Points Singuliers Réguliers, Lecture Notes Math. 163, Springer (1970).

[16] J. Dieudonne, Éléments d'analyse. 1. Fondements de l'analyse moderne, Gauthier-Villars (1969).

[17] M. Eichler, Eine Verallgemeinerung der Abelschen Integrale, Mat. Z., 67 (1957), 267-298.

[18] L. Euler, Opusculis varii argumenti, Berlin, 1746.

[19] P. Gordan, Invariantentheorie, Teubner (1887).

[20] J. Gray, Linear Differential Equations and Group Theory from Riemann to Poincaré. Birkhäuser (1986).

[21] P. Griffiths and J. Harris, Principles of Algebraic Geometry, Wiley (1978).

[22] R.G. Gunning, Lectures on Riemann Surfaces, Princeton University Press, (1966).

[23] R.G. Gunning, Special coordinate coverings of Riemann surfaces, Math. Ann., 170 (1967), 67-86.

[24] R.G. Gunning, On uniformization of complex manifolds: the role of connections. Princeton University Press (1980).

[25] B. Gustafsson and J. Peetre, Notes on projective structures on complex manifolds, Nagoya Math. J. (to appear).

[26] B. Gustafsson and J. Peetre, Hankel forms on multiply connected plane domains. Part two. The case of higher connectivity. Complex Variables (to appear).

[27] P. Henrici, Applied and Computational Complex Analysis III, Wiley (1986).

[28] E. Hille, Ordinary Differential Equations in the Complex Domain, Wiley (1976).

[29] R.-P. Holzapfel, Geometry and Arithmetic Around Euler Partial Differential Equations, Reidel (1986).

[30] C. Houzel, Fonctions elliptiques et intégrales abéliennes, In J. Dieudonné (ed.), Abregé d'Histoire des Mathematiques 1700-1900 pp. 1-133, Hermann (1978).

[31] C.G.J. Jacobi, Fundamenta nova theoriae functiorum ellipticum. In Werke I, pp. 49-239.

[32] C.G.J. Jacobi, Über die Differentialgleichung, welcher die Reihen $1 \pm 2q \pm 2q^4 \pm 2q^9 + \text{etc.}$, $2\sqrt[4]{q} + 2\sqrt[4]{q^9} + 2\sqrt[4]{q^{25}} + \text{etc.}$ genüge leisten, J. reine angew. Math., 36 (1847), 97-112. Also Werke II, pp. 97-112.

[33] S. Janson and J. Peetre, A new generalization of Hankel operators (the case of higher weight), Math. Nachr., 132 (1987), 313-328.

[34] S. Janson, J. Peetre and R. Wallstén, A new look on Hankel forms over Fock space. Technical report, Lund (1988) (submitted).

[35] S. Janson, J. Peetre and R. Rochberg, Hankel forms and the Fock space, Revista Mat. Iberoamer., 3 (1987), 61-138.

[36] I. Kra, Automorphic Functions and Kleinian Groups, Benjamin (1972).

[37] E.E. Kummer, De generali quadam aequatione differentiali tertii ordinis. In Propgramm des evangelischen Königl. und Stadtgümnasiums in Liegnitz, 1834. Also J. reine angew. Math., 100 (1887), 1-9; Werke pp. 33-41.

[38] E. Ligocka, On the Forelli-Rudin construction and weighted Bergman projections, Institute Mittag-Leffler, Report no. 14 (1987).

[39] H. Morikawa, Some analytic and geometric applications of the invariant theoretic methods, Nagoya Math. J., 80 (1980), 1-47.

[40] N.K. Nikol'skiĭ, Treatise on the Shift Operator. Spectral function theory. Grundlehren, Berlin (1986).

[41] J. Peetre, Some unsolved problems. In Colloquie Mathematica Societatis Janos Bolyai 49, Alfred Haar Memorial Conference, Budapest, Hungary (1985), pp.711-735. North-Holland (1986).

[42] J. Peetre, Invariant function spaces and Hankel operators - a rapid survey, Expositiones Mathematicae, 5 (1986), 3-16.

[43] J. Peetre, Hankel kernels of higher weight for the ball. Technical report, Lund (1988) (submitted).

[44] J. Peetre, Generalizing the arithmetic-geometric mean - a hapless computer experiment, Int. J. Math. Math. Sci. (to appear).

[45] J. Peetre, Some calculations related to Fock space and the Shale-Weil representation, Integral Equations Operator Theory (to appear).

[46] J. Peetre, Invariant function spaces connected with the holomorphic discrete series. In P.L. Butzer et al (eds.), Anniversary Volume on Approximation and Functional Analysis, pp. 119-134, Birkhäuser, (1984).

[47] J. Peetre and J. Karlsson, Rational approximation - analysis of the work of Pekarskiĭ. Rocky Mountain J. Math. (to appear).

[48] J. Pöschel and E. Trubowitz, Inverse Spectral Theory, Academic Press (1987).

[49] L. Rubel, Some research problems about algebraic differential equations, Trans. Amer. Math. Soc., 280 (1983), 43-52.

[50] R. Sachs, Review of [48], Bull. Amer. Math. Soc., 19 (1988), 356-362.

[51] M. Schiffer, Personal communication, Jan. 1985.

[52] Y. Teranishi, The variational theory of higher-order linear differential equations. Nagoya Math. J., 95 (1984), 137-161.

[53] A.N. Tyurin, On periods of quadratic differentials, Usp. Mat. Nauk, 33:6 (1978), 149-195 (Russian); Russian Math. Surveys, 33:6 (1978), 169-221.

[54] A. Unterberger, Symbolic calculi and the duality of homogeneous spaces, Contemporary Math., 27 (1984), 237-252.

[55] A. Unterberger, L'origine et les buts de diverses méthodes de quantification, Preprint (to appear).

[56] A. Unterberger and J. Unterberger, La série discrète de SL(2,R) et les opérateurs pseudo-différentielles sur une demi-droite, Ann. Sci. École Norm. Sup. (4), 17 (1984), 83-116.

[57] H. Upmeier, Jordan algebras in analysis and quantum mechanics, C.B.S. Regional Conf. Series in Math. 67, Amer. Math. Soc. (1987).

[58] H. Weber, Elliptische Functionen und algebraische Zahlen, Vieweg (1891).

[59] H. Weyl, Gruppentheorie und Quantenmechanik, Hirzel (1931).

[60] E.J. Wilczynski, Projective differential geometry of curves and ruled surfaces, Chelsea (1961).

Björn Gustafsson
Matematiska Institutionen
KTH
S-100 44 Stockholm
Sweden

Jaak Peetre
Matematiska Institutionen
Stockholms Universitet
Box 6701
S-113 85 Stockholm
Sweden

2. FUNCTION SPACES II

M. KRBEC

Weighted norm inequalities in Orlicz spaces

1. INTRODUCTION

The 70's and 80's witnessed a real flood of papers on weighted inequalities triggered by the pioneering results due to B. Muckenhoupt and R.L. Wheeden [17], [18] dealing with necessary and sufficient conditions on weight functions in terms of the A_p or the Muckenhoupt classes. The results are very well known; the membership of the weight function and/or the weight functions in appropriate A_p class guarantees boundedness of various integral operators as the fractional maximal operator, the Riesz potential in Lebesgue and/or Marcinkiewicz spaces. It should be noted that the study of these weighted problems goes back to Hardy, Littlewood, and Pólya [6], Stein and Weiss [23], Lizorkin [14], Walsh [25] to name at least some fine results which are various generalizations of the maximal operator theorem and the original Sobolev inequality.

Here we shall concentrate our attention on one of the offshoots of the theory, namely, on Orlicz spaces.

Let us briefly recall some basic concepts. Let be a Young function, i.e. $\phi:\langle 0,\infty) \to \langle 0,\infty)$ is convex, $\lim_{x\to 0} \phi(x)/x = \lim_{x\to\infty} x/\phi(x) = 0$. Let $\Omega \subset R^n$ be measurable and ρ a weight in Ω, i.e. a measurable a.e. positive function. We define the modular

$$m_\rho(f,\phi) = \int_\Omega \phi(|f(x)|)\, \rho(x)\, dx$$

and define the weighted Orlicz space $L_{\phi,\rho}$ as the linear hull of all f for which $m_\rho(f,\phi) < \infty$, endowing it with the (Luxemburg) norm

$$\|f\|_{\phi,\rho} = \inf\ \{\lambda > 0;\ \ m_\rho(f/\lambda,\phi) \leq 1\}.$$

If $\rho \equiv 1$ in Ω we have the usual Orlicz space and we simply write $m(f,\phi)$, $L_\phi(\Omega)$, $\|f\|_\phi$.

A Young function ϕ is said to satisfy the Δ_2 condition (we write $\phi \in \Delta_2$)

if there exist constants c and T such that $\phi(2t) \leq c\phi(t)$ for $t \geq T$. If $T = 0$ the function ϕ is said to satisfy the Δ_2 condition globally.

Let $\phi \in \Delta_2$ globally and put

$$h(\lambda) = h_\phi(\lambda) = \sup_{t>0} \phi(\lambda t)/\phi(t), \quad \lambda \geq 0.$$

The lower index of ϕ is

$$i(\phi) = \lim_{\lambda\to 0_+} \frac{\log h(\lambda)}{\log \lambda}$$

and the upper index of ϕ is

$$I(\phi) = \lim_{\lambda\to\infty} \frac{\log h(\lambda)}{\log \lambda} .$$

The existence of the above limits is e.g. a consequence of properties of submultiplicative functions; see Matuszewska and Orlicz [16], Gustavson and Peetre [4], Maligranda [15]. The condition $I(\phi) < \infty$ is equivalent to $\phi \in \Delta_2$, and if ψ is the complementary function to ϕ, i.e.

$$\psi(t) = \sup_{\tau\geq 0} (t\tau - \phi(\tau)), \quad t \geq 0,$$

then $i(\phi) > 1 \iff \psi \in \Delta_2$.

The standard reference to the theory of non-weighted Orlicz spaces is the monograph [11] by Krasnosel'skii and Rutitskii and [19] by Musielak. Basic properties of weighted Orlicz spaces in the above sense can easily be verified going along the lines of proofs of nonweighted theorems.

2. REFLEXIVE ORLICZ SPACES CASE

One of the first papers in this area is that by A. Torchinsky [24] who studied the boundedness of the Riesz potential in nonweighted Orlicz spaces. It was shown that the Riesz potential

$$T_\gamma f(x) = \int_{R^n} \frac{f(y)}{|x - y|^{n-\gamma}}\, dy, \quad 0 < \gamma < n$$

is bounded between $L_{\phi_1}(R^n)$ and $L_{\phi_2}(R^n)$ if $\phi_2^{-1}(t) \sim t^{-\gamma/n}\phi_1^{-1}(t)$.

The first very interesting and deep result about weighted Orlicz spaces appeared in 1982 and belongs to R. Kerman and A. Torchinsky. They considered the maximal operator

$$Mf(x) = \sup_{Q \ni x} \frac{1}{|Q|} \int_Q |f(y)|\, dy$$

between weighted Orlicz spaces $L_{\phi,\rho}(R^n)$ with $i(\phi) > 1$, $I(\phi) < \infty$. One could except that the necessary and sufficient condition for M to be bounded in $L_{\phi,\rho}$ will be

$$(A_\phi) \quad \sup_Q \left(\frac{1}{|Q|} \int_Q \rho\right) \phi'\left(\frac{1}{|Q|} \int_Q (\phi')^{-1}\left(\frac{1}{\rho}\right)\right) < \infty.$$

Kerman and Torchinsky showed much more, their necessary and sufficient condition for

$$\int_{R^n} \phi(Mf(x))\rho(x)dx \leqq C \int_{R^n} \phi(f(x))\, \rho(x)\, dx$$

is very surprising and simply reads $\rho \in A_{i(\phi)}$ which is also much more convenient than A_ϕ.

In [9], Kokilashvili and the author considered the fractional maximal operator

$$M_\gamma f(x) = \sup_{Q \ni x} \frac{1}{|Q|^{1-\gamma/n}} \int_Q |f(y)|\, dy, \quad 0 < \gamma < n,$$

and the corresponding Riesz potential

$$T_\gamma f(x) = \int_{R^n} \frac{f(y)}{|x - y|^{n-\gamma}}\, dy, \quad 0 < \gamma < n.$$

Using the result of [8], we get

<u>THEOREM 1</u>: Let ϕ_1 and ϕ_2 be Young functions, $\phi_2^{-1}(t) \sim t^{-\gamma/n}\phi_1^{-1}(t)$, $1 < i(\phi_1) = p \leqq I(\phi_1) < \infty$, $1 < i(\phi_2) = q \leqq I(\phi_2) < \infty$. Then the following conditions are equivalent

(i) $\|T_\gamma(f\rho^{\gamma/n})\|_{L_{\phi_2,\varepsilon\rho}} \leqq c\,\|f\|_{L_{\phi_1,\varepsilon\rho}}, \quad \varepsilon > 0,$

(ii) $\|M_\gamma(f\rho^{\gamma/n})\|_{L_{\phi_2,\varepsilon\rho}} \leqq c\,\|f\|_{L_{\phi_1,\varepsilon\rho}}, \quad \varepsilon > 0,$

(iii) $\rho \in A_{1+q/p'}$ $\quad (p' = p/(p - 1))$.

Immediately, if $\rho \in A_{1+q/p'}$, we have the imbedding inequality

$$\|f\|_{L_{\phi_2,\rho}} \leq c[\,\|f\rho^{-1/n}\|_{L_{\phi_1,\rho}} + \sum_{i=1}^{n} \|\frac{\partial f}{\partial x_i}\rho^{-1/n}\|_{L_{\phi_1,\rho}}].$$

More generally, one can consider anisotropic operators. Let $\alpha_j > 0$, $j = 1,\dots,n$, $|\alpha| = \alpha_1 + \dots + \alpha_n$,

$$\|x\| = (\sum_{j=1}^{n} |x_j|^{2/\alpha_j})^{1/2},$$

be the anisotropic norm, and, for $x \in R^n$ and $t > 0$, let

$$E = E(x,t) = \{y \in R^n;\ |x_j - y_j| < 2^{-1}t^{\alpha_j}\}$$

be the anisotropic cube centred at x. Then we define the anisotropic potential of order s,

$$T_s f(x) = \int_{R^n} \frac{f(y)}{\|x - y\|^s}\,dy, \quad 0 < s < |\alpha|,$$

and the anisotropic fractional maximal function

$$M_\gamma f(x) = \sup_{t>0} |E(x,t)|^{-\gamma} \int_{E(x,t)} |f(y)|\,dy, \quad 0 < \gamma < 1.$$

The following theorem was proved by Kokilashvili and the author in [10].

THEOREM 2: Let ϕ_1, ϕ_2 be Young functions,

$$1 < i(\phi_1) = p \leqq I(\phi_1) = P < \infty,$$

$$1 < i(\phi_2) = q \leqq I(\phi_2) = Q < \infty.$$

Further, let

$$0 < s < |\alpha| \ , \quad r = p^{-1} - q^{-1} > 0,$$

$$q = |\alpha|p/(|\alpha| - (|\alpha| - s)p),$$

$$Q = |\alpha|P/(|\alpha| - (|\alpha| - s)P).$$

Then the following conditions are equivalent

(i) $\|T_s(f(\varepsilon\rho)^r)\|_{L_{\phi_2,\varepsilon\rho}} \leqq c \ \|f\|_{L_{\phi_1,\varepsilon\rho}}, \ \varepsilon > 0,$

(ii) $\|M_{s/|\alpha|}(f(\varepsilon\rho)^r)\|_{L_{\phi_2,\varepsilon\rho}} \leqq c \ \|f\|_{L_{\phi_1,\varepsilon\rho}}, \ \varepsilon > 0,$

(iii) $\rho \in A_{1+q/p'}$.

This theorem and its proof only appeared in Russian; a sketch of the proof might be useful.

STEP 1: Let (ii) hold for each $f \in L_{\phi_1,\varepsilon\rho}$ and $\varepsilon > 0$. Then

$$\|\chi_E(\varepsilon\rho)^{r-1}\|_{L_{\psi_1,\varepsilon\rho}} \cdot \|\chi_E\|_{L_{\phi_2,\varepsilon\rho}} \leqq c|E|^{1/|\alpha|}$$

for any anisotropic cube $E = E(x,t)$ with ψ_1 complementary to ϕ_1. Indeed,

$$M_{s|\alpha|}(f(\varepsilon\rho)^r)(x) \geqq |E|^{-s/|\alpha|} \int_E |f(y)|(\varepsilon\rho(y))^{r-1}\varepsilon\rho(y)dy\cdot\chi_E(x).$$

As ψ_1 is the Young function complementary to ϕ_1, choose $f_o \geqq 0$, $f_o \in L_{\phi_1,\varepsilon\rho}$, $\|f_o\|_{L_{\phi_1,\varepsilon\rho}} = 1$ in such a way that

$$\int_E f_o(y)\,(\varepsilon\rho(y))^{r-1}\,\varepsilon\rho(y)\,dy = \|\chi_E/(\varepsilon\rho)^{r-1}\|_{L_{\psi_1,\varepsilon\rho}}.$$

It follows that

$$\|M_{s/|\alpha|}(f_o(\varepsilon\rho)^r)\|_{L_{\phi_2,\varepsilon\rho}} \geqq \|\chi_E(\varepsilon\rho)^{r-1}\|_{L_{\psi_1,\varepsilon\rho}} \cdot \|\chi_E\|_{L_{\phi_2,\varepsilon\rho}},$$

so that

$$\|\chi_E(\varepsilon\rho)^{r-1}\|_{L_{\psi_1,\varepsilon\rho}} \cdot \|\chi_E\|_{L_{\phi_2,\varepsilon\rho}} \leqq c\,\|f_o\|_{L_{\phi_1,\varepsilon\rho}} |E|^{s/|\alpha|} \leqq c|E|^{s/|\alpha|}. \tag{2.1}$$

Step 2: If ϕ and ψ are complementary Young functions and

$$\|\chi_E/\varepsilon\rho\|_{L_{\psi,\varepsilon\rho}} \cdot \|\chi_E\|_{L_{\phi,\varepsilon\rho}} \leqq c|E|, \tag{2.2}$$

for each $E = E(x,t)$ and each $\varepsilon > 0$, then $\rho \in A_{i(\phi)}$. If E are isotropic cubes this is essentially the crucial part of [8]. The proof of our case is rather lengthy and it is an amalgam of [8] and the following anisotropic decomposition theorem of the Calderon-Zygmund type (it can be found in [2]):

Let $\lambda > 0$ and $f \in L^1(R^n)$. Then there exist $\lambda_o \geqq 1$ and a sequence $\{R_i\}$ of nonoverlapping parallepipeds such that

(i) for each $i \in N$ there exist anisotropic cubes U_i and V_i such that $V_i \subset R_i \subset U_i$, $|U_i| = \lambda_o^{|\alpha|}|V_i|$,

(ii) $\lambda < |R_i|^{-1} \int_{R_i} |f(y)|\,dy \leqq 2^{|\alpha|}\,\lambda_o^{|\alpha|}\,\lambda$

(iii) $|f(y)| \leqq \lambda$, $y \in R^n \setminus \bigcup_j R_j$.

STEP 3: Put $\phi_3^{-1} = (\phi_2^{-1}(t))^{|\alpha|/s}$, $\psi_3(t) = \psi_1(t^{s/|\alpha|})$. Then ϕ_3 and ψ_3 are mutually complementary Young functions and

$$\|\chi_E\|_{L_{\phi_2,\varepsilon\rho}}^{|\alpha|/s} = \|\chi_E\|_{L_{\phi_3,\varepsilon\rho}} ,$$

$$\|\chi_E(\varepsilon\rho)^{r-1}\|_{L_{\psi_1,\varepsilon\rho}}^{|\alpha|/s} = \|\chi_E/\varepsilon\rho\|_{L_{\psi_3,\varepsilon\rho}} .$$

Thus we can combine (2.1) with (2.2) and get $\rho \in A_{i(\phi_3)}$. But

$$i(\phi_3) = 1/I(\phi_3^{-1}) = s/|\alpha|\ I(\phi_2^{-1}) = 1 + q/p'.$$

STEP 4: (iii) ⇒ (i) follows by interpolation as $\rho \in A_{1+q/p'}$ implies $\rho \in A_\beta$ for $\beta > 1 + q/p' - \varepsilon$ with some $\varepsilon > 0$.

The implication (i) ⇒ (ii) holds in virtue of the pointwise estimate

$$M_{s/|\alpha|}(f)(x) \leq c\ T_s(|f|)(x), \quad x \in \mathbb{R}^n.$$

REMARK: There is a lot of likely unsolved problems. The prominent place is occupied by the related one and/or more dimensional Hardy inequality in weighted Orlicz spaces. At this state there is established a generalized Carleman inequality (a forthcoming paper by Heinig, Kerman and the author).

3. NONREFLEXIVE ORLICZ SPACES CASE

The situation changes dramatically if we consider nonreflexive spaces L_ϕ and $L_{\phi,\rho}$, i.e. $i(\phi) = 1$ or $I(\phi) = \infty$. The picture is far from being complete, as well.

First, let us look at the maximal operator. The well known fact is that $M : L_p \to L_p$ is bounded for $p > 1$ and its behaviour deteriorates as we are approaching $p = 1$. M maps L_1 into the Marcinkiewicz space L_1^*. It is also a classical fact that

$$\int_0^a Mf(x)\,dx \leq C(1 + \int_0^a |f(x)|\,(1 + \log^+|f(x)|)\,dx$$

(see Hardy, Littlewood, and Pólya [6], de Guzmán [5] or, more generally, a more dimensional inequality holds with respect to general measure μ and ν and any $K \subset \mathbb{R}^n$ with $\nu(K) < \infty$,

$$\int_K Mf(x)\,d\nu \leq C[\nu(K) + \int_{\mathbb{R}^n} |f(x)|(1 + \log^+|f(x)|)d\mu]$$

(see, e.g. [21]).

It is natural to study two weights weak type inequalities in Orlicz spaces. Two weights problem was solved by Muckenhoupt [17] and Sawyer [22] in Lebesgue spaces. The basic fact is that

$$\sup_Q \left(\frac{1}{|Q|}\int_Q \rho\right)\left(\frac{1}{|Q|}\int_Q \sigma^{-1/(p-1)}\right)^{p-1} < \infty \tag{3.1}$$

is equivalent to

$$\rho(\{Mf > \lambda\}) \leq C\lambda^{-p}\int_{\mathbb{R}^n} |f(x)|^p\,\sigma(x)\,dx.$$

Moreover, the lower upper bound in (3.1) is $\leq C^{1/p}$.

The corresponding problem turns out to be easier near L_1 rather than in L_ϕ, $i(\phi) > 1$. Let $1 < p < \infty$, $p' = p/(p - 1)$. A simple calculation shows that ρ and σ satisfy (3.1) iff

$$A_p(\rho,\sigma) = \sup_{Q\ni x}\left(\int_Q \left(\frac{\rho(Q)}{|Q|\,\sigma(x)}\right)^{p'}\frac{\sigma(x)}{\rho(Q)}\,dx\right)^{1/p'} < \infty\ ;$$

where $\rho(Q) = \int_Q \rho(x)dx$. Therefore, it is natural to consider the limiting condition

there exists $\eta > 0$ such that

$$\sup_Q \int_Q \exp\left(\frac{\eta(Q)}{|Q|\,\sigma(x)}\right)\frac{\sigma(x)}{\rho(Q)}\,dx = A_{\log}(\rho,\sigma) < \infty. \tag{3.2}$$

We have

<u>THEOREM 3</u> ([12]): The inequality

$$\rho(\{Mf > \lambda\}) \leq C(\rho,\sigma)\lambda^{-1}\int_{\mathbb{R}^n} |f(x)|(1 + \log^+|f(x)|/\lambda)\sigma(x)\,dx \tag{3.3}$$

holds iff ρ and σ satisfy (3.2).

SKETCH OF THE PROOF: Let us write $\phi(t) = |t|(1 + \log^+|t|)$. If (3.3) holds, then for any positive integer j_0 the integral in (3.2) can be written as

$$\sum_{j=0}^{j_0-1} \ldots + \sum_{j=j_0}^{\infty} \frac{\eta^j}{j!} (A_{j'}(\rho,\sigma))^j \quad (j' = j/(j-1)).$$

As $\phi(t) \leq C(p-1)^{-1} t^p$ for $t > 2^{-1}$ and p near 1, we can arrive at

$$\rho(\{Mf > \lambda\}) \leq \frac{C \cdot C(\rho,\sigma)}{p-1} \int_{\mathbb{R}^n} |f(x)|/\lambda)^p \sigma(x)\, dx.$$

The best constant in the last inequality is $\geq (A_p(\rho,\sigma))^p$, therefore $A_p(\rho,\sigma)(p-1) = O(1)$ for p near 1, so that $A_{j'}(\rho,\sigma)(p-1) \leq \tilde{C}$. In particular, $A_{j'}(\rho,\sigma) \leq \tilde{C}(j'-1) < \tilde{C}j$, giving

$$\sum_{j=j_0}^{\infty} \frac{\eta^j}{j!} (A_{j'}(\rho,\sigma))^j \leq \sum_{j=j_0}^{\infty} \frac{\eta^j}{j!} \tilde{C}^j j^j$$

which is finite for small η.

On the other hand, if the condition holds, then

$$\frac{1}{|Q|} \int_Q |f(x)|dx \leq C\, A_{\log}(\rho,\sigma) + C(\eta) \int_Q \phi(f(x)) \frac{\sigma(x)}{\rho(Q)}\, dx.$$

Now, we can estimate $\rho(\{Mf > C\, A_{\log}(\rho,\sigma) + C(\eta)\})$, covering the set inside with cubes Q_j in such a manner that

$$\sum \rho(Q_j) \leq C_n \int_{\mathbb{R}^n} \phi(f(x))\, \sigma(x)\, dx.$$

The estimate for $\rho(\{Mf > \lambda\})$ follows then by the Δ_2 condition. A more general result is due to Pick [22].

A similar limit type approach can be applied to obtain a weighted imbedding theorem $L_{\phi,\rho} \hookrightarrow L_{1,\sigma}$. Let us briefly describe it. It is known that

$$\int |f(x)|\sigma(x)\, dx \leq C_p(\int_\Omega |f(x)|^p \rho(x)\, dx)^{1/p} \tag{3.4}$$

iff

$$B_p(\sigma,\rho) = \int_\Omega (\sigma^p(x)\rho^{-1}(x))^{1/(p-1)}\, dx < \infty. \tag{3.5}$$

(Avantaggiati [1] for continuous ρ and σ and Kabaila [7] for general measures). Moreover, it can be established (see [13]) that the best C_p in (3.4) equals $(B_p(\sigma,\rho))^{1/p'}$. The limit exponential form of (3.5) is

$$\text{there exists } \mu > 0 \text{ such that} \quad \int_\Omega \exp(\mu \frac{\sigma(x)}{\rho(x)})\, \rho(x)\, dx < \infty. \tag{3.6}$$

A procedure analogous to that in the proof of the preceding theorem gives us the following

THEOREM 4: Let Ω be bounded and $\rho \in L_1(\Omega)$. Then

$$\|f\|_{L_{1,\sigma}} \leq c\, \|f\|_{L_{\phi,\rho}}$$

iff (3.6) is true.

The details can be found in Krbec and Pick [13]. Also, a more general theorem holds:

THEOREM 5 ([13]): Let ϕ_1 and ϕ_2 be Young functions such that $\phi_2\phi_1^{-1}$ is a Young function, let $\phi_1 \in \Delta_2$ globally, and let ψ be the Young function complementary to $\phi_2\phi_1^{-1}$. Then

$$\|f\|_{L_{\phi_1,\sigma}} \leq c\, \|f\|_{L_{\phi_2,\rho}}$$

iff

there exists $\eta > 0$ such that

$$\int_\Omega \psi(\eta \frac{\sigma(x)}{\rho(x)})\ \rho(x)\ dx < \infty.$$

References

[1] A. Avantaggiati, On compact embedding theorems in weighted Sobolev spaces. Czechoslovak Math. J. 29 (104) (1979), 635-648.

[2] O.V. Besov, V.P. I'lin and S.M. Nikol'skii, Integral representation of functions and imbedding theorems. Nauka, Moscow (1975) (Russian).

[3] D.W. Boyd, Indices of function spaces and their relationship to interpolation. Canad. J. Math. 21 (1969), 1245-1254.

[4] J. Gustavsson and J. Peetre, Interpolation of Orlicz spaces. Studia Math. 60 (1977), 33-59.

[5] M. de Guzmán, Differentiation in R^n. Springer-Verlag, Berlin-Heidelberg-New York 1975.

[6] G.H. Hardy, J.E. Littlewood and G. Pólya, Inequalities. Cambridge Univ. Press, Cambridge 1934.

[7] V.P. Kabaila, On the imbedding of $L^p(\mu)$ into $L^r(\nu)$ (Russian, English summary). Lit. Mat. Sb. 21 (1981), 143-148.

[8] R.A. Kerman and A. Torchinsky, Integral inequalities with weights for the Hardy maximal function. Studia Math. 71 (1982), 277-284.

[9] V.M. Kokilashvili and M. Krbec, Weighted norm inequalities for Riesz potentials and fractional maximal functions in Orlicz spaces (Russian). Dokl. Akad. Nauk SSSR 283 (1985), 280-283. English transl. in Soviet Math. Dokl. 32 (1985), 70-73.

[10] V.M. Kokilashvili and M. Krbec, On boundedness of anisotropic fractional order maximal functions and potentials in weighted Orlicz spaces (Russian). Trudy Tbilis. Mat. Inst. Razmadze 82 (1986), 106-115.

[11] M.A. Krasnosel'skii and J.B. Rutitskii, Convex functions and Orlicz spaces. Noordhof, Groningen 1961.

[12] M. Krbec, Two weights weak type inequalities for the maximal function in the Zygmund class. In: Function Spaces and Applications. Proceedings of the US-Swedish Seminar held in Lund, June 1986. Springer-Verlag, Berlin-Heidelberg-New York 1988.

[13] M. Krbec and L. Pick, On imbeddings between weighted Orlicz spaces. Preprint No. 37/1988, Math. Inst. Czech. Acad. Sci.

[14] P.I. Lizorkin, Multipliers of Fourier integrals and bounds on convolution in spaces with mixed norms. Applications (Russian). Izv. Akad. Nauk, Ser. Mat. 34 (1970), 225-255.

[15] L. Maligranda, Indices and interpolation. Dissertationes Mathematicae No. 234, 1-54. Polish Sci. Publ., Warsaw 1985.

[16] W. Matuszewska and W. Orlicz, On certain properties of ϕ-functions. Bull Acad. Polon. Sci. 7 (1960), 439-443.

[17] B. Muckenhoupt, Weighted norm inequalities for the Hardy maximal function. Trans. Amer. Math. Soc. 165 (1972), 207-226.

[18] B. Muckenhoupt and R.L. Wheeden, Weighted norm inequalities for fractional integrals. Trans. Amer. Math. Soc. 190 (1974), 261-274.

[19] J. Musielak, Modular spaces. Springer-Verlag, Berlin-Heidelberg-New York 1983.

[20] L. Pick, Two weights weak type inequality for the maximal function in $L(1 + \log^+ L)^K$. To appear in Proceedings of the Conference on Constructive Theory of Function, Varna 1987.

[21] C. Sadosky, Interpolation of operators and singular integrals. Marcel Dekker, Inc., New York and Basel 1979.

[22] E.T. Sawyer, Two weight norm inequalities for certain maximal and integral operators. In: Harmonic Analysis, Proceedings, Minneapolis 1981, pp. 102-127. Springer-Verlag, Berlin-Heidelberg-New York 1982.

[23] E.M. Stein and G. Weiss, Fractional integrals on n-dimensional Euclidean space. J. Math. Mech. 7 (1958), 503-514.

[24] A. Torchinsky, Interpolation of operations and Orlicz classes. Studia Math. 59 (1976), 177-207.

[25] T. Walsh, On weighted norm inequalities for fractional and singular integrals. Canad. J. Math. 23 (1971), 907-928.

M. Krbec
Mathematical Institute
Czech. Acad. Sci.
Žitná 25
115 67 Prague 1
Czechoslovakia

L. NIKOLOVA AND L.E. PERSSON

On interpolation between X^p-spaces

0. INTRODUCTION

The X^p-spaces have been used and studied in some recent papers (see e.g. [1], [20], [23] and [25]). The complex interpolation method by Calderon (see [3]) was extended to the case with families of Banach spaces by Coifman, Cwikel, Rochberg, Sagher and Weiss (see e.g. [5], [6]) and Krein and Nikolova (see e.g. [11]). In this paper we study complex interpolation between families of X^p-spaces. Moreover, we introduce the more general $X^p(\tilde{A},\omega)$-spaces ($\tilde{A}$ denotes a family of Banach spaces and ω a positive weight function) and study real interpolation between this kind of spaces. We prove some new exact descriptions, estimates and imbedding results. Several concrete examples are given and the connections to classical results of this kind are pointed out.

The paper is organized in the following way. In section 1 we collect some basic definitions, notations and other preliminaries. In section 2 we prove a useful estimate (of interpolation type), which may be regarded as an infinite-dimensional version of Hölder's inequality for X^p-spaces (see [25]). In section 3 we give an exact description of the spaces $[X^{p(e^{it})}]_\theta$. Section 4 is used to characterize the spaces $[X_0^{p_0}(\omega_0), X_1^{p_1}(\omega_1)]_\theta$. In section 5 we prove some imbedding results in connection with the spaces $(X^{p_0}(\tilde{A}_0,\omega_0), X^{p_1}(\tilde{A}_1,\omega_1))_{\theta,q}$. We do not exclude the troublesome off-diagonal case $q \neq p_\theta$. Finally, in section 6 we present some concluding remarks and examples.

In this paper we use the following conventions.

CONVENTIONS: For given Banach (or quasi-Banach) spaces A and B the notations $A \subset B$, $A = B$ and $A \equiv B$ mean that the imbedding is continuous, A is equal to B as sets, and the norms are equivalent or the norms are even equal (A coincides with B isometrically), respectively. The equivalence notation $f(t) \approx g(t)$ means that $c_0 f(t) \le g(t) \le c_1 f(t)$ for some positive constants c_0 and c_1 and all $t \in D_f = D_g$. C denotes an immaterial constant, not necessarily the same in different appearances. We let $p, p_0, p_1, \theta, p_\theta$ and q denote parameters satisfying $0 < p_0, p_1, p < \infty$, $0 < \theta < 1$, $1/p_\theta = (1-\theta)p_0 + \theta/p_1$ and $0 < q \le \infty$.

1. PRELIMINARIES

1.1 BANACH FUNCTION SPACES: Let (Ω,Σ,μ) be a σ-finite complete measure space and let $L^0(\Omega)$ denote the space of all complex-valued μ-measurable functions on Ω. A Banach subspace of $L^0(\Omega)$ is a Banach function space (on (Ω,Σ,μ)), if, for all $x \in L^0(\Omega)$, the following implication holds:

$$y \in X, \ |x| \leq |y| \quad \mu\text{-.a.e} \Rightarrow x \in X \text{ and } \|x\|_X \leq \|y\|_X.$$

1.2 GENERALIZED X^p-SPACES: Let X be a Banach function space, let $\underset{\sim}{A} = \{A_s\}_{s\in\Omega}$ be a family of Banach function spaces on Ω and let $\omega = \omega(s)$ be a positive weight function on Ω. The space $X^p(\underset{\sim}{A},\omega)$ consists of all strongly measurable functions (cross-sections) $x = x(t)$ satisfying

$$\|x\|_{X^p(\underset{\sim}{A},\omega)} = (\|(\|x(s)\|_{A_s}\omega(s))^p\|_X)^{1/p} < \infty.$$

For the case $X = L^1(\mu)$ and $A_s = A$ for all $s \in \Omega$ where A denotes a fixed Banach space (constant fibres) we have the usual space $L^p(A,\omega)$. For the cases $A_s = R$ and $\omega \equiv 1$ we use the abbreviated notations $X^p(\omega)$ and $X^p(\underset{\sim}{A})$, respectively. We obviously have $X^p(R,1) = X^p$. It is easy to prove that the X^p-spaces are Banach function spaces if $p \geq 1$ and at least quasi-Banach function spaces for all $p > 0$ (see e.g. [25]). Sometimes the space X^1 is also called the complexification of X.

1.3 FURTHER DEFINITIONS: We say that the Banach function space X is p-*convex* or q-*concave* if there exists a positive constant M such that, for every finite set $x_1,x_2,\ldots,x_n$ of elements in X we have

$$\|(\Sigma|x_i|^p)^{1/p}\|_X \leq M\,(\Sigma(\|x_i\|_X)^p)^{1/p},$$

or

$$(\Sigma(\|x_i\|_X)^q)^{1/q} \leq M\|(\Sigma|x_i|^q)^{1/q}\|_X,$$

respectively (the summation is carried out from 1 to n).

The Banach function space X has the *dominated convergence property* if, for every $x \in X$, the assumptions $|x_n(s)| \leq |x(s)|$ and $x_n(s) \to 0$ a.e. as $n \to \infty$ imply that $\|x_n\|_X \to 0$. In this case we also say that the norm of X is *absolutely continuous*.

Let X' denote the associate space of X. It is well known that X" = X if X has the Fatou property (see e.g. [12, p.45]).

1.4 THE COMPLEX INTERPOLATION METHOD FOR FAMILIES OF BANACH SPACES: First we recall some definitions from [5], [6]. A family $\underset{\sim}{X} = \{X(t)\}_{t\in[0,2\pi)}$ of Banach spaces is called an interpolation family if each X(t) is continuously imbedded in a large Banach space $(U, \|\cdot\|_U)$, if the function $t \to \|x\|_{X(t)}$ is measurable for each $x \in \bigcap_{t\in[0,2\pi)} X(t)$ and if $\|x\|_U \leq K(t)\, \|x\|_{X(t)}$ for $x \in \beta$, where

$$\beta = \{x \in \bigcap_{t\in[0,2\pi)} X(t), \int_0^{2\pi} \log^+ \|x\|_{X(t)}\, dt < \infty\}$$

and $\log^+ K(t) \in L[0,2\pi)$. (The space β is called the log-intersection space of the given family and U is called the containing space.)

Let $N^+(X(t))$ denote the space of all β-valued analytic functions of the form $g(z) = \Sigma\Psi_i(z)x_i$, where $\Psi_j \in N^+$ (N^+ denotes the usual positive Nevanlinna class on the unit disc D), for which

$$\|g\|_\infty = \sup_{t\in[0,2\pi)} \|g(t)\|_{X(t)} < \infty.$$

The space $[X(t)]_z$, $z \in D$, consists of all elements x of the form f(z) for $f \in \mathbf{F}$, where $\mathbf{F}$ denotes the completion of $N^+(X(t))$ with respect to the norm $\|\cdot\|_\infty$. It can be proved that $[X(t)]_z$ is a Banach space with the norm

$$\|x\|_{[X(t)]_z} = \inf \{\|f\|_\infty,\ f \in \mathbf{F},\ f(z) = x\}.$$

1.5 THE CALDERON CONSTRUCTION FOR FAMILIES OF BANACH FUNCTION SPACES: For a family $\underset{\sim}{X} = \{X(t)\}_{t\in[0,2\pi)}$ of Banach function spaces, the space $[X(t)]^z$, $z \in D$, consists of all $x = x_z(s) \in L^0(\Omega)$ for which there exists a $\lambda > 0$ and a function $F(t,s)$: $[0,2\pi)X\Omega \to R$, measurable on $[0,2\pi) \times \Omega$, $\|F(t,\cdot)\|_{X(t)} \leq 1$ a.e. and such that

$$|x_z(s)| \leq \lambda \exp[\frac{1}{2\pi} \int_0^{2\pi} \log|F(t,s)|\ P(z,t)dt],\ z \in D. \qquad (1.1)$$

Here and in the sequel $P(z,t)$ denotes the Poisson kernel. The norm $\|x\|_{[X(t)]^z}$ is equal to the infimum of all λ satisfying (1.1) (see [10]). This construction generalizes Calderon's original definition of the spaces $X_0^{1-\theta}X_1^{\theta}$ between two given Banach function spaces X_0 and X_1 (see [3].)

1.6 THE REAL INTERPOLATION METHOD: Let $\bar{A} = (A_0, A_1)$ denote a compatible Banach couple. For $x \in A_0 + A_1$ Peetre's K-functional is defined as follows:

$$K(t,x) = K(t,x,A_0,A_1) = \inf_{x=x_0+x_1} (\|x_0\|_{A_0} + t\|x_1\|_{A_1}),\ t > 0.$$

The Lions-Peetre real interpolation spaces $\bar{A}_{\theta,q}$ consist of all x satisfying

$$\|x\|_{\bar{A}_{\theta,q}} = [\int_0^\infty (t^{-\theta}K(t,x))^q \frac{dt}{t}]^{1/q} < \infty$$

with the usual (supremum) interpretation for the case $q = \infty$. For technical reasons we also consider the functionals

$$K^*(t,x) = K^*(t,x,A_0,A_1) = \inf_{x=x_0+x_1} (\max(\|x_0\|_{A_0}, t\|x_1\|_{A_1})),$$

$$L(t,x) = L(t,x,A_0,A_1) = \inf_{x=x_0+x_1} ((\|x_0\|_{A_0})^{p_0} + t(\|x_1\|_{A_1})^{p_1}),$$

and

$$L^*(t,x) = L^*(t,x,A_0,A_1) = \inf_{x=x_0+x_1} (\max((\|x_0\|_{A_0})^{p_0}, t(\|x_1\|_{A_1})^{p_1})).$$

We note that

$$K^*(t,x) \leq K(t,x) \leq 2K^*(t,x) \text{ and } L^*(t,x) \leq L(t,x) \leq 2L^*(t,x). \qquad (1.2)$$

The following useful relation holds:

LEMMA 1: If $u = t^{p_1}(K^*(t,x))^{p_0-p_1}$, then $L^*(u,x) = (K^*(t,x))^{p_0}$ and

$du/u = \alpha(t)dt/t$, where $\min(p_0,p_1) \leq \alpha(t) \leq \max(p_0,p_1)$.

An elementary proof can be found in [2, p.68] and the original proof is due to Peetre [21, p. 28]. □

In this paper we consider two families $\underset{\sim}{A}_0 = \{A_{0;s}\}_{s\in\Omega}$ and $\underset{\sim}{A}_1 = \{A_{1;s}\}_{s\in\Omega}$ of Banach function spaces over Ω. We assume that $\underset{\sim}{A}_0$ and $\underset{\sim}{A}_1$ are compatible families, i.e. that all the fibres $A_{0;s}$ and $A_{1;s}$, $s \in \Omega$, are continuously imbedded in a large Hausdorff topological vector space.

Finally we remark that if the assumptions above hold, then the spaces $[X(t)]_z$ and $(X^{p_0}(\underset{\sim}{A}_0,\omega_0), X^{p_1}(\underset{\sim}{A}_1,\omega_1))_{\theta,q}$ have all appropriate interpolation properties. Moreover, also the spaces $[X(t)]^z$ are interpolation spaces if we impose some appropriate additional assumptions.

2. A GENERALIZED VERSION OF HÖLDER'S INEQUALITY

Let $p(e^{it})$ be a measurable function on $[0,2\pi)$ such that $0 < p(e^{it}) < \infty$ and $1/p(e^{it}) \in L[0,2\pi)$. Let $p(z)$, $z \in D$, be defined by

$$\frac{1}{p(z)} = \int_0^{2\pi} \frac{1}{p(e^{it})} P(z,t)dt.$$

Let us consider a family of weight functions $\{\omega_t(s)\}$, $t \in [0,2\pi)$, $s \in \Omega$, such that $\omega_t(s)$ is measurable on $[0,2\pi)X\Omega$, $\log \omega_t(s) \in L[0,2\pi)$ a.e. $s \in \Omega$, and define

$$\omega = \omega_z(s) = \exp\left(\frac{1}{2\pi}\int_0^{2} \log \omega_t(s)P(z,t)dt\right), \quad z \in D, \ s \in \Omega.$$

<u>THEOREM 1</u>: Let X be a Banach function space such that $X'' = X$ and put $Y_t = X^{p(e^{it})}(\omega_t)$. If $x_t \in Y_t$ and $\log |x_t(s)| \in L[0,2\pi)$ for $s \in \Omega$ $X_t(s)$ is measurable on $[0,2\pi) \times \Omega$ and if $\log \|x_t\|_{Y_t} \in L[0,2\pi)$, then

$$x = x_z(s) = \exp\left[\frac{1}{2\pi}\int_0^{2\pi} \log|x_t(s)|\, P(z,t)dt\right]$$

belongs to $X^{p(z)}(\omega)$ and

$$\|x\|_{X^{p(z)}(\omega)} \leq \exp\left[\frac{1}{2\pi}\int_0^{2\pi} \log \|x_t\|_{Y_t} P(z,t)dt\right].$$

PROOF: Let $y_t = x_t / \|x_t\|_{Y_t}$. Then

$$(\|y_t(s)|\omega_t(s))^{p(e^{it})}\|_X = \| \frac{(|x_t(s)|\omega_t(s))^{p(e^{it})}}{(\|x_t(s)\|_{Y_t})^{p(e^{it})}} \|_X = 1$$

and $\log |y_t(s)| \in L[0,2\pi)$ for every $s \in \Omega$. Since

$$\left\{\exp\left[\frac{1}{2\pi}\int_0^{2\pi} \log|y_t(s)|\ P(z,t)dt\right]\omega(z)\right\}^{p(z)}$$

$$= \left\{\exp\left[\frac{1}{2\pi}\int_0^{2\pi} \log(|y_t(s)|\omega_t(s))P(z,t)dt\right]\right\}^{p(z)}$$

$$= \exp\left[\frac{1}{2\pi}\int_0^{2\pi} \log(|y_t(s)|\omega_t(s))^{p(e^{it})}\ \frac{p(z)}{p(e^{it})}\ P(z,t)dt\right],$$

the function exp is convex and

$$\frac{1}{2}\int_0^{2\pi} \frac{p(z)}{p(e^{it})}\ P(z,t)dt = 1$$

we can use Jensen's inequality to obtain that

$$\left\{\exp\left[\frac{1}{2}\int_0^{2\pi} \log|y_t(s)|\ P(z,t)dt\right]\omega(z)\right\}^{p(z)} \tag{2.1}$$

$$\leq \frac{1}{2\pi}\int_0^{2\pi} \frac{p(z)}{p(e^{it})}\ (|y_t(s)|\omega_t(s))^{p(e^{it})}P(z,t)dt.$$

We consider

$$U(t,s) = \frac{1}{2\pi}\ (|y_t(s)|\omega_t(s))^{p(e^{it})}\ \frac{p(z)}{p(e^{it})}\ P(z,t)$$

and note that

$$\|U(t,s)\|_X = \frac{1}{2\pi}\ \frac{p(z)}{p(e^{it})} P(z,t).$$

Therefore

$$\int_0^{2\pi} \|U(t,s)\|_X dt = 1. \tag{2.2}$$

Using (2.1), (2.2) and a generalization of Minkovski's inequality (see e.g. [12, p.45]) we find that

$$\left\{ \left\|\exp\left[\frac{1}{2\pi}\int_0^{2\pi} \log|y_t(s)| \ P(z,t)dt\right]\right\|_{X^{p(z)}(\omega)}\right\}^{p(z)}$$

$$\leq \left\|\int_0^{2\pi} U(t,s)dt\right\|_{X''} \leq \int_0^{2} \|U(t,s)\|_X \ dt = 1.$$

We insert $y_t = x_t/ \|x_t\|_{Y_t}$ into this estimate and the proof follows. □

REMARK: Let $(x_t\omega_t)^{p(e^{it})}$, $t \in [0,2\pi)$, be strongly measurable. Then $U(t,s)$ is Bochner integrable and

$$\left\|\int_0^{2} U(t,s)dt\right\|_X \leq \int_0^{2} \|U(t,s)\|_X \ dt.$$

(See e.g. [13, p. 110].) Therefore, in view of the proof above, we find that if $(x_t\omega_t)^{p(e^{it})}$ is strongly measurable, then Theorem 1 holds also without the assumption $X'' = X$.

Let $0 \leq a_0 < a_1 < \ldots < a_N = 1$, $E_j = [a_{j-1},a_j)$, $p(e^{it}) = p_j$, $t \in 2\pi E_j$, $q_j = p_j/m(E_j)$, $j = 0,1,\ldots,N$. Then, by using Theorem 1 and the remark above with $z = 0$ and $x_t(s) = |y_j(s)|^{1/m(E_j)}$, $t \in 2\pi E_j$, $j = 0,1,\ldots,N$, we obtain the following corollary:

COROLLARY 1: Let $y_i \in X^{q_i}$, $0 < q_i < \infty$, $i = 0,1,\ldots,N$. Then $\prod_0^N y_i \in X^r$, where

$$\frac{1}{r} = \sum_0^N \frac{1}{q_i} \quad \text{and} \quad \left\|\prod_0^N y_i\right\|_{X^r} \leq \prod_0^N \|y_i\|_{X^{q_i}}.$$

REMARK: Another proof of this generalization of Hölder's inequality can be found in [25].

3. THE SPACES $[X^{p(e^{it})}]_z$

If $p(e^{it}) \geq 1$, $t \in [0,2\pi)$, then the identity

$$[X^{p(e^{it})}]^z \equiv X^{p(z)}$$

follows from our Theorem 1 under some conditions on X. Using this fact together with [10, Theorem 6.1] we obtain an interpolation theorem of the following type:

THEOREM 2: Let $p(e^{it}) \geq 1$, $t \in [0,2\pi)$, and assume that the Banach function space X has the dominated convergence property. Then

$$[X^{p(e^{it})}]_z \equiv X^{p(z)}.$$

REMARK 1: THEOREM 2 holds also if the assumption that X has the dominated convergence property is replaced by the condition that X is q-concave for some $q < \infty$.

For the sake of completeness we include an independent proof of our statements.

PROOF: Let $\alpha_t = 1-1/p(e^{it})$. If X has the dominated convergence property (or satisfies the condition in remark 1) then the complexification of the Calderon space $(X)^{1-\alpha_t}(L^\infty)^{\alpha_t}$ coincides isometrically with

$$[X^1,(L^\infty)^1]_{\alpha_t} \equiv [(L^\infty)^1,X^1]_{1/p(e^{it})},$$

see e.g. [3], [10] and [12]. Using the reiteration theorem (see [5, Theorem 5.1]) and the well-known estimate $X^p \equiv (X)^{1/p}(L^\infty)^{1-1/p}$ (see e.g. [20]) we find that $\{X^{p(e^{it})}\}$, $t \in [0,2\pi)$, is an interpolation family of Banach function spaces and

$$[X^{p(e^{it})}]_z \equiv [((X)^{1-\alpha_t}(L^\infty)^{\alpha_t})^1]_z \equiv [[(L^\infty)^1,X^1]_{1/p(e^{it})}]_z$$

$$\equiv [(L^\infty)^1,X^1]_{1/p(z)} \equiv ((X)^{1/p(z)}(L^\infty)^{1-1/p(z)})^1 \equiv X^{p(z)}.$$

The proof is complete. □

COROLLARY 2: Let X satisfy the assumptions in Theorem 2 (or in Remark 1). Then

(a) $[X^{p_0}, X^{p_1}]_\theta \equiv X^{p_\theta}$ and (b) $[X^{p_0}, L^\infty]_\theta \equiv X^{p_0/(1-\theta)}$.

PROOF: Obviously (a) is the special case when $p(e^{it})$ only takes two values and the proof of (b) follows by inserting $Y = X^p$ into the identity

$$Y^{1/(1-\theta)} \equiv (Y)^{1-\theta}(L^\infty)^\theta \equiv [Y, L^\infty]_\theta .$$

REMARK: The identity (a) has also been pointed out in [26] (see Remark 4.3).

4. THE SPACES $[X_0^{p_0}(\omega_0), X_1^{p_1}(\omega_1)]_\theta$

Let X_i and ω_i, $i = 0,1$, denote Banach function spaces and positive weight functions over (Ω,Σ,μ), respectively. First we state the following lemma which is of independent interest:

LEMMA 2: Let $p_0, p_1 \geq 1$, $\mu = \theta p_\theta/p_1$. If $x_0 \in X_0^{p_0}(\omega_0)$ and $x_1 \in X_1^{p_1}(\omega_1)$, then $|x_0|^{1-\theta}|x_1|^\theta \in X_\mu^{p_\theta}(\omega_\theta)$, where $X_\mu = (X_0)^{1-\mu}(X_1)^\mu$, $\omega_\theta = \omega_0^{1-\theta}\omega_1^\theta$ and

$$\| |x_0|^{1-\theta}|x_1|^\theta \|_{X_\mu^{p_\theta}(\omega_\theta)} \leq (\|x_0\|_{X_0^{p_0}(\omega_0)})^{1-\theta}(\|x_1\|_{X_1^{p_1}(\omega_1)})^\theta .$$

PROOF: For $i = 0,1$ we define $y_i = y_i(s) = (|x_i(s)|\omega_i(s))^{p_i}/(\|x_i\|_{X_i^{p_i}(\omega_i)})^{p_i}$. Then $\|y_i\|_{X_i} = 1$, $y_0^{1-\mu}y_1^\mu \in X_\mu$ and $\|y_0^{1-\mu}y_1^\mu\|_{X_\mu} \leq 1$. We note that $1-\mu = (1-\theta)p_\theta/p_0$ and

$$y_0^{1-\mu}y_1^\mu = \frac{(|x_0|\omega_0)^{p_0(1-\mu)}(|x_1|\omega_1)^{p_1\mu}}{(\|x_0\|_{X_0^{p_0}(\omega_0)})^{p_0(1-\mu)}(\|x_1\|_{X_1^{p_1}(\omega_1)})^{p_1\mu}} =$$

$$= \left(\frac{|x_0|^{1-\theta}|x_1|^\theta\omega_\theta}{(\|x_0\|_{X_0^{p_0}(\omega_0)})^{1-\theta}(\|x_1\|_{X_1^{p_1}(\omega_1)})^\theta}\right)^{p_\theta} .$$

This means that

$$\left\| \frac{|x_0|^{1-\theta}|x_1|^{\theta}}{(\|x_0\|_{X_0^{p_0}(\omega_0)})^{1-\theta}(\|x_1\|_{X_1^{p_1}(\omega_1)})^{\theta}} \right\|_{X^{p_\theta}(\omega_\theta)} \leq 1$$

and the lemma is proved. □

THEOREM 3: Let $p_0, p_1 \geq 1$ and $\mu = \theta p_\theta/p_1$. Assume that one of the Banach function spaces X_0 and X_1 has absolutely continuous norm. Then

$$[X_0^{p_0}(\omega_0), X_1^{p_1}(\omega_1)]_\theta \equiv X_\mu^{p_\theta}(\omega_\theta),\ \omega_\theta = (\omega_0)^{1-\theta}(\omega_1)^\theta.$$

PROOF: Let $x \in [X_0^{p_0}(\omega_0), X_1^{p_1}(\omega_1)]_\theta$. Then x belongs to (the complexification of) the Calderon space $(X_0^{p_0}(\omega_0))^{1-\theta}(X_1^{p_1}(\omega_1))^\theta$, i.e. there exists $\lambda > 0$ and $x_i \in X_i^{p_i}(\omega_i)$ with $\|x_i\|_{X_i^{p_i}(\omega_i)} \leq 1$, $i = 0,1$, such that $|x(s)| \leq \lambda|x_0(s)|^{1-\theta}|x_1(s)|^\theta$. Therefore, according to Lemma 1 and the lattice property of $X_\mu^{p_\theta}(\omega_\theta)$, we conclude that $|x| \in X_\mu^{p_\theta}(\omega_\theta)$ and

$$\|x\|_{X_\mu^{p_0}(\omega_\theta)} \leq \lambda(\|x_0\|_{X_0^{p}(\omega_0)})^{1-\theta}(\|x_1\|_{X_1^{p_1}(\omega_1)})^\theta \leq \lambda.$$

Hence it follows that

$$\|x\|_{X_\mu^{p_0}(\omega_\theta)} \leq \|x\|_{(X_0^{p_0}(\omega_0))^{1-\theta}(X_1^{p_1}(\omega_1))^\theta} \leq \|x\|_{[X_0^{p_0}(\omega_0),X_1^{p_1}(\omega_1)]_\theta}.$$

On the other hand, let $x \in X_\mu^{p_\theta}(\omega_\theta)$. Then $|x\omega_\theta|^{p_\theta} \in X_\mu = (X_0)^{1-\mu}(X_1)^\mu$, i.e. there exists $\lambda > 0$ and $x_i \in X_i$ with $\|x_i\|_{X_i} \leq 1$, $i = 0,1$, such that

$$|x(s)\omega_\theta(s)|^{p_\theta} \leq \lambda|x_0(s)|^{1-\mu}|x_1(s)|^\mu.$$

We put $y_i = |x_i|^{1/p_i}/\omega_i$ and find that $\|y_i\|_{X_1^{p_i}(\omega_1)} = (\|x_i\|_{X_i})^{1/p_i} \leq 1$, $i = 0,1$, and

$$|x(s)| \leq \lambda^{1/p_\theta} \frac{|x_0(s)|^{(1-\mu)/p_\theta}|x_1(s)|^{\mu/p_\theta}}{(\omega_0(s))^{1-\theta}(\omega_1(s))^{\theta}} = \lambda^{1/p_\theta}|y_0(s)|^{1-\theta}|y_1(s)|^{\theta}.$$

We conclude that x belongs to (the complexification of) the space $((X_0^{p_0}(\omega_0))^{1-\theta}(X_1^{p_1}(\omega_1))^{\theta}$ and, thus, $x \in [X_0^{p_0}(\omega_0), X_1^{p_1}(\omega_1)]_\theta$ and the corresponding norms are even equal (see [12, pp. 244-245]). Therefore

$$\|x\|_{[X_0^{p_0}(\omega_0),X_1^{p_1}(\omega_1)]} \leq \inf \lambda^{1/p_0} = (\| \, |x\omega_\theta|^{p_\theta} \, \|_{X_\mu})^{1/p_\theta} = \|x\|_{X^{p_\theta}(\omega_\theta)}.$$

The proof is complete. □

5. THE SPACES $(X^{p_0}(\underset{\sim}{A}_0,\omega_0),X^{p_1}(\underset{\sim}{A}_1,\omega_1))_{\theta,q}$

Let ω_0 and ω_1 be positive weight functions on Ω and let $\underset{\sim}{A}_0 = \{A_{0;s}\}_{s\in\Omega}$ and $\underset{\sim}{A}_1 = \{A_{1;s}\}_{s\in\Omega}$ be compatible families of Banach function spaces over Ω.

<u>THEOREM 4</u>: Let $\underset{\sim}{A}_\theta = \{A_{\theta;s}\}_{s\in\Omega}$, where $A_{\theta;s} = (A_{0;s}, A_{1;s})_{\theta,p_\theta}$ and $\omega_\theta = (\omega_0)^{1-\theta}(\omega_1)^{\theta}$. Then for any Banach function space X,

$$(X^{p_0}(\underset{\sim}{A}_0,\omega_0), X^{p_1}(\underset{\sim}{A}_1,\omega_1))_{\theta,p_\theta} \subset X^{p_\theta}(\underset{\sim}{A}_\theta,\omega_\theta). \tag{5.1}$$

<u>PROOF</u>: Let $\mu = \theta p_\theta/p_1$. Using (1.2) and Lemma 1 we find that

$$\|x\|_{(X^{p_0}(\underset{\sim}{A}_0,\omega_0),X^{p_1}(\underset{\sim}{A}_1,\omega_1))_{\theta,p_\theta}} = \left[\int_0^\infty (t^{-\eta}L^*(t,x,X^{p_0}(\underset{\sim}{A}_0,\omega_0),X^{p_1}(\underset{\sim}{A}_1,\omega_1))) \frac{dt}{t}\right]^{1/p_\theta} \tag{5.2}$$

Moreover, for every t > 0, we have

$$L^*(t,x,X^{p_0}(\underset{\sim}{A}_0,\omega_0),X^{p_1}(\underset{\sim}{A}_1,\omega_1))$$

$$= \inf_{x=x_0+x_1} (\max(\| (\|x_0\omega_0\|_{A_{0;s}})^{p_0} \|_X, t\| (\|x_1\omega_1\|_{A_{1;s}})^{p_1} \|_X)) \tag{5.3}$$

$$\geq 0.5 \|L(t,x,A_{0;s}(\omega_0),A_{1;s}(\omega_1))\|_X .$$

According to (5.2), (5.3) and a generalized version of the triangle inequality (see e.g. [13, p. 110]) we obtain that

$$\left[\| \int_0^\infty t^{-\mu} L(t,x(s),A_{0;s}(\omega_0),A_{1;s}(\omega_1)) \frac{dt}{t} \|_X\right]^{1/p_\theta}$$

$$\leq C(\|x\|_{(X^{p_0}(\underset{\sim}{A}_0,\omega_0),X^{p_1}(\underset{\sim}{A}_1,\omega_1))_{\theta,p_\theta}}). \tag{5.4}$$

Furthermore, using Lemma 1 and(1.2) once more and making some elementary calculations, we obtain that

$$\int_0^\infty u^{-\mu} L(u,x(s),A_{0;s}(\omega_0),A_{1;s}(\omega_1)) \frac{du}{u}$$

$$\approx \int_0^\infty t^{-\mu p_1}(K(t,x(s),A_{0;s}(\omega_0),A_{1;s}(\omega_1))^{p_\theta} \frac{dt}{t}$$

$$= (\omega_\theta(s))^{p_\theta} \int_0^\infty t^{-\theta p_\theta}(K(t,x(s),A_{0;s},A_{1;s}))^{p_\theta} \frac{dt}{t} = (\omega_\theta(s) \|x(s)\|_{A_{0;s}})^{p_\theta}.$$

Inserting this estimate into (5.4) we get

$$\|\omega_\theta(s)\| x(s)\|_{A_{\theta;s}}\|_{X^{p_\theta}} \leq C \|x\|_{(X^{p_0}(\underset{\sim}{A}_0,\omega_0),X^{p_1}(\underset{\sim}{A}_1,\omega_1))_{\theta,p_\theta}}$$

and the proof is complete. □

Any description of the spaces $(X^{p_0}(\underset{\sim}{A}_0,\omega_0), X^{p_1}(\underset{\sim}{A}_1,\omega_1))_{\theta,q}$, $q \neq p_\theta$, must in general be fairly complicated. For the special case when $A_{0;s} = A_0$ and $A_{1;s} = A_1$ for all $s \in \Omega$ and $X = L^1(\mu)$ see the descritpions in [23] and the counter-example by Cwikel [7]. We state the following partial result for the case $p_0 = p_1 = p$.

THEOREM 5: Let $\gamma = 1/q - 1/p$ and $\omega_\theta = (\omega_0)^{1-\theta}(\omega_1)^\theta$. Then for any Banach function space X and any small $\varepsilon > 0$,

$$(X^p(\underset{\sim}{A}_0,\omega_0),X^p(\underset{\sim}{A}_1,\omega_0))_{\theta,q} \subset \begin{cases} \bigcap_{\psi\in Q_\varepsilon} X^p(\underset{\sim}{B},\omega_\theta) \text{ if } p < q, \\ \bigcup_{\psi\in Q_\varepsilon} X^p(\underset{\sim}{B},\omega_\theta) \text{ if } p > q, \end{cases} \tag{5.5}$$

where $B = \{B_{\psi;s}\}_{s\in\Omega}$, $B_{\psi;s} = (A_{0;s},A_{1;s})_{t^\theta\psi^\gamma(t\omega_0/\omega_1),p}$.

Here Q_ε denotes the class of positive functions ψ on $[0,\infty)$ such that $\psi(t)t^\varepsilon$ is non-decreasing, $\psi(t)t^{-\varepsilon}$ is non-increasing and $\int_0^\infty \psi(t)\,dt/t = 1$.

Moreover, $B_{\psi,s}$ is a parameter function space defined by replacing t^θ by $t^\theta\psi^\gamma(t\omega_0/\omega_1)$ in the definition of the space $A_{\theta,p}$ (see e.g. [24] and the references given there).

PROOF: The proof of Theorem 5 can be carried out by first making quasi-linearization with aid of Lemma 2.3 in [23] and by using similar arguments as in the proof of Theorem 4, so we omit the details. □

REMARK: Analysing our proofs in this section we see that '$\subset$' in (5.1) and (5.5) can be replaced by '=' if, in addition, X satisfies that

$$\int_0^\infty \|f(t,s)\|_X \, dt \leq C \left\|\int_0^\infty f(t,s)dt\right\|_X \tag{5.6}$$

for every non-negative $f(t,s)$ on $R\times\Omega$. In particular, if $A_{0;s} = A_0$, $A_{1;s} = A_1$ for every $s \in \Omega$ and $X = L^1(\Omega)$ we get the usual Lions-Peetre formula

$$(L^{p_0}(A_0,\omega_0),\ L^{p_1}(A_1,\omega_1))_{\theta,p_\theta} = L^{p_\theta}((A_0,A_1)_{\theta,p_\theta},\omega_\theta)$$

and a special case of the descriptions given in [23] for the off-diagonal case $q \neq p_\theta$.

We complete this section by stating the following corollary:

COROLLARY 3: Let $\underset{\sim}{A} = \{A_{\theta,q;s}\}_{s\in\Omega}$, where $A_{\theta,q;s} = (A_{0;s},A_{1;s})_{\theta,q}$ and let X denote a Banach function space.

(a) If $0 < q \leqq p < \infty$, then

$$(X^p(\underset{\sim}{A_0}),\ X^p(\underset{\sim}{A_1}))_{\theta,q} \subset X^p(\underset{\sim}{A}).$$

(b) If $0 < p \leqq q < \infty$ and (5.6) holds, then

$$(X^p(\underset{\sim}{A_0}),\ X^p(\underset{\sim}{A_1}))_{\theta,q} \supset X^p(\underset{\sim}{A}).$$

PROOF: Let $0 < q < p < \infty$ and assume that $x = x(s) \in (X^p(A_0),X^p(A_1))_{\theta,q}$. Using Theorem 5 with $\omega_0 \equiv \omega_1 \equiv 1$, Hölder's inequality and the lattice property we find that

$$(||x||_{(X^p(\underset{\sim}{A_0}),X^p(\underset{\sim}{A_1}))_{\theta,q}})^p \geqq C\ ||\int_0^\infty \left(\frac{K(t,x(s),A_{0,s},A_{1,s})}{t^\theta \psi^\gamma(t)}\right)^p \frac{dt}{t}||_X$$

$$\geqq C||\left[\int_0^\infty \left(\frac{K(t,x(s),A_{0,s},A_{1,s})}{t^\theta}\right)^q \frac{dt}{t}\right]^{p/q}||_X = C(||\,\|x(s)\|_{A_{\theta,q;s}}||_{X^p})^p$$

$$= C(||x||_{X^p(\underset{\sim}{A})})^p.$$

For the case q = p Corollary 3 is a special case of Theorem 4. The proof of (a) is complete. The proof of (b) is similar, so we omit the details. □

REMARK: For the case $q \geqq 1$, $A_0 = \{A_0\}$, $A_1 = \{A_1\}$ and $X = L^1(\mu)$ or $X = \ell^1$ another proof of Corollary 3 can be found in [14, pp. 197-198].

6. CONCLUDING REMARKS AND EXAMPLES

First we remark that it is well known that the space $[X(t)]^z$ need not necessarily be an interpolation space even in the case with only two spaces X_0 and X_1. In fact, Lozanovskii [15] has proved that there exists two Banach function spaces X and Y and an operator T acting continuously from

X into Y and from L^∞ into L^∞, but not from $(X)^{1-\theta}(L^\infty)^\theta$ into $(Y)^{1-\theta}(L^\infty)^\theta$, i.e. not from X^p into Y^p, $p = 1/(1-\theta)$.

EXAMPLE 1: We note that, for any Banach function spaces X_0 and X_1,

$$K^*(t,x,X_0^p,X_1^p) = (K^*(t^p,|x|^p,X_0,X_1))^{1/p}.$$

Thus, by (1.2), we have the estimate

$$K(t,x,X_0^p,X_1^p) \approx (K(t^p,|x|^p,X_0,X_1))^{1/p}. \tag{6.1}$$

For the case with symmetric spaces and $p \geqq 1$ see also Maligranda [16, p. 176]. In particular (6.1) implies the following convexification (concavification) formula

$$((X_0,X_1)_{\theta,q})^p = (X_0^p,X_1^p)_{\theta,pq}.$$

We give another example concerning interpolation between martingale spaces and refer to the notations used in [19].

EXAMPLE 2: It is easy to verify that

$$(MHX)^p = MHX^p,\ (MX)^p = MX^p \text{ and } (Mx)^p = M|x|^p.$$

Therefore, according to (6.1) and the estimate

$$K(t,x,MH^1,ML^\infty) \approx \int_0^t (Mx)^*(u)du$$

(see [18, Theorem 3.2]) we have

$$K(t,x,MH^p,ML^\infty) \approx \left[\int_0^{t^p} (Mx^*(u))^p \, du\right]^{1/p}$$

and, thus, if $p > 1$, then

$$K(t,x,ML^p,ML^\infty) \approx \left[\int_0^{t^p} (Mx^*(u))^p \, du\right]^{1/p}. \tag{6.2}$$

We conclude that the crucial estimate in [19, p. 315] in fact can be replaced by the equivalence (6.2) at least for the case $p > 1$.

In a similar way we find that Kree's estimate

$$K(t,x,L^p,L^\infty) = \left[\int_0^{t^p} (x^*(u))^p \, du\right]^{1/p}$$

follows at once from (6.1) and Peetre's original identity for $p = 1$ (see [2]).

EXAMPLE 3 (on the coincidence between real and complex interpolation spaces): It is well known that

$$[X_0,X_1]_\theta \supset (X_0,X_1)_{\theta,p} \tag{6.3}$$

if X_i is of Fourier type p_i, $i = 1,2$ (see [22]). According to our Theorems 3 and 4 we see that (6.3) holds also if $X_0 = X^{p_0}(\omega_0)$, $X_1 = X^{p_1}(\omega_1)$, where X is any Banach function space with absolutely continuous norm. Moreover, we find that the identity

$$(X^{p_0}(\omega_0),\ X^{p_1}(\omega_1))_{\theta,p_\theta} = [X^{p_0}(\omega_0),\ X^{p_1}(\omega_1)]_\theta$$

holds if X has absolutely continuous norm and (5.6) is satisfied (e.g. if $X = L^1(\mu)$). On the other hand Cwikel and Nilsson [8] have proved that if the spaces $[Y(\omega_0),\ Y(\omega_1)]_\theta$ are, to within equivalence of norms, spaces generated by the K-functional of $(Y(\omega_0),\ Y(\omega_1))$ for all ω_0,ω_1 and the constants of equivalence are independent of ω_0 and ω_1, then $X = L^p(\mu)$ for some p.

We remark that the results in section 5 can also be formulated and proved for the more general function parameter case (for basic definitions, notations and results in this connection we refer to [24]). We only point out the following generalization of Corollary 3 and a recent statement in [4].

EXAMPLE 4: Let $\underset{\sim}{A} = \{A_{\lambda,q;s}\}_{s\in\Omega}$, where $A_{\lambda,q;s} = (A_{0;s},A_{1;s})_{\lambda,q}$ and $\lambda = \lambda(t)$ is a parameter function and let X denote a Banach function space.

(a) If $0 < q \leqq p < \infty$, then

$$(X^p(\underset{\sim}{A}_0),\ X^p(\underset{\sim}{A}_1))_{\lambda,q} \subset X^p(\underset{\sim}{A}).$$

(b) If $0 < p \leq q < \infty$ and (5.6) holds, then

$$(X^p(\underset{\sim}{A}_0),\ X^p(\underset{\sim}{A}_1))_{\lambda,q} \supset X^p(\underset{\sim}{A}).$$

For the constant fibre case $\underset{\sim}{A}_0 = \{A_0\}$, $\underset{\sim}{A}_1 = \{A_1\}$, and when $X = L^1(\mu)$ and $q = p$ another proof of the statement above can be found in [4].

In this paper we have assumed that X is a Banach function space, but in many cases the corresponding results can be proved also if X is only a c-quasi-Banach function space, i.e. if the usual triangle inequality is replaced by the c-triangle inequality $\|x+y\| \leq c(\|x\| + \|y\|)$, $c > 1$. We only give the following example.

EXAMPLE 5: Let X be a c-quasi-function space and let $x_0 \in X^{p_0}$, $x_1 \in X^{p_1}$. Then, by arguing as in the proof of Lemma 2, we find that $|x_0|^{1-\theta}|x_1|^{\theta} \in X^{p_\theta}$ and

$$\|\,|x_0|^{1-\theta}|x_1|^{\theta}\|_{X^{p_\theta}} \leq c^{1/p_\theta}(\|x_0\|_{X^{p_0}})^{1-\theta}(\|x_1\|_{X^{p_1}})^{\theta}.$$

It follows that $(X^{p_0})^{1-\theta}(X^{p_1})^{\theta} = X^{p_\theta}$ and

$$\|x\|_{(X^{p_0})^{1-\theta}(X^{p_1})^{\theta}} \leq \|x\|_{X^{p_\theta}} \leq c^{1/p_\theta}\|x\|_{(X^{p_0})^{1-\theta}(X^{p_1})^{\theta}},\quad p_0,p_1 \geq 1.$$

EXAMPLE 6: Cwikel and Reisner [9] proved that if at least one of the spaces X_0 and X_1 is uniformly convex, then also $[X_0,X_1]_\theta$ is uniformly convex and we have also a corresponding estimate of the modules of convexity. Using this fact we find that if X satisfies the assumptions in Corollary 2 (or in Remark 1) and if X^p is uniformly convex for some p, then X^q is uniformly convex for every $q > p$. Moreover, for every $p > 1$, $\delta_{X^p}(\varepsilon) \succ \delta_X(\varepsilon^p)$ ($f \succ g$ if there exist $a,b > 0$ such that $af(bt) \geq g(t)$ for all $t \in R_+$). Another result concerning uniform convexity of X^p-spaces can be found in [25].

References

[1] J. Arazy, The K-functional of certain pairs of rearrangement invariant spaces, Bull. Austral. Math. Soc., 27 (1983), 249-257.

[2] J. Bergh and J. Löfström, Interpolation Spaces, An Introduction, Grundlehren der Matematischen Wissenschaften 223, Springer-Verlag (1976).

[3] A.P. Calderon, Intermediate spaces and interpolation, the complex method, Studia Math., 24 (1964), 114-190.

[4] F. Cobos, Some spaces in which martingale difference sequences are unconditional, Bull. Acad. Pol. Sci. Math., 34 (1986), 695-703.

[5] R. Coifman, M. Cwikel, R. Rochberg, Y. Sagher and G. Weiss, A theory of complex interpolation for families of Banach spaces, Adv. Math., 43 (1982), 203-229.

[6] R. Coifman, M. Cwikel, R. Rochberg, Y. Sagher and G. Weiss, The complex method of interpolation of operators acting on families of Banach spaces, Lecture Notes in Math. 779, Springer-Verlag (1980), 123-153.

[7] M. Cwikel, On $(L^{p_0}(A_0), L^{p_1}(A_1))_{\theta,q}$, Proc. Amer. Math. Soc., 44:2 (1974), 286-292.

[8] M. Cwikel and P. Nilsson, The coincidence of real and complex interpolation methods for couples of weighted Banach lattices, Lecture Notes in Math. 1070, Springer-Verlag (1984), 54-65.

[9] M. Cwikel and S. Reisner, Interpolation of uniformly convex Banach spaces, Proc. AMS, 84 (1982), 55-59.

[10] E. Hernandes, Intermediate spaces and the complex method of interpolation for families of Banach spaces, Ann. Scuola Norm. Sup. Pisa, 13 (1986), 245-266.

[11] S.G. Krein and L.I. Nikolova, Holomorphic functions in families of Banach spaces and interpolation, Dokl. Akad. Nauk SSSR, 250 (1980), 547-550; Sov. Math. Dokl., 21 (1980), 131-134.

[12] S.G. Krein, Ju.I. Petunin and E.M. Semenov, Interpolation of linear operators, Nauka (1978) (Russian); English translation, Amer. Math. Soc. (1982).

[13] A. Kufner, O. John and S. Fucik, Function Spaces, Noordhoff, (1977).

[14] H. König, Eigenvalue Distribution of Compact Operators, Birkhäuser (1986).

[15] G.Ja. Lozanovskii, Remark of an interpolation theorem of Calderon, Funkcional. Anal. Prilozen., 6 (1972), 89-90; Functional Anal. Appl., 6 (1972).

[16] L. Maligranda, The K-functional for symmetric spaces, Lecture Notes in Math. 1070, Springer-Verlag (1984), 169-182.

[17] L. Maligranda and L.E. Persson, Generalized duality of some Banach function spaces, Research report 4, Dept of Math., Luleå University (1987); Proc. Netherlands Acad. Sci. (to appear).

[18] M. Milman, Interpolation of some concrete scales of spaces, Research report 7, Dept of Math., Lund University (1982).

[19] M. Milman, On interpolation of martingale L^p-spaces, Indiana Math. J., 30 (1981), 313-318.

[20] P. Nilsson, Interpolation of Banach lattices, Studia Math., 82 (1985), 135-154.

[21] J. Peetre, A new approach in interpolation spaces, Studia Math., 34 (1970), 23-42.

[22] J. Peetre, Sur la transformation de Fourier des fonctions a valeurs vectorielles, Rend. Sem. Mat. Padova, 42 (1969), 15-26.

[23] L.E. Persson, Descriptions of some interpolation spaces in off-diagonal cases, Lecture Notes in Math. 1070, Springer-Verlag (1984), 213-231.

[24] L.E. Persson, Interpolation with a parameter function, Math. Scand. 59 (1986), 199-222.

[25] L.E. Persson, Some elementary inequalities in connection with X^p-spaces, Proceedings from the conference 'Constructive Theory of Functions' Varna, June 1987, Publishing House of the Bulgarian Academy of Sciences (to appear).

[26] G. Pisier, Some applications of the complex interpolation method to Banach lattices, J. d'Anal. Math., 35 (1979), 264-281.

Ljudmila I. Nikolova
Department of Mathematics
Sofia University
A. Ivanov 5
1126 Sofia
Bulgaria

Lars Erik Persson
Department of Mathematics
Lulea University
S-95187 Lulea
Sweden

B. OPIC AND P. GURKA

N-dimensional Hardy inequality and imbedding theorems for weighted Sobolev spaces on unbounded domains

1. INTRODUCTION

The inequality

$$\left[\int_\Omega |u(x)|^q\, w(x)\, dx\right]^{1/q} \le C\left[\sum_{i=1}^N \int_\Omega \left|\frac{\partial u(x)}{\partial x_i}\right|^p v_i(x)\, dx\right]^{1/p},\quad u \in T(\Omega) \tag{1.1}$$

where $1 \le p,q < \infty$, $w, v_1, \ldots, v_N$ are weight functions on the domain Ω, $T(\Omega)$ is a certain class of functions containing $C_0^\infty(\Omega)$ and C is a positive constant independent of $u \in T(\Omega)$, is called the *N-dimensional Hardy inequality* (see [11]).

General necessary and sufficient conditions for the validity of the inequality (1.1) on the class of functions $C_0^\infty(\Omega)$ were found by V.G. Maz'ja (see [7]). These conditions involving capacities can be difficult to verify because there are no good means of computing the capacity of an arbitrary measurable set.

In 1985 Wheeden, Gatto and Guitiérrez [2] proved that for $1 \le p \le q < \infty$, $\Omega = R^N \setminus \{0\}$, $w(x) = |x|^\alpha$, $v_i(x) = |x|^\beta$, $i = 1,\ldots,N$, the inequality (1.1) holds on the class $C_0^\infty(\Omega)$ if and only if

$$N\left(\frac{1}{q} - \frac{1}{p}\right) + 1 \ge 0,\quad \frac{\alpha}{q} - \frac{\beta}{p} + N\left(\frac{1}{q} - \frac{1}{p}\right) + 1 = 0,\ \beta \ne p - N. \tag{1.2}$$

This result is a consequence of the theorems on continuity of fractional integral operators in weighted H^p spaces.

In [4], [10] the authors found simple necessary and sufficient conditions for the validity of the N-dimensional Hardy inequality

$$\left[\int_\Omega |u(x)|^q \operatorname{dist}^\alpha(x,\partial\Omega)dx\right]^{1/q} \le C\left[\sum_{i=1}^N \int_\Omega \left|\frac{\partial u(x)}{\partial x_i}\right|^p \operatorname{dist}^\beta(x,\partial\Omega)dx\right]^{1/p},$$
$$u \in C_0^\infty(\Omega),$$

assuming $\Omega \in C^{0,1}$, $p,q \in \langle 1,\infty)$, $\beta < p - 1$. These conditions read as follows:

either

$$1 \leq p \leq q < \infty, \; N(\frac{1}{q} - \frac{1}{p}) + 1 \geq 0, \; \frac{\alpha}{q} - \frac{\beta}{p} + N(\frac{1}{q} - \frac{1}{p}) + 1 \geq 0 \tag{1.3}$$

or

$$1 \leq q < p < \infty, \; \frac{\alpha}{q} - \frac{\beta}{p} + \frac{1}{q} - \frac{1}{p} + 1 > 0. \tag{1.4}$$

The first part of this result (the condition (1.3)) is very similar to (1.2). Nevertheless, our method differs from that of Wheeden, Gatto and Gutiérrez [2]. To obtain (1.3) and (1.4) we make use of necessary and sufficient conditions for imbeddings of weighted Sobolev spaces $W_0^{1,p}(\Omega;d^\beta(x),d^\beta(x))$ into weighted Lebesgue spaces $L^q(\Omega;d^\alpha(x))$, and of theorems on equivalent norms on spaces $W_0^{1,p}(\Omega;d^\beta(x),d^\beta(x))$ (here $d(x) = \text{dist}(x,\partial\Omega)$).

The aim of this paper is to establish simple necessary and sufficient conditions for the validity of the N-dimensional inequality provided that the set Ω is an unbounded domain of a special type. We will use the method from our papers [4], [10]. Similarly as in [4], [10], here the principal role is also played by imbedding theorems for weighted Sobolev spaces.

2. NOTATION

Throughout this paper we will suppose that Ω is an *unbounded domain* in $\mathbb{R}^N$ with a boundary $\partial\Omega$. For $n \in \mathbb{N}$ we set $\Omega_n = \{x \in \Omega;\ |x| < n\}$, $\Omega^n = \text{int}(\Omega \setminus \Omega_n)$. The open ball with a radius R and a centre x is denoted by $B(x,R)$. We use $\mathcal{W}(\Omega)$ to denote the set of *weight functions* on Ω, i.e. the set of all measurable, a.e. in Ω positive and finite functions. For $w \in \mathcal{W}(\Omega)$, $1 \leq q < \infty$ the *weighted Lebesgue space* $L^q(\Omega;w)$ is the set of all measurable functions u defined on Ω with a finite norm

$$\|u\|_{q,\Omega,w} = \left[\int_\Omega |u(x)|^q \, w(x) \, dx\right]^{1/q}.$$

Further we assume that (see [6])

$$v_0, v_1 \in \mathcal{W}(\Omega) \cap L^1_{loc}(\Omega), \quad v_0^{-1/p}, v_1^{-1/p} \in L^{p'}_{loc}(\Omega) \quad (\frac{1}{p} + \frac{1}{p'} = 1). \quad (2.1)$$

We define the *weighted Sobolev space* $W^{1,p}(\Omega;v_0,v_1)$ as the set of all functions $u \in L^p(\Omega;v_0)$ which have distributional derivatives $\partial u/\partial x_i \in L^p(\Omega;v_1)$, $i = 1,\dots,N$. The space $W^{1,p}(\Omega;v_0,v_1)$ with the norm

$$\|u\|_{1,p,\Omega,v_0,v_1} = \left(\|u\|^p_{p,\Omega,v_0} + \sum_{i=1}^{N} \left\|\frac{\partial u}{\partial x_i}\right\|^p_{p,\Omega,v_1} \right)^{1/p} \quad (2.2)$$

is a Banach space. Finally, we define the space $W_0^{1,p}(\Omega;v_0,v_1)$ as the closure of $C_0^\infty(\Omega)$ with respect to the norm (2.2).

For two Banach spaces X, Y, we write $X \hookrightarrow Y$ or $X \hookrightarrow\hookrightarrow Y$ if $X \subset Y$ and the natural injection of X into Y is continuous or compact, respectively.

Let $I = (n,\infty)$, $n \in \mathbb{N}$, $r : I \to (0,\infty)$. We write $r \in V(n)$ if

(i) r is continuous and non-decreasing on I, $r(x) \leq x/2$ for $x \in I$;

(ii) $x - r(x)$ is non-decreasing on I, $\lim_{x\to\infty} [x - r(x)] = +\infty$;

(iii) there exists a constant $c \geq 1$ such that

$$c^{-1} \leq \frac{r(y)}{r(x)} \leq c$$

for all $x \in I$ and all $y \in (x - r(x), x + r(x)) \cap I$.

3. IMBEDDING THEOREMS

In this section we present some imbedding theorems for Sobolev weighted spaces on unbounded domains Ω of a special type. These theorems are based on the following two lemmas, the proofs of which can be found in [8].

3.1 LEMMA[1]: Suppose $p, q \in \langle 1,\infty)$. If

$$W^{1,p}(\Omega_n;v_0,v_1) \hookrightarrow L^q(\Omega_n;w) \quad \forall n \in \mathbb{N}$$

and

---

[1] Let us remark that an analogous assertion also holds for spaces $W_0^{1,p}(\Omega;v_0,v_1)$ (see [8]).

$$\lim_{n\to\infty} \sup_{\|u\|_{1,p,\Omega,v_0,v_1} \leq 1} \|u\|_{q,\Omega^n,w} < \infty, \tag{3.1}$$

then

$$W^{1,p}(\Omega;v_0,v_1) \hookrightarrow L^q(\Omega;w). \tag{3.2}$$

Conversely, if (3.2) holds, then the condition (3.1) is satisfied.

3.2 LEMMA[1]: Suppose $p,q \in \langle 1,\infty)$. If

$$W^{1,p}(\Omega_n;v_0,v_1) \hookrightarrow\hookrightarrow L^q(\Omega_n;w) \quad \forall n \in \mathbb{N}$$

and

$$\lim_{n\to\infty} \sup_{\|u\|_{1,p,\Omega,v_0,v_1} \leq 1} \|u\|_{q,\Omega^n,w} = 0, \tag{3.3}$$

then

$$W^{1,p}(\Omega;v_0,v_1) \hookrightarrow\hookrightarrow L^q(\Omega;w). \tag{3.4}$$

Conversely, if (3.4) holds, then the condition (3.3) is satisfied.

Now we are going to formulate our main results concerning the above-mentioned imbeddings.

(A) THE CASE $1 \leq p \leq q < \infty$

3.3 THEOREM (the continuous imbedding): Let $1 \leq p \leq q < \infty$, $N(q^{-1} - p^{-1}) + 1 \geq 0$. Suppose that the following conditions are fulfilled:

(D1) There exists $n_0 \in \mathbb{N}$ such that $\Omega^{n_0} = \{x \in \mathbb{R}^N;\ |x| > n_0\}$.

(D2) $W^{1,p}(\Omega_n;v_0,v_1) \hookrightarrow L^q(\Omega_n;w)$, $n \geq n_0$.

(D3) There exist positive constants $c_0 \leq C_0$, $c_1 \leq C_1$, positive measurable functions a_0, a_1 defined on Ω^{n_0} and a function $r \in V(n_0)$ such that

$$c_0\, a_0(x) \leq w(y) \leq C_0\, a_0(x), \quad c_1\, a_1(x) \leq v_1(y) \leq C_1\, a_1(x)$$

for all $x \in \Omega^{n_0}$ and for a.e. $y \in B(x, r(|x|))$.

(D4) There exist positive constants $k_0 \leq K_0$ such that

$$k_0\, v_0(x) \leq v_1(x)\, r^{-p}(|x|) \leq K_0\, v_0(x) \quad \text{for a.e.} \quad x \in \Omega^{n_0}.$$

Then

$$W^{1,p}(\Omega; v_0, v_1) \hookrightarrow L^q(\Omega; w)$$

(and also $W_0^{1,p}(\Omega; v_0, v_1) \hookrightarrow L^q(\Omega; w)$) if and only if

$$\text{(D5)} \quad \lim_{n\to\infty} \sup_{x \in \Omega^n} \frac{a_0^{1/q}(x)}{a_1^{1/p}(x)}\, r^{N/q - N/p + 1}(|x|) < \infty .$$

3.4 THEOREM (the compact imbedding): Let $1 \leq p \leq q < \infty$, $N(q^{-1} - p^{-1}) + 1 > 0$. Suppose that (D1), (D3), (D4) and the following condition are fulfilled:

$$\text{(D2*)} \quad W^{1,p}(\Omega_n; v_0, v_1) \hookrightarrow\hookrightarrow L^q(\Omega_n; w), \quad n \geq n_0.$$

Then

$$W^{1,p}(\Omega; v_0, v_1) \hookrightarrow\hookrightarrow L^q(\Omega; w)$$

(and also $W_0^{1,p}(\Omega; v_0, v_1) \hookrightarrow\hookrightarrow L^q(\Omega; w)$ if and only if

$$\text{(D5*)} \quad \lim_{n\to\infty} \sup_{x \in \Omega^n} \frac{a_0^{1/q}(x)}{a_1^{1/p}(x)}\, r^{N/q - N/p + 1}(|x|) = 0.$$

For the proofs of Theorems 3.3 and 3.4 see [3]. □

If a domain Ω satisfies the condition (D1), we write $\Omega \in D1$. For such domains Ω we put $\underline{a} = \inf\{|x|;\ x \in \Omega\}$. If $\Omega \in D1$ and each Ω_n, $n \geq n_0$, has the *cone property* (in the sense of [1]), we write $\Omega \in C1$.

From Theorems 3.3 and 3.4 we obtain:

3.5 EXAMPLE: Suppose $\Omega \in C1$, $1 \leqq p \leqq q < \infty$, $\alpha, \beta \in \mathbb{R}$.

(I) Let $\underline{a} > 0$. Then

$$W^{1,p}(\Omega;|x|^{\beta-p},|x|^{\beta}) \hookrightarrow L^{q}(\Omega;|x|^{\alpha})$$

or

$$W^{1,p}(\Omega;|x|^{\beta-p},|x|^{\beta}) \hookrightarrow\hookrightarrow L^{q}(\Omega;|x|^{\alpha})$$

if and only if

$$N(\frac{1}{q} - \frac{1}{p}) + 1 \geqq 0, \quad \frac{\alpha}{q} - \frac{\beta}{p} + N(\frac{1}{q} - \frac{1}{p}) + 1 \leqq 0$$

or

$$N(\frac{1}{q} - \frac{1}{p}) + 1 > 0, \quad \frac{\alpha}{q} - \frac{\beta}{p} + N(\frac{1}{q} - \frac{1}{p}) + 1 < 0,$$

respectively[2].

(II) Let $\Omega = \mathbb{R}^N \setminus \{0\}$ or $\Omega = \mathbb{R}^N$. Then

$$W^{1,p}(\Omega;|x|^{\beta-p},|x|^{\beta}) \hookrightarrow L^{q}(\Omega;|x|^{\alpha})$$

if and only if

$$N(\frac{1}{q} - \frac{1}{p}) + 1 \geqq 0, \quad \frac{\alpha}{q} - \frac{\beta}{p} + N(\frac{1}{q} - \frac{1}{p}) + 1 = 0.$$

The space $W^{1,p}(\Omega;|x|^{\beta-p},|x|^{\beta})$ is not compactly imbedded into $L^{q}(\Omega;|x|^{\alpha})$ for any $\alpha, \beta \in \mathbb{R}$.

(Let us remark that in the case $\Omega = \mathbb{R}^N$ the spaces $W_0^{1,p}(\Omega;|x|^{\beta-p},|x|^{\beta})$, $W^{1,p}(\Omega;|x|^{\beta-p},|x|^{\beta})$ are defined (due to the condition (2.1)) only for $p - N < \beta < N(p - 1)$.)

---

2 In Theorems 3.3 and 3.4 we take $r(x) = x/3$ for $x > n_0$ and $a_0(x) = |x|^{\alpha}$, $a_1(x) = |x|^{\beta}$ for $x \in \Omega^{n_0}$.

3.6 EXAMPLE: Suppose $\Omega \in C1$, $\underline{a} > 1$, $1 \leqq p \leqq q < \infty$. For $x \in \Omega$ we put

$$w(x) = |x|^{\alpha} \log^{\gamma}|x|, \quad v_0(x) = |x|^{\beta-p} \log^{\delta}|x|, \quad v_1(x) = |x|^{\beta} \log^{\delta}|x|.$$

(I) Then

$$W^{1,p}(\Omega;v_0,v_1) \hookrightarrow L^q(\Omega;w)$$

if and only if

$$N(\frac{1}{q} - \frac{1}{p}) + 1 \geqq 0, \quad \frac{\alpha}{q} - \frac{\beta}{p} + N(\frac{1}{q} - \frac{1}{p}) + 1 < 0$$

or

$$\frac{\alpha}{q} - \frac{\beta}{p} + N(\frac{1}{q} - \frac{1}{p}) + 1 = 0, \quad \frac{\gamma}{q} - \frac{\delta}{p} \leqq 0.$$

(II) Then

$$W^{1,p}(\Omega;v_0,v_1) \hookrightarrow\hookrightarrow L^q(\Omega;w)$$

if and only if

$$N(\frac{1}{q} - \frac{1}{p}) + 1 > 0, \quad \frac{\alpha}{q} - \frac{\beta}{p} + N(\frac{1}{q} - \frac{1}{p}) + 1 < 0$$

or

$$\frac{\alpha}{q} - \frac{\beta}{p} + N(\frac{1}{q} - \frac{1}{p}) + 1 = 0, \quad \frac{\gamma}{q} - \frac{\delta}{p} < 0.$$

(In Theorems 3.3 and 3.4 we take $r(x) = x/3$ for $x > n_0$ and $a_0(x) = w(x)$, $a_1(x) = v_1(x)$ for $x \in \Omega^{n_0}$.)

3.7 EXAMPLE: Suppose $\Omega \in C1$, $1 \leqq p \leqq q < \infty$. Then

$$W^{1,p}(\Omega;\, e^{\beta|x|},\, e^{\beta|x|}) \hookrightarrow L^q(\Omega;\, e^{\alpha|x|})$$

or

$$W^{1,p}(\Omega;\ e^{\beta|x|},\ e^{\beta|x|}) \hookrightarrow\hookrightarrow L^q(\Omega;\ e^{\alpha|x|})$$

if and only if

$$N(\frac{1}{q} - \frac{1}{p}) + 1 \geqq 0, \quad \frac{\alpha}{q} - \frac{\beta}{p} \leqq 0$$

or

$$N(\frac{1}{q} - \frac{1}{p}) + 1 > 0, \quad \frac{\alpha}{q} - \frac{\beta}{p} < 0,$$

respectively. (In Theorems 3.3 and 3.4 we take $r(x) \equiv 1$ for $x > n_0$, $a_0(x) = e^{\alpha|x|}$, $a_1(x) = e^{\beta|x|}$ for $x \in \Omega^{n_0}$.)

(B) THE CASE $1 \leqq q < p < \infty$

The following lemma plays the principal role in the proofs of the imbedding theorems with $1 \leqq q < p < \infty$.

3.8 LEMMA (the one-dimensional Hardy inequality): Let $1 \leqq q < p \leqq \infty$, $1/r = 1/q - 1/p$, $-\infty \leqq a < b \leqq +\infty$, $\omega_0, \omega_1 \in W((a,b))$. Then there exists a positive constant C such that the inequality

$$\left[\int_a^b |u(t)|^q\, \omega_0(t)\, dt\right]^{1/p} \leqq C\left[\int_a^b |u'(t)|^p\, \omega_1(t)\, dt\right]^{1/p} \tag{3.5}$$

holds for all functions

$$u \in T_1(a,b) = \{f \in AC((a,b));\ \lim_{t\to a_+} f(t) = 0\} \tag{3.6}$$

or

$$u \in T_2(a,b) = \{f \in AC((a,b));\ \lim_{t\to b_-} f(t) = 0\} \tag{3.7}$$

or

$$u \in T(a,b) = T_1(a,b) \cap T_2(a,b) \tag{3.8}$$

if and only if

$$A_1(a,b;\omega_0,\omega_1)$$

$$= \left\{\int_a^b \left[\left(\int_x^b \omega_0(t)dt\right)^{1/q} \left(\int_a^x \omega_1^{1-p'}(t)\, dt\right)^{1/q'}\right]^r \omega_1^{1-p'}(x)\, dx\right\}^{1/r} < \infty \tag{3.9}$$

or

$$A_2(a,b;\omega_0,\omega_1)$$

$$= \left\{\int_a^b \left[\left(\int_a^x \omega_0(t)\, dt\right)^{1/q} \left(\int_x^b \omega_1^{1-p'}(t)\, dt\right)^{1/q'}\right]^r \omega_1^{1-p'}(x)\, dx\right\}^{1/r} < \infty \tag{3.10}$$

or

$$A(a,b;\omega_0,\omega_1)$$

$$= \inf_{c \in \langle a,b \rangle} \max \{A_1(a,c;\omega_0,\omega_1),\ A_2(c,b;\omega_0,\omega_1)\} < \infty \ , \tag{3.11}$$

respectively[3].

For the proof see [7] (the cases (3.6) and (3.7)) and [9] (the case (3.8)).

3.9 REMARK: If $C > 0$ is the least constant such that the inequality (3.5) holds on the class $T_i(a,b)$ ($i = 1,2$) or $T(a,b)$, then there is an estimate of C in terms of $A_i(a,b;\omega_0,\omega_1)$ or $A(a,b;\omega_0,\omega_1)$, respectively (see [7], [9]).

We introduce the following notation:

For an unbounded open interval $I \subset R$ we define $\mathcal{W}_B(I)$ as the set of all weight functions $\rho \in \mathcal{W}(I)$ bounded from below and from above by positive constants on each bounded interval $J \subset I$.

---

3 For $a = c$ we set $A_1(a,c;\omega_0,\omega_1) = 0$ and similarly we put $A_2(c,b;\omega_0,\omega_1) = 0$ if $c = b$.

In the rest of the paper we suppose that the weight functions[4] w, v_0, v_1 are *radial*, i.e.

$$w(x) = \bar{w}(|x|), \quad v_i(x) = \bar{v}_i(|x|), \quad i = 0,1, \quad x \in \Omega.$$

3.10 THEOREM: Suppose $\Omega \in C1$, $1 \leq q < p < \infty$, $\bar{w}, \bar{v} \in \mathcal{W}_B((\underline{a},\infty))$,[5] $v \in C(\Omega)$. Let there exist $R \in \langle \underline{a},\infty)$ such that

$$A_1(R,\infty;\bar{w}(t)t^{N-1},\bar{v}(t)t^{N-1}) < \infty.$$

Then

$$W^{1,p}(\Omega;v,v) \hookrightarrow\hookrightarrow L^q(\Omega;w).$$

3.11 THEOREM: Suppose $\Omega \in C1$, $1 \leq q < p < \infty$, $\bar{w}$, $\bar{v}_0$, $\bar{v}_1 \in \mathcal{W}_B((\underline{a},\infty))$. Let the following conditions be fulfilled:

There exist $k > 0$ and $t_0 \in (\underline{a},\infty)$ such that (3.12)

$$\bar{v}_0(t) \geq k\,\bar{v}_1(t)t^{-p} \quad \text{for all} \quad t \geq t_0.$$

There exists $R \in \langle \underline{a},\infty)$ such that (3.13)

$$A_2(R,\infty;\bar{w}(t)t^{N-1},\bar{v}_1(t)t^{N-1}) < \infty .$$

Then

$$W^{1,p}(\Omega;v_0,v_1) \hookrightarrow\hookrightarrow L^q(\Omega;w).$$

3.12 THEOREM: Suppose $\Omega \in D1$, $1 \leq q < p < \infty$, $\bar{w}$, $\bar{v}_0,\bar{v}_1$, $\bar{\lambda} \in \mathcal{W}_B((\underline{a},\infty))$,

---

[4] If $v_0 \equiv v_1$, then we write v instead of v_0 and v_1.

[5] This means that the weight functions w, v may have singularities or degenerations only at infinity.

$$A(\underline{a},\infty;\bar{w}(t)t^{N-1},\bar{v}_1(t)t^{N-1}) < \infty.$$

Let the function $\bar{\lambda}$ satisfy:

$$\bar{\lambda} \text{ is decreasing in some interval } (R,\infty) \subset (\underline{a},\infty); \tag{3.14}$$

$$\lim_{t\to\infty} \bar{\lambda}(t) = 0. \tag{3.15}$$

Then

$$W_0^{1,p}(\Omega;v_0,v_1) \hookrightarrow\hookrightarrow L^q(\Omega;w\cdot\lambda)$$

where $\lambda(x) = \bar{\lambda}(|x|)$, $x \in \Omega$.

3.13 <u>THEOREM</u>: Suppose $\Omega \in D1$, $1 \leq q < p < \infty$, $\bar{w}$, $\bar{v}$, $\bar{\lambda} \in \mathcal{W}_B((\underline{a},\infty))$. Let the following conditions be fulfilled:

(i) there exists $R \in \langle\underline{a},\infty)$ such that

$$A(R,\infty;\bar{w}(t)t^{N-1}, \bar{v}(t)t^{N-1}) < \infty ;$$

(ii) the function $\bar{\lambda}$ satisfies (3.14) and (3.15).

Then

$$W_0^{1,p}(\Omega;v,v) \hookrightarrow\hookrightarrow L^q(\Omega;w\cdot\lambda)$$

where $\lambda(x) = \bar{\lambda}(|x|)$, $x \in \Omega$.

For $1 \leq p < \infty$, $-\infty \leq a < b \leq +\infty$ and weight functions ω_0, $\omega_1 \in \mathcal{W}((a,b))$ let us define

$$B(a,b;\omega_0,\omega_1) = \inf_{c\in\langle a,b\rangle} \max \{B_1(a,c;\omega_0,\omega_1), B_2(c,b;\omega_0,\omega_1)\}$$

where

$$B_1(a,a;\omega_0,\omega_1) = 0 = B_2(b,b;\omega_0,\omega_1)$$

and

$$B_1(a,c;\omega_0,\omega_1) = \sup_{a<x<c} \|\omega_0^{1/p}\|_{p,(x,c)} \cdot \|\omega_1^{-1/p}\|_{p',(a,x)} \quad \text{if } a < c \leqq b,$$

$$B_2(c,b;\omega_0,\omega_1) = \sup_{c<x<b} \|\omega_0^{1/p}\|_{p,(c,x)} \cdot \|\omega_1^{-1/p}\|_{p',(x,b)} \quad \text{if } a \leqq c < b.$$

Sufficient conditions for non-existence of imbeddings are given by the following theorem.

3.14 THEOREM: Suppose $\Omega \in D1$, $1 \leqq q < p < \infty$. Let there exist $R \geqq \max\{0; \sup\{|x|; x \in \mathbb{R}^N \setminus \Omega\}\}$ such that

$$B(R,\infty;\bar{v}_0(t)t^{N-1},\bar{v}_1(t)t^{N-1}) < \infty, \quad A(R,\infty;\bar{w}(t)t^{N-1},\bar{v}_1(t)t^{N-1}) = \infty.$$

Then the space $W_0^{1,p}(\Omega;v_0,v_1)$ is not continuously imbedded into the space $L^q(\Omega;w)$.

The proofs of Theorems 3.10 - 3.14 can be found in [5].

3.15 EXAMPLE: Suppose $\Omega \in C1$, $1 \leqq q < p < \infty$.

(I) Let $\underline{a} > 0$, $\beta \neq p - N$. Then the following three conditions are equivalent:

(i) $W_0^{1,p}(\Omega;|x|^{\beta-p},|x|^{\beta}) \hookrightarrow\hookrightarrow L^q(\Omega;|x|^{\alpha})$,

(ii) $W_0^{1,p}(\Omega;|x|^{\beta-p},|x|^{\beta}) \hookrightarrow L^q(\Omega;|x|^{\alpha})$,

(iii) $\frac{\alpha}{q} - \frac{\beta}{p} + N(\frac{1}{q} - \frac{1}{p}) + 1 < 0$.

(II) Let $\underline{a} > 0$, $\beta > p - N$. Then the following three conditions are equivalent:

(i) $W^{1,p}(\Omega;|x|^{\beta-p},|x|^{\beta}) \hookrightarrow\hookrightarrow L^q(\Omega;|x|^{\alpha})$,

(ii) $W^{1,p}(\Omega;|x|^{\beta-p},|x|^{\beta}) \hookrightarrow L^q(\Omega;|x|^{\alpha})$,

(iii) $\frac{\alpha}{q} - \frac{\beta}{p} + N(\frac{1}{q} - \frac{1}{p}) + 1 < 0.$

(III) Let $\Omega = \mathbb{R}^N \setminus \{0\}$ or $\Omega = \mathbb{R}^N$, $\beta \neq p - N$. Then the space $W_0^{1,p}(\Omega;|x|^{\beta-p},|x|^{\beta})$ is not continuously imbedded into the space $L^q(\Omega;|x|^{\alpha})$ for any $\alpha \in \mathbb{R}$.

3.16 EXAMPLE: Suppose $\Omega \in C1$, $\underline{a} > 1$, $1 \leqq q < p < \infty$. For $x \in \Omega$ we define

$$w(x) = |x|^{\alpha} \log^{\gamma}|x|, \quad v_0(x) = |x|^{\beta-p} \log^{\delta}|x|, \quad v_1(x) = |x|^{\beta} \log^{\delta}|x|.$$

(I) Let $\beta \neq p - N$. Then the following conditions are equivalent:

(i) $W_0^{1,p}(\Omega;v_0,v_1) \hookrightarrow\hookrightarrow L^q(\Omega;w)$,

(ii) $W_0^{1,p}(\Omega;v_0,v_1) \hookrightarrow L^q(\Omega;w)$,

(iii) $\frac{\alpha}{q} - \frac{\beta}{p} + N(\frac{1}{q} - \frac{1}{p}) + 1 < 0$ or

$$\frac{\alpha}{q} - \frac{\beta}{p} + N(\frac{1}{q} - \frac{1}{p}) + 1 = 0, \quad \frac{\gamma}{q} - \frac{\delta}{p} + \frac{1}{q} - \frac{1}{p} < 0.$$

(II) Let $\beta > p - N$. Then the following conditions are equivalent:

(i) $W^{1,p}(\Omega;v_0,v_1) \hookrightarrow\hookrightarrow L^q(\Omega;w)$,

(ii) $W^{1,p}(\Omega;v_0,v_1) \hookrightarrow L^q(\Omega;w)$,

(iii) $\frac{\alpha}{q} - \frac{\beta}{p} + N(\frac{1}{q} - \frac{1}{p}) + 1 < 0$ or

$$\frac{\alpha}{q} - \frac{\beta}{p} + N(\frac{1}{q} - \frac{1}{p}) + 1 = 0, \quad \frac{\gamma}{q} - \frac{\delta}{p} + \frac{1}{q} - \frac{1}{p} < 0.$$

3.17 EXAMPLE: Suppose $\Omega \in C1$, $1 \leqq q < p < \infty$, $\beta \neq 0$. Then the following five conditions are equivalent:

(i) $W_0^{1,p}(\Omega;e^{\beta|x|},e^{\beta|x|}) \hookrightarrow\hookrightarrow L^q(\Omega;e^{\alpha|x|})$,

(ii) $W^{1,p}(\Omega;e^{\beta|x|},e^{\beta|x|}) \hookrightarrow\hookrightarrow L^q(\Omega;e^{\alpha|x|})$,

(iii) $W_0^{1,p}(\Omega;e^{\beta|x|},e^{\beta|x|}) \hookrightarrow L^q(\Omega;e^{\alpha|x|})$,

(iv) $W^{1,p}(\Omega;e^{\beta|x|},e^{\beta|x|}) \hookrightarrow L^q(\Omega;e^{\alpha|x|})$,

(v) $\frac{\alpha}{q} - \frac{\beta}{p} < 0$.

4. N-DIMENSIONAL HARDY INEQUALITY

As a consequence of the imbedding theorems and theorems on equivalent norms on the spaces $W_0^{1,p}(\Omega;v_0,v_1)$ or $W^{1,p}(\Omega;v_0,v_1)$ we obtain conditions for the validity of the N-dimensional Hardy inequality. The corresponding result is formulated in Lemma 4.1. (As concerns theorems on equivalent norms the reader is referred to [5].)

4.1 LEMMA: Let the norms $\|\cdot\|_{1,p,\Omega,v_0,v_1}$ and $|||\cdot|||_{1,p,\Omega,v_1}$, where

$$|||u|||_{1,p,\Omega,v_1} = \left[\sum_{i=1}^{N} \int_\Omega \left|\frac{\partial u}{\partial x_i}(x)\right|^p v_1(x)\, dx \right]^{1/p},$$

be equivalent on the space $W_0^{1,p}(\Omega;v_0,v_1)$ (or $W^{1,p}(\Omega;v_0,v_1)$). Then

$$W_0^{1,p}(\Omega;v_0,v_1) \hookrightarrow L^q(\Omega;w)$$

(or $W^{1,p}(\Omega;v_0,v_1) \hookrightarrow L^q(\Omega;w)$) if and only if there exists a positive constant C such that the N-dimensional Hardy inequality

$$\left[\int_\Omega |u(x)|^q\, w(x)\, dx \right]^{1/q} \le C \left[\sum_{i=1}^{N} \int_\Omega \left|\frac{\partial u}{\partial x_i}(x)\right|^p v_1(x)\, dx \right]^{1/p}$$

holds for all $u \in W_0^{1,p}(\Omega;v_0,v_1)$ (or $u \in W^{1,p}(\Omega;v_0,v_1)$, respectively).

The proof is trivial. □

We shall write $\Omega \in \overline{C1}$ if the following conditions are fulfilled:

(i) $\Omega \in C1$

(ii) $x \in \Omega,\ t > 1 \Rightarrow tx \in \Omega$.

Lemma 4.1 and the examples from section 3 imply:

4.2 EXAMPLE:[6] Suppose $\Omega \in \overline{CT}$, $1 \leq p,q < \infty$, $\alpha, \beta \in \mathbb{R}$. For $x \in \Omega$ we define

$$w(x) = |x|^{\alpha}, \quad v_0(x) = |x|^{\beta-p}, \quad v_i(x) = |x|^{\beta}, \quad i = 1,\dots,N.$$

(I) Let $\underline{a} > 0$, $\beta \neq p - N$ (or $\beta > p - N$). Then the inequality (1.1) with $T(\Omega) = W_0^{1,p}(\Omega;v_0,v_1)$ (or $T(\Omega) = W^{1,p}(\Omega;v_0,v_1)$, respectively) holds if and only if

(i) $1 \leq p \leq q < \infty$, $N(\frac{1}{q} - \frac{1}{p}) + 1 \geq 0$, $\frac{\alpha}{q} - \frac{\beta}{p} + N(\frac{1}{q} - \frac{1}{p}) + 1 \leq 0$

or

(ii) $1 \leq q < p < \infty$, $\frac{\alpha}{q} - \frac{\beta}{p} + N(\frac{1}{q} - \frac{1}{p}) + 1 < 0$.

(II) Let $\Omega = \mathbb{R}^N \setminus \{0\}$, $\beta \neq p - N$ (or $\beta > p - N$). Then the inequality (1.1) with $T(\Omega) = W_0^{1,p}(\Omega;v_0,v_1)$ (or $T(\Omega) = W^{1,p}(\Omega;v_0,v_1)$, respectively) holds if and only if

$$1 \leq p \leq q < \infty, \; N(\frac{1}{q} - \frac{1}{p}) + 1 \geq 0, \; \frac{\alpha}{q} - \frac{\beta}{p} + N(\frac{1}{q} - \frac{1}{p}) + 1 = 0. \tag{4.1}$$

(III) Let $\Omega = \mathbb{R}^N$, $p - N < \beta < N(p - 1)$. Then the inequality (1.1) with $T(\Omega) = W^{1,p}(\Omega;v_0,v_1)$ or $T(\Omega) = W_0^{1,p}(\Omega;v_0,v_1)$ holds if and only if the condition (4.1) is fulfilled.

4.3 EXAMPLE: Suppose $\Omega \in \overline{CT}$, $\underline{a} > 1$, $1 \leq p,q < \infty$, $\beta \neq p - N$ (or $\beta > p - N$). For $x \in \Omega$ we define

$$w(x) = |x|^{\alpha} \log^{\gamma}|x|, \quad v_0(x) = |x|^{\beta-p} \log^{\delta}|x|, \quad v_i(x) = |x|^{\beta} \log^{\delta}|x|,$$
$$i = 1,\dots,N.$$

---

[6] If we consider the inequality (1.1) only on the space $W_0^{1,p}(\Omega;v_0,v_1)$, it suffices to assume only $\Omega \in D1$. Similarly in other examples.

Then the inequality (1.1) with $T(\Omega) = W_0^{1,p}(\Omega;v_0,v_1)$ (or $T(\Omega) = W^{1,p}(\Omega;v_0,v_1)$, respectively) holds if and only if

(i) $1 \leqq p \leqq q < \infty$, $N(\frac{1}{q} - \frac{1}{p}) + 1 \geqq 0$,

$$\frac{\alpha}{q} - \frac{\beta}{p} + N(\frac{1}{q} - \frac{1}{p}) + 1 < 0$$

or

$$\frac{\alpha}{q} - \frac{\beta}{p} + N(\frac{1}{q} - \frac{1}{p}) + 1 = 0 \text{ and } \frac{\gamma}{q} - \frac{\delta}{p} \leqq 0;$$

or

(ii) $1 \leqq q < p < \infty$, $\frac{\alpha}{q} - \frac{\beta}{p} + N(\frac{1}{q} - \frac{1}{p}) + 1 < 0$

or

$$\frac{\alpha}{q} - \frac{\beta}{p} + N(\frac{1}{q} - \frac{1}{p}) + 1 = 0 \text{ and } \frac{\gamma}{q} - \frac{\delta}{p} + \frac{1}{q} - \frac{1}{p} < 0.$$

4.4 EXAMPLE: Suppose $\Omega \in D1$, $1 \leqq p,q < \infty$, $\alpha, \beta \in \mathbb{R}$. For $x \in \Omega$ we define

$$w(x) = e^{\alpha|x|}, \quad v_i(x) = e^{\beta|x|}, \quad i = 0,1,\ldots,N.$$

(I) Let one of the following two conditions be satisfied:

(i) $\underline{a} > 0$, $\beta \neq 0$;

(ii) $\underline{a} = 0$, $\beta > 0$ or $\beta < 0$ and $p < N$.

Then the inequality (1.1) with $T(\Omega) = W_0^{1,p}(\Omega;v_0,v_1)$ holds if and only if

$$1 \leqq p \leqq q < \infty, \quad N(\frac{1}{q} - \frac{1}{p}) + 1 \geqq 0, \quad \frac{\alpha}{q} - \frac{\beta}{p} \leqq 0 \tag{4.2}$$

or

$$1 \leqq q < p < \infty, \quad \frac{\alpha}{q} - \frac{\beta}{p} < 0. \tag{4.3}$$

(II) Let $\Omega \in \overline{CT}$, $\beta > 0$. Then the inequality (1.1) with $T(\Omega) = W^{1,p}(\Omega;v_0,v_1)$ holds if and only if (4.2) or (4.3) is fulfilled.

References

[1] R.A. Adams, Sobolev spaces, Academic Press (1975).

[2] A.E. Gatto, C.E. Gutiérrez and R.L. Wheeden, Fractional integrals on weighted H^p spaces, Trans. Amer. Math. Soc., 289 (1985), 575-589.

[3] P. Gurka and B. Opic, Continuous and compact imbeddings of weighted Sobolev spaces I, Czech. Math. J., 38 (1988), 730-744.

[4] P. Gurka and B. Opic, A note on N-dimensional Hardy's inequality, Constructive Theory of Functions '87, Publishing House of the Bulgarian Academy of Sciences, Sofia (1988), 194-196.

[5] P. Gurka and B. Opic, Continuous and compact imbeddings of weighted Sobolev spaces III, Czech. Math. J. (to appear).

[6] A. Kufner and B. Opic, How to define reasonably weighted Sobolev spaces, Comment. Math. Univ. Carolinae, 25 (1984), 537-554.

[7] V.G. Maz'ja, Sobolev Spaces, Springer-Verlag (1985).

[8] B. Opic, Necessary and sufficient conditions for imbeddings in weighted Sobolev spaces, Časopis Pěst. mat. (to appear).

[9] B. Opic, Hardy's inequality for absolutely continuous functions with zero limits on both ends of the interval (to appear).

[10] B. Opic and P. Gurka, Continuous and compact imbeddings of weighted Sobolev spaces II, Czech. Math. J. (to appear).

[11] B. Opic and A. Kufner, Weighted Sobolev spaces and the N-dimensional Hardy inequality, Trudy Sem. S.L. Soboleva, 1 (1983), 108-117 (Russian).

Bohumír Opic & Petr Gurka
Mathematics Institute
Academy of Science
Žitná 25
115 67 Prague 1
Czechoslovakia

J. PEETRE AND L.E. PERSSON

General Beckenbach's inequality with applications

0. INTRODUCTION

Let x_j, y_j, $j = 1,2,\ldots,n$, denote positive numbers and write $\sum = \sum_1^n$. The well-known inequality

$$\left(\frac{\sum (x_j + y_j)^\alpha}{\sum (x_j + y_j)^\beta}\right)^{1/(\alpha-\beta)} \leq \left(\frac{\sum x_j^\alpha}{\sum x_j^\beta}\right)^{1/(\alpha-\beta)} + \left(\frac{\sum y_j^\alpha}{\sum y_j^\beta}\right)^{1/(\alpha-\beta)} \tag{0.1}$$

holds if $0 \leqq \beta \leqq 1 \leqq \alpha$, $\alpha \neq \beta$. For the case $\beta = \alpha - 1$ this inequality was proved by Beckenbach [3]. Using moment space techniques Dresher [14] proved an integral version of (0.1). Different elementary proofs have been presented by Danskin [12] and Daroczy [13] (see also [4]). The inequality (0.1) has been generalized and complemented in various ways (see e.g. [6], [22], [23], [24] and [25]).

In this paper we prove a general inequality (see Theorem 2.1), which in particular generalizes (0.1) and all generalizations known to us of this inequality. Moreover, this inequality can also be seen as a generalization of some other classical inequalities.

Furthermore, the reversed inequality

$$\left(\frac{\sum (x_j + y_j)^\alpha}{\sum (x_j + y_j)^\beta}\right)^{1/(\alpha-\beta)} \geq \left(\frac{\sum x_j^\alpha}{\sum x_j^\beta}\right)^{1/(\alpha-\beta)} + \left(\frac{\sum y_j^\alpha}{\sum y_j^\beta}\right)^{1/(\alpha-\beta)} \tag{0.2}$$

holds if $\beta \leqq 0 \leqq \alpha \leqq 1$, $\alpha \neq \beta$.

We note that the inequalities (0.1) and (0.2) obviously hold also in the symmetric cases $0 \leqq \alpha \leqq 1 \leqq \beta$, $\alpha \neq \beta$, and $\alpha \leqq 0 \leqq \beta \leqq 1$, $\alpha \neq \beta$, respectively.

In this paper we also prove a reversed version of our general inequality (see Theorem 2.2), which generalizes both (0.2) and reversed versions of some classical inequalities. We also investigate the limiting cases and present some new applications.

The paper is organized in the following way. In section 1 we collect some preliminaries. In section 2 the above-mentioned general inequalities are proved. In section 3 we apply our general inequalities to obtain generalized versions of the Beckenbach-Dresher inequality, including some recent results of Beesack and Pečarić [6] and Persson [25]. In section 4 we discuss the limiting cases including the case $\alpha = \beta = 1$ in (0.1) and the case $\alpha = \beta = 0$ in (0.2). Section 5 is used to discuss the close connections between some of the inequalities presented and inequalities for generalized entropies, which, in its turn, are important for applications in information theory, statistics, compression of data, etc. (see e.g. [7], [11] and [18]). In particular we generalize and complement some of the inequalities obtained in [11], [26] and [27]. Section 6 is used to give some concluding remarks.

1. PRELIMINARIES

Let D denote an additive Abelian semigroup. Let $u = (u_1, u_2, \ldots, u_n)$ and $v = (v_1, v_2, \ldots, v_n) \in R_+^n$. We write $u \leqq v$ if $u_1 \leqq v_1$, $u_2 \leqq v_2, \ldots, u_n \leqq v_n$. We say that $f: D \to R_+^n$, is *subadditive* if

$$f(x + y) \leqq f(x) + f(y) \tag{1.1}$$

for all $x, y \in D$. If (1.1) holds in the reversed direction, then we say that f is *superadditive*.

Let X denote a Banach lattice of measurable functions on a σ-finite measure space (Ω, Σ, μ). The space X^p, $-\infty < p < \infty$, $p \neq 0$, consists of all μ-measurable functions x satisfying

$$\|x\|_{X^p} = (\| |x|^p \|_X)^{1/p} < \infty,$$

see e.g. [25] and the references given there.

$P(\Omega)$ denotes the *power set* of Ω, i.e. the set of subsets of Ω.

Following Beesack and Pečarić [6] we also consider a linear class L of real-valued functions $x = x(t)$ defined on a non-empty set E. We also consider an isotonic linear functional $A: L \to R$ ('isotonic' means that if $x = x(t) \geqq 0$ on E, then $A(x) \geqq 0$) and define the *generalized Gini mean* $G(\alpha, \beta; x)$, $-\infty < \alpha, \beta < \infty$, $\alpha \neq \beta$, by

$$G(\alpha,\beta;x) = G_A(\alpha,\beta;x) = (A(x^{\alpha})/A(x^{\beta}))^{1/(\alpha-\beta)}, \tag{1.2}$$

whenever x^{α}, $x^{\beta} \in L$ and $A(x^{\beta}) \neq 0$.

We need the following version of Hölder's inequality:

LEMMA 1.1: Let x and y be non-negative and $x, y, x^{\theta}y^{1-\theta} \in L$.

(a) If $0 < \theta < 1$, then

$$A(x^{\theta}y^{1-\theta}) \leqq (A(x))^{\theta}(A(y))^{1-\theta}.$$

(b) If $\theta > 1$ or $\theta < 0$, then

$$A(x^{\theta}y^{1-\theta}) \geqq (A(x))^{\theta}(A(y))^{1-\theta}.$$

For the case $0 < \theta < 1$ see [6, p. 134] and [5] but for the sake of completeness we give the proof here.

PROOF: Let $0 < \theta < 1$. Assume that $A(x) = A(y) = 1$. Since

$$(x(t))^{\theta}(y(t))^{1-\theta} \leqq \theta x(t) + (1-\theta)y(t)$$

we find that

$$A(x^{\theta}y^{1-\theta}) \leqq \theta A(x) + (1-\theta)A(y) = 1.$$

For the general case we replace x by $x/A(x)$ and y by $y/A(y)$ and the proof of (a) follows. The proof of (b) is similar. □

LEMMA 1.2: Let $x, x^r \in L$. Then the functional $A_r(x) = (A(x^r))^{1/r}$ is subadditive if $r > 1$ and superadditive if $r < 1$, $r \neq 0$.

PROOF: Write $(x+y)^r = (x+y)^{r-1}x + (x+y)^{r-1}y$ and use Lemma 1.1. □

2. A GENERAL INEQUALITY

THEOREM 2.1: Let $f:D \to R_+^n$ be subadditive and let $F:R_+^n \to R_+$ be a convex and increasing function such that $F(0) = 0$. Assume that $G:D \to P(R_+)$ is such that, for all $a \in G(x)$ and $b \in G(y)$, there exists $c \in G(x + y)$ such that $c \geqq a + b$. Then the function $f_1 : D \to R_+$, defined by

$$f_1(x) = \inf_{a \in G(x)} (aF(f(x)/a)), \tag{2.1}$$

is subadditive.

PROOF: Let $x,y \in D$, $a \in G(x)$ and $b \in G(y)$. We choose $c \in G(x+y)$ such that $c \geqq a + b$. Using the subadditivity assumption we obtain

$$\frac{f(x+y)}{c} \leqq \frac{f(x) + f(y)}{c} = \frac{a}{c}\frac{f(x)}{a} + \frac{b}{c}\frac{f(y)}{b} .$$

Therefore, according to our assumptions on F, we get

$$F(\frac{f(x+y)}{c}) \leqq \frac{a}{c} F(\frac{f(x)}{a}) + \frac{b}{c} F(\frac{f(y)}{b}).$$

We conclude that for any $\varepsilon > 0$ there exists $c \in G(x+y)$ such that

$$F(\frac{f(x+y)}{c}) \leqq \frac{1+\varepsilon}{c} f_1(x) + \frac{1+\varepsilon}{c} f_1(y).$$

It follows that

$$f_1(x+y) \leqq f_1(x) + f_1(y)$$

and the proof is complete. □

REMARK: If G is additive, i.e. if the condition '$c \geqq a + b$' can be replaced by the condition '$c = a + b$', then Theorem 2.1 holds also if the condition '$F(0) = 0$' is removed.

EXAMPLE 2.1: Assume that $D = R_+^n$, $f(x) = x = (x_1,x_2,\ldots,x_n)$, $G(x) = R_+$ and $F(u) = \max(1,\Sigma u_j^p)$, $p \geqq 1$. In this case we have

$$f_1(x) = \inf_{a\in R_+} (a \max(1, \Sigma x_j^p/a^p)) = (\Sigma x_j^p)^{1/p}.$$

Thus, using Theorem 2.1 and the remark, we obtain Minkowski's inequality

$$(\Sigma(x_j + y_j)^p)^{1/p} \leqq (\Sigma x_j^p)^{1/p} + (\Sigma y_j^p)^{1/p}. \tag{2.2}$$

More generally taking

$$f_1(x) = \inf_{a\in R_+} (a[1 + \Sigma\phi(x_j/a)]),$$

where ϕ is any convex function, we have the well-known generalization of this inequality to an Orlicz environment.

We also state the following reversed version of Theorem 2.1:

THEOREM 2.2: Let $f : D \to R_+^n$ be superadditive and let $F : R_+^n \to R_+$ be a concave and increasing function. Assume that $G : D \to P(R_+)$ satisfies that, for all $a \in G(x)$ and $b \in G(y)$, $c \geqq a + b$ for every $c \in G(x + y)$. Then the function f_1 (see (2.1)) is superadditive.

PROOF: The proof of Theorem 2.2 is quite similar to that of Theorem 2.1 so we omit the details. □

EXAMPLE 2.2: By applying Theorem 2.2 for the case considered in Example 2.1 with $0 < p < 1$ we obtain the reversed inequality (2.2).

3. SOME GENERAL VERSIONS OF BECKENBACH-DRESHER'S INEQUALITY

In Theorems 2.1 and 2.2 G was a set-valued function (a 'many-valued' function). Now we restate these results for 'single-valued' functions.

PROPOSITION 3.1: Let $F : R_+^n \to R_+$ be an increasing function and let $g:D \to R_+$ be superadditive.

(a) If F is convex and $f : D \to R_+^n$, is subadditive, then

$$g(x+y)\, F(\frac{f(x+y)}{g(x+y)}) \leqq g(x)F(\frac{f(x)}{g(x)}) + g(y)F(\frac{f(y)}{g(y)}). \tag{3.1}$$

(b) If F is concave and $f : D \to R^n_+$, is superadditive, then (3.1) holds in the opposite direction.

PROOF: Apply Theorems 2.1 and 2.2 with $G(x) = \{g(x)\}$ (singleton). □

In the rest of this paper we are going to work with this case only. The authors really regret that they do not have any more substantial applications of the general case except of the rather simple Examples 2.1 and 2.2.

EXAMPLE 3.1: Let $D = R^n_+$, $f(x) = (\Sigma x_j^\alpha)^{1/\alpha}$, $g(x) = (\Sigma x_j^\beta)^{1/\beta}$ and $F(u) = u^p$. Then, according to Proposition 3.1, we find that

$$\frac{(\sum(x_j + y_j)^\alpha)^{p/\alpha}}{(\sum(x_j + y_j)^\beta)^{(p-1)/\beta}} \leq \frac{(\sum x_j^\alpha)^{p/\alpha}}{(\sum x_j^\beta)^{(p-1)/\beta}} + \frac{(\sum y_j^\alpha)^{p/\alpha}}{(\sum y_j^\beta)^{(p-1)/\beta}}$$

holds if $p \geqq 1$, $\beta \leqq 1 \leqq \alpha$, $\beta \neq 0$, and the reversed inequality holds if $0 < p \leqq 1$, $\alpha, \beta \leqq 1$, $\alpha, \beta \neq 0$. In particular if $p = \alpha/(\alpha-\beta)$ we obtain (0.1)-(0.2).

Now we pass to continuous versions of our inequalities.

COROLLARY 3.2: Let x and y be μ-measurable functions such that, for some real number β,

$$\|x+y\|_{X^\beta} \geqq \|x\|_{X^\beta} + \|y\|_{X^\beta}.$$

(a) If $0 < \beta \leqq 1 \leqq \alpha$, $\alpha \neq \beta$, then

$$\left(\frac{(\|x+y\|_{X^\alpha})^\alpha}{(\|x+y\|_{X^\beta})^\beta}\right)^{1/(\alpha-\beta)} \leqq \left(\frac{(\|x\|_{X^\alpha})^\alpha}{(\|x\|_{X^\beta})^\beta}\right)^{1/(\alpha-\beta)} + \left(\frac{(\|y\|_{X^\alpha})^\alpha}{(\|y\|_{X^\beta})^\beta}\right)^{1/(\alpha-\beta)} \tag{3.2}$$

(b) If $\alpha < 0 < \beta \leqq 1$, then (3.2) holds in the opposite direction.

PROOF: Let $0 < \beta \leqq 1 \leqq \alpha$. Then the functionals

$$f(\cdot) = \|\cdot\|_{X^\alpha} \text{ and } g(\cdot) = \|\cdot\|_{X^\beta}$$

are subadditive (see [25]) and superadditive (by assumption), respectively. Therefore the proof of (a) follows by using Proposition 3.1(a) with $F(u) = u^{\alpha/(\beta-\alpha)}$. In a similar way we find that the statement in (b) is a special case of Proposition 3.1(b). □

REMARK: Another proof of Corollary 3.2(a) can be found in [25].

COROLLARY 3.3: Let $A,B : L \to R$ be isotonic linear functionals and let the functions $z_i : E \to R_+$ satisfy that z_i^α, z_i^β, $(\Sigma z_i)^\alpha$, $(\Sigma z_i)^\beta \in L$, $i = 1,2,\ldots,n$.

(a) If $\beta \leq 1 \leq \alpha$, $\beta \neq 0$ and $p \geq 1$, then

$$\frac{[A((\Sigma z_i)^\alpha)]^{p/\alpha}}{[B((\Sigma z_i)^\beta)]^{(p-1)/\beta}} \leq \sum \frac{(A(z_i^\alpha))^{p/\alpha}}{(B(z_i^\beta))^{(p-1)/\beta}} . \tag{3.3}$$

(b) If α, $\beta \leq 1$, $\alpha,\beta \neq 0$ and $0 < p \leq 1$, then (3.3) holds in the reversed direction.

PROOF: Let $f(z) = (A(z^\alpha))^{1/\alpha}$, $g(z) = (B(z^\beta))^{1/\beta}$, $D = L$ and $F(u) = u^p$, $p \geq 1$. Then, according to Lemma 1.2, it follows from Proposition 3.1 that

$$\frac{[A((z_1+z_2)^\alpha)]^{p/\alpha}}{[B((z_1+z_2)^\beta)]^{(p-1)/\beta}} \leq \frac{(A(z_1^\alpha))^{p/\alpha}}{(B(z_1^\beta))^{(p-1)/\beta}} + \frac{(A(z_2^\alpha))^{p/\alpha}}{(B(z_2^\beta))^{(p-1)/\beta}} .$$

Thus (3.3) is proved for the case $n = 2$ and the proof of the general case follows by using induction. The proof of (b) is similar. □

REMARK: For the case $\beta < 1$, $p = \alpha/(\alpha-\beta)$, $\alpha \neq \beta$, another proof of Corollary 3.3(a) can be found in [6, p. 137].

4. SOME LIMITING CASES

If $G(\alpha,\beta;x)$ is the functional defined by (1.2), then, according to Corollary

3.3, the following generalization of (0.1) and (0.2) holds:

(a) If $0 < \beta \leqq 1 \leqq \alpha$ or if $0 < \alpha \leqq 1 \leqq \beta$, $\alpha \neq \beta$, then

$$G(\alpha,\beta;\Sigma z_j) \leqq \Sigma G(\alpha,\beta;z_j). \tag{4.1}$$

(b) If $\beta < 0 < \alpha \leqq 1$ or if $\alpha < 0 < \beta \leqq 1$, then (4.1) holds in the reversed direction.

We shall now investigate the limiting cases $\alpha = \beta = 1$ and $\alpha = \beta = 0$. First we consider $H(a;x) = \ln A(x^a)$ and note that

$$\ln G(\alpha,\beta;x) = \frac{H(\alpha;x) - H(\beta;x)}{\alpha - \beta}, \quad \alpha \neq \beta. \tag{4.2}$$

We assume that $H(a;x)$ is continuously differentiable in a and define

$$\ln G(\alpha,\alpha;x) = H'(\alpha;x), \tag{4.3a}$$

i.e.

$$G(\alpha,\alpha;x) = \exp[A(x^{\alpha}\ln x)/A(x^{\alpha})]. \tag{4.3b}$$

PROPOSITION 4.1: Let $G(a,a;x)$ be defined by (4.3). Then

$$G(\alpha,\beta;x) = \exp[\frac{1}{\alpha-\beta}\int_{\beta}^{\alpha} \ln G(a,a;x)\, da]. \tag{4.4}$$

Moreover

$$G(1,1;\Sigma z_j) \leqq \Sigma G(1,1;z_j) \tag{4.5}$$

and

$$G(0,0;\Sigma z_j) \geqq \Sigma G(0,0;z_j). \tag{4.6}$$

PROOF: Using (4.2) and (4.3a) we obtain the representation formula (4.4). We use (4.1) and (4.4) and an obvious continuity argument and the proof of

(4.5) follows. The proof of (4.6) is similar. □

EXAMPLE 4.1: For the case n = 2, $A(x) = \sum x_j$ and $A(y) = \sum y_j$ Proposition 4.1 implies the following complements of the estimates (0.1)-(0.2):

$$(\Pi(x_j + y_j)^{x_j+y_j})^{1/\Sigma(x_j+y_j)} \leq (\Pi x_j^{x_j})^{1/\Sigma x_j} + (\Pi y_j^{y_j})^{1/\Sigma y_j},$$

and

$$(\Pi(x_j + y_j))^{1/n} \geq (\Pi x_j)^{1/n} + (\Pi y_j)^{1/n},$$

respectively ($\Pi = \Pi_1^n$). Other proofs of the last inequality can be found in [4, p. 26] and [19, p. 81].

REMARK: Other limiting cases of the estimates (0.1) and (0.2) are the trivial estimates

$$\sup_{1\leq j\leq n} (x_j + y_j) \leq \sup_{1\leq j\leq n} x_j + \sup_{1\leq j\leq n} y_j \quad \text{(the case } \alpha = \infty \text{ or } \beta = \infty\text{)},$$

$$\inf_{1\leq j\leq n} (x_j + y_j) \geq \inf_{1\leq j\leq n} x_j + \inf_{1\leq j\leq n} y_j \quad \text{(the case } \alpha = -\infty \text{ or } \beta = -\infty\text{)}.$$

We note that the derivative of the function $g(a) = \ln G(a,a;x)$ is a function with the numerator

$$A(x^a)A(x^a(\ln x)^2) - (A(x^a \ln x))^2.$$

Therefore, according to Lemma 1.1 ($\theta = 1/2$), we find that $g(a)$ is an increasing function. Using this fact together with the representation formula (4.4) we obtain the following useful information.

COROLLARY 4.2: Let $G(\alpha,\beta;x)$, $-\infty < \alpha, \beta < \infty$, be defined by (1.2) and (4.3). Then $G(\alpha,\beta;x)$ is an increasing function of α (β fixed) and of β (α fixed).

REMARK: Another (more complicated) proof of Corollary 4.2 can be found in [6, p. 135]. For special cases see also [9, p. 177] and [20, p. 89].

5. GENERALIZED GINI MEANS AND ENTROPIES

Assume that $x = (x_j)_1^n$, $x_j \geqq 0$, and

$$E(\alpha,\beta) = E(\alpha,\beta;x) = (\Sigma x_j^\alpha / \Sigma x_j^\beta)^{1/(\alpha-\beta)}, \quad \alpha \neq \beta. \tag{5.1}$$

(That is $E(\alpha,\beta;x)$ is just $G(\alpha,\beta;x)$ in a special situation.) Writing $\alpha = a + \beta$ we see that $E(\alpha,\beta)$ may be regarded as the self-weighted means introduced by Gini [16]. The limiting cases are the following:

$$\begin{gathered} E(\infty,\beta) = E(\alpha,\infty) = \max_{1\leqq j\leqq n} x_j, \quad E(-\infty,\beta) = E(\alpha,-\infty) = \min_{1\leqq j\leqq n} x_j, \\ E(\alpha,\alpha) = \exp\,(\Sigma x_j^\alpha \ln x_j / \Sigma x_j^\alpha), \quad \alpha \neq 0, \quad E(0,0) = (\prod_1^n x_j)^{1/n}. \end{gathered} \tag{5.2}$$

<u>EXAMPLE 5.1</u>: If $\Sigma x_j = 1$, α, $\beta > 0$, then

$$h_{\alpha,\beta} = h_{\alpha,\beta}(x) = -\ln E(\alpha,\beta;x)$$

is called the entropy of order (α,β). In particular $I_1 = h_{1,1}$ and $I_\alpha = h_{\alpha,1}$, $\alpha \neq 1$, are the usual Shannon's entropy and Renyi's entropy, respectively. The most important properties and applications (e.g. in information theory, statistics, compression of data, etc.) have been discussed and surveyed in [1], [2] and [11].

Let X be a continuous random variable with the probability density function $x = x(t)$. The *differential entropy*, denoted $H(X)$, is defined by

$$H(X) = -\int_{-\infty}^{\infty} x(t) \log x(t)\, dt,$$

see e.g. [7]. (As we are using natural logarithms instead of logarithms to the base 2 the units of information, $-\log(x(t))$, are called 'nats' instead of 'bits', and $H(X)$ is the expected amount of information in nats.) Using the functional

$$A(x) = \int_{-\infty}^{\infty} x(t)\, dt$$

we can generalize the differential entropy in a similar way as the original entropy of Shannon has been generalized in Example 5.1. The properties and possible applications of these entropies are, as far as we know, far from being fully investigated (compare with [7] and [18]). In this paper we introduce the *generalized entropy of order* (α,β), denoted $H_{\alpha,\beta}(x)$, and the *generalized entropy of degree* (α,β), denoted $H^{\alpha,\beta}(x)$, in the following way:

If $G(\alpha,\beta;x)$ is defined by (1.2) and (4.3), α, $\beta > 0$ and $A(x) = 1$, then

$$H_{\alpha,\beta}(x) = -\ln G(\alpha,\beta;x),$$

and

$$H^{\alpha,\beta}(x) = (A(x^{\alpha})-A(x^{\beta}))/(\beta-\alpha), \ \alpha \neq \beta.$$

For the discrete case compare with [1, p. 344]. We note that

$$H_{\alpha,\beta}(x) = \lambda H_{\alpha}(x) + (1-\lambda)H_{\beta}(x), \tag{5.3}$$

and

$$H^{\alpha,\beta}(x) = \lambda H^{\alpha}(x) + (1-\lambda)H^{\beta}(x), \tag{5.4}$$

where $\lambda = (\alpha-1)/(\alpha-\beta)$, $H_{\alpha}(x) = H_{\alpha,1}(x)$ and $H^{\alpha}(x) = H^{\alpha,1}(x)$.

PROPOSITION 5.1: Let $\alpha,\beta > 0$. Then

(a) $H_{\alpha,\beta}(x)$ is a decreasing function of α (β fixed) and of β (α fixed).

(b) $H^{\alpha}(x)$ is a decreasing function of α.

If, in addition, $\alpha < 1 \leqq \beta$ or $\beta < 1 \leqq \alpha$, then

(c) $H_{\alpha,\beta}(x) \leqq H^{\alpha,\beta}(x)$.

(d) $H^{\alpha,\beta}(\cdot)$ is a concave functional.

PROOF: (a) This statement follows from Corollary 4.2.

(b) We consider

$$g_\alpha(t) = (1-\alpha)^{-1}(e^{(1-\alpha)t}-1)$$

and note that

$$H^\alpha(x) = g_\alpha(H_\alpha(x)). \tag{5.5}$$

Thus the proof follows by using (a) and the fact that $g_\alpha(t)$ is increasing.

(c) Since $e^t - 1 \geqq t$ we have

$$H^\alpha(x) = g_\alpha(H_\alpha(x)) \geqq H_\alpha(x)$$

and the proof follows by using (5.3)-(5.4).

(d) Let $\beta = 1$, $0 < \alpha < 1$, $x = \lambda_1 x_1 + \lambda_2 x_2$, $\lambda_1 \geqq 0$, $\lambda_2 \geqq 0$ and $\lambda_1 + \lambda_2 = 1$. Then

$$H^{\alpha,\beta}(x) = H^\alpha(x) = \frac{1}{1-\alpha}\,(A(x^\alpha) - 1) \geqq \frac{1}{1-\alpha}(A(\lambda_1 x_1^\alpha + \lambda_2 x_2^\alpha) - 1)$$

$$= \frac{1}{1-\alpha}(\lambda_1 A(x_1^\alpha) + \lambda_2 A(x_2^\alpha) - \lambda_1 - \lambda_2) = \lambda_1 H^{\alpha,\beta}(x_1) + \lambda_2 H^{\alpha,\beta}(x_2).$$

In a similar way we find that also $H^{1,\beta}(\cdot) = H^\beta(\cdot)$, $\beta > 1$, is a concave functional. Thus, in view of (5.4), the proof is complete for the case $0 < \alpha < 1 \leqq \beta$. The proof of the case $0 < \beta < 1 \leqq \alpha$ is similar. □

REMARK: Proposition 5.1(a) (together with (5.2)) generalizes Satz 6 in [13]. Moreover, Proposition 5.1 generalizes several of the estimates in [11, pp. 347-349]. The statement in (b) seems to be new also for the discrete case.

REMARK: By replacing $g_\alpha(t)$ in (5.5) by

$$h_\alpha(t) = (2^{1-\alpha}-1)^{-1}(e^{(1-\alpha)t}-1),\ \alpha \neq 1,$$

we obtain generalized versions of the entropies studied in [26] and [27]. Exactly as in the proof of Proposition 5.1(d) we find that the corresponding generalized entropy is concave for all $\alpha \neq 1$. In particular we conclude that Theorem 3 in [26] is not correct (see also [27]).

6. CONCLUDING REMARKS

1. Pales [24] has proved that (0.1) holds *only if* $0 < \beta \leqq 1 \leqq \alpha$ (or $0 < \alpha \leqq 1 \leqq \beta$), $\alpha \neq \beta$, and (0.2) holds *only if* $\beta \leqq 0 \leqq \alpha \leqq 1$ (or $\alpha \leqq 0 \leqq \beta \leqq 1$), $\alpha \neq \beta$ (see also [22] and [23]). Thus if $E(\alpha,\beta)$ is defined by (5.1) and (5.2) we find, by using the representation formula (4.3) and a continuity argument, that the estimates

$$E(\alpha,\alpha;x+y) \leqq E(\alpha,\alpha;x) + E(\alpha,\alpha;y)$$

and

$$E(\beta,\beta;x+y) \geqq E(\beta,\beta;x) + E(\beta,\beta;y)$$

hold iff $\alpha = 1$ or $\beta = 0$, respectively (compare with Example 4.1). We conclude that also Proposition 4.1 is in a way best possible.

2. Some geometric interpretations of special cases of the Gini means $E(\alpha,\beta;x)$ can be found in [15]. More information about $E(\alpha,\beta;x)$ are given in [20] and [21]. Concerning classical means and their interpretations and inequalities we refer to [17] and the excellent new books [10] and [8, ch. 8].

3. Let A be an isotonic linear functional (see section 1) and define $B_a = A(x^a)$, $a \in R$. If $0 < \gamma < \beta < \alpha$, $\theta = (\alpha-\beta)/(\alpha-\gamma)$, then, according to Corollary 4.2, we get

$$(B_\alpha/B_\beta)^{1/(\alpha-\beta)} \geqq (B_\alpha/B_\gamma)^{1/(\alpha-\gamma)}, \text{ i.e. } B_\beta \leqq (B_\gamma)^\theta (B_\alpha)^{1-\theta}.$$

We conclude that the *generalized Liapounoff's inequality*

$$A(x^\beta) \leqq (A(x^\gamma))^{(\alpha-\beta)/(\alpha-\gamma)}(A(x^\alpha))^{(\beta-\gamma)/(\alpha-\gamma)}$$

follows from Proposition 4.2. Compare with [17, p. 27], [6, p. 133] and the (partly incomplete) statement in [9, p. 178].

4. The estimate (4.1) and Proposition 4.1 may also be regarded as complements of the usual 'entropy power inequality' (see e.g. [7, p. 287]). Moreover it follows from (4.1) (and a homogeneity argument) that the

generalized Gini mean $G(\alpha,\beta;x)$ is a convex functional if $0 < \alpha \leqq 1 \leqq \beta$. Some further complements and applications of these and other estimates obtained in this paper can be found in the forthcoming paper [18].

ACKNOWLEDGEMENT: We want to thank Dr Timo Koski, Luleå, for interesting discussions concerning the possibilities to apply some of the inequalities presented in information thoery, statistics, compression of data, etc.

References

[1] J. Aczel and Z. Daroczy, On Measures of Information and Their Characterizations, Mathematics in Science and Engineering, Academic Press (1975).

[2] J. Aczel and Z. Daroczy, Über verallgemeinerte quasilineare Mittelwerte die mit Gewichtsfunktionen gebildet sind, Publ. Math. Debrechen, 10 (1963), 171-190.

[3] E.F. Beckenbach, A class of mean-value functions, Amer. Math. Monthly, 57 (1950), 1-6.

[4] E.F. Beckenbach and R. Bellman, Inequalities, Springer (1983).

[5] P.R. Beesack and J.E. Pečarić, On Jessen's inequality for convex functions, J. Math. Anal. Appl., 110 (1985), 536-552.

[6] P.R. Beesack and J.E. Pečarić, On Jessen's inequality for convex functions 2, J. Math. Anal. Appl., 118 (1986), 125-144.

[7] R.E. Blahut, Principles and Practice of Information Theory, Addison-Wesley (1987).

[8] J.M. Borwein and P.B. Borwein, Pi and the AGM, Wiley (1987).

[9] J.L. Brenner, A unified treatment and extension of some means of classical analysis - comparison theorems, J. Combinatoric Inf. Syst. Sci., 3:4 (1978), 175-199.

[10] P.S. Bullen, D.S. Mitrinovic and P.M. Vasic, Means and Their Inequalities, Reidel (1988).

[11] R.M. Capocelli and I.J. Tanja, On some inequalities and generalized entropies: a unified approach, Cybern. Syst., 16 (1985), 341-376.

[12] J.M. Danskin, Dresher's inequality, Amer. Math. Monthly, 59 (1952), 687-688.

[13] Z. Daroczy, Einige Ungleichungen über die mit Gewichtsfunktionen gebildeten Mittelwerte, Monatsh. Math., 68 (1964), 102-112.

[14] M. Dresher, Moment spaces and inequalities, Duke Math. J., 20, (1953), 261-271.

[15] D. Farnsworth and P. Orr, Gini Means, Amer. Math. Monthly, 93: 8 (1986), 603-607.

[16] C. Gini, Di una formula comprensiva delle medie, Metron, 13 (1938), 3-22.

[17] G.H. Hardy, J.E. Littlewood and G. Polya, Inequalities, Cambridge University Press (1978).

[18] T. Koski and L.E. Persson, Some properties of generalized entropies with applications to compression of data, in preparation.

[19] A. Kovacec, On an algorithmic method to prove inequalities, General Inequalities 3, Proceedings of the Third International Conference on General Inequalities in Oberwolfach (1981), 69-89.

[20] E. Leach and M. Sholander, Extended mean values, Amer. Math. Monthly, 85 (1978), 84-90.

[21] E. Leach and M. Sholander, Multi-variable extended mean values, J. Math. Anal. Appl., 104 (1984), 390-407.

[22] L. Losonczi, Subadditive Mittelwerte, Arch. Math., 22 (1971), 168-174.

[23] L. Losonczi, Inequalities for integral mean values, J. Math. Anal. Appl., 61 (1977), 586-606.

[24] Z. Pales, A generalization of the Minkowski inequality, J. Math. Anal. Appl., 90 (1982), 456-462.

[25] L.E. Persson, Some elementary inequalities in connection with X^p-spaces, Research report, Dept. of Math., Luleå University (1987); Proceedings of the conference in Varna in June 1987, Publishing House of the Bulgarian Academy of Sciences (1988), 367-376.

[26] B.D. Sharma and R. Autar, An inversion theorem and generalized entropies for continuous distributions, SIAM J. Appl. Math., 25:2 (1973), 125-131.

[27] B.D. Sharma and R. Autar, On characterization of a generalized inaccuracy measure in information theory, J. Appl. Probability, 10 (1973),

Jaak Peetre
Department of Mathematics
Stockholm University
Box 6701, S-113 85, Stockholm
Sweden

and

Lars Erik Persson
Department of Mathematics
Luleå University
S-951 87 Luleå
Sweden

J. RÁKOSNÍK

On embeddings of anisotropic spaces of Sobolev type

1. INTRODUCTION

Let B_1 and B_2 be two Banach spaces, $B_1 \subset B_2$, and let the natural embedding operator $I : B_1 \to B_2$ be continuous. The n-th entropy number $e_n(I)$ of I, $n \in \mathbb{N}$, is the infimum of all numbers $\varepsilon > 0$ such that there exist 2^{n-1} balls in B_2 of radius ε which cover the unit ball in B_1.

The notion of entropy numbers was introduced by A. Pietsch [13] (he calls it outer entropy numbers). For the case when B_1 is the usual Sobolev space $W^{k,p}(\Omega)$, where $k \in \mathbb{N}$, $1 < p < \infty$ and Ω is a smooth bounded domain in $\mathbb{R}^N$, and $B_2 = L^q(\Omega)$, $p < q < \infty$, $\frac{k}{N} > \frac{1}{p} - \frac{1}{q}$, there is a well-known estimate

$$e_n(I) \leqq c\, n^{-k/N},\ n \in \mathbb{N}, \tag{1.1}$$

due to M.S. Birman and M.Z. Solomjak [4, 5]. V.V. Borzov [6, 7] used the same method of piecewise-polynomial approximation to extend this result for the anisotropic Sobolev-Slobodeckij space in place of B_1. In fact, an estimate is proved for ε-entropy of the unit ball of $W^{k,p}(\Omega)$ in $L^q(\Omega)$, a quantity in certain sense inverse to entropy numbers. The corresponding estimate for entropy numbers follows immediately.

If Ω is unbounded, then in general there is no compact embedding of $W^{k,p}(\Omega)$ into $L^q(\Omega)$ and so an estimate of type (1.1) cannot hold. The domain Ω has to be quasibounded, i.e. such that

$$\lim_{\substack{|x|\to\infty \\ x\in\Omega}} \operatorname{dist}(x,\partial\Omega) = 0 \tag{1.2}$$

(see [2; VI.6]). The rate of decay of entropy numbers then, generally, depends on the rate of convergence (1.2).

M. Namasivayam [10, 11] obtained estimates for entropy numbers of the embedding of anisotropic Sobolev space $W^{k,p}(\Omega)$, $k = (k_1,\dots,k_N) \in \mathbb{N}^N$, in $L^q(\Omega)$, when Ω is a certain type of quasibounded domain. However, he used an isotropic characterization of domains which yielded some loss in estimates and some of his steps were not careful enough.

Our aim is to suggest an anisotropic characterization of domains which more closely correponds to the nature of anisotropic spaces and to prove better estimates for entropy numbers. At the same time we shall deal with the anisotropic "fractional-order" Sobolev-Slobodeckij spaces. The method is based on integral representations of functions (see [3]) and on inequalities of Poincaré type (cf. [1], [8, 9] and [10, 11]).

2. DEFINITIONS AND PRELIMINARIES

2.1. Let $\mathbb{R}^N$ be Euclidean N-space. Let $k = (k_1,\dots,k_N) \in \mathbb{R}^N$ be an N-tuple of positive real numbers. We set $\omega = \sum_{i=1}^{N} \frac{1}{k_i}$, $\lambda = (\lambda_1,\dots,\lambda_N)$ with $\lambda_i = \omega k_i/N$ and in $\mathbb{R}^N$ we define the quasi-norm

$$|x|_k = \max_{1\leqq i\leqq N} |x_i|^{\lambda_i}, \quad x \in \mathbb{R}^N.$$

The ball of radius $h > 0$ with centre $x \in \mathbb{R}^N$ is the N-dimensional interval

$$B_h(x) = \{y \in \mathbb{R}^N;\ |x_i - y_i|^{\lambda_i} < h, \quad i = 1,\dots,N\}.$$

2.2. Given $0 < h < \infty$, $0 < b < \infty$ and $a_i \in \mathbb{R}\setminus\{0\}$, $i = 1,\dots,N$, the set

$$V(\lambda) = V(\lambda,h) = \bigcup_{0<v<h} \{x \in \mathbb{R}^N;\ x_i/a_i > 0,\ v < (x_i/a_i)^{\lambda_i} < (1+b)v,\quad i = 1,\dots,N\}$$

is called the λ-horn (of radius h and opening b).

2.3. Throughout the whole paper we shall suppose that Ω is a (non-empty) open set in $\mathbb{R}^N$ and $1 \leqq p < \infty$.

2.4. DEFINITION: Let $L^p(\Omega)$ be the usual Lebesgue space normed by $\|\cdot\|_{p,\Omega}$. The anisotropic Sobolev-Slobodeckij space $W^{k,p}(\Omega)$ is the collection of all functions $f \in L^p(\Omega)$ which have generalized derivatives $D_i^{[k_i]} f \in L^p(\Omega)$ and such that

$$\|f\|_{k,p,\Omega} = \|f\|_{p,\Omega} + \sum_{i=1}^{N} M_i(f;k_i,p,\Omega) < \infty, \tag{2.1}$$

where

$$M_i(f;k_i,p,\Omega) = \|D_i^{k_i} f\|_{p,\Omega} \text{ if } k_i \in \mathbb{N} \tag{2.2}$$

and

$$M_i(f;k_i,p,\Omega) = \left[\int_\Omega \int_{-\infty}^{\infty} \frac{|\Delta_i(\Omega) D_i^{[k_i]} f(x,y_i)|^p}{|x_i - y_i|^{1+\kappa_i p}} dy_i dx\right]^{1/p} \text{ if } k_i \notin \mathbb{N}. \tag{2.3}$$

Here $[k_i]$ stands for the integer part of k_i, $\kappa_i = k_i - [k_i]$ and $\Delta_i(\Omega)g(x,y_i) = (x_1,\dots,x_{i-1},y_i,x_{i+1},\dots,x_N) - g(x)$ if $(x_1,\dots,x_{i-1}, tx_i + (1-t)y_i, x_{i+1},\dots,x_N) \in \Omega$ for all $t \in [0,1]$, otherwise $\Delta_i(\Omega)g(x,y_i) = 0$.

$W^{k,p}(\Omega)$ becomes a Banach space when endowed by the norm (2.1).

2.5. We shall utilise the following easy fact: Let $a, b \in R^N$, $h > 0$, $H > 0$ and let $f \in L^p(B_h(a))$ and $g \in W^{k,p}(B_h(a))$. Define the functions $\tilde{f}$, $\tilde{g}$ by $\tilde{f}(x) = f(y)$ and $\tilde{g}(x) = g(y)$ where $y_i = a_i + (x_i - b_i)(h/H)^{1/\lambda_i}$, $i = 1,\dots,N$. Then $\tilde{f} \in L^p(B_H(b))$ and $\tilde{g} \in W^{k,p}(B_H(b))$, and

$$\|\tilde{f}\|_{p,B_H(b)} = (h/H)^{-N/p} \|f\|_{p,B_h(a)}, \tag{2.4}$$

$$M_i(\tilde{g};k_i,p,B_H(b)) = (h/H)^{N/\omega - N/p} M_i(g;k_i,p,B_h(a)). \tag{2.5}$$

2.6. DEFINITION: The space $W_0^{k,p}(\Omega)$ is defined as the closure of $C_0^\infty(\Omega)$ with respect to the norm $\|\cdot\|_{k,p,R^N}$. (More precisely, we take the closure of $\{f \in C^\infty(R^N);\ f$ has a compact support in $\Omega\}$ and then restrict the functions from the resulting space to Ω.) If $f \in W_0^{k,p}(\Omega)$ and g is the extension of f by zero outside Ω, then $g \in W^{k,p}(R^N)$ and we define

$$\|f\|_{k,p} = \|g\|_{k,p,R^N}. \tag{2.6}$$

Obviously, (2.6) defines a norm on $W_0^{k,p}(\Omega)$.

2.7. REMARK: If $k \in \mathbb{N}^N$, then $W_0^{k,p}(\Omega)$ coincides with the usual space defined as the closure of $C_0^\infty(\Omega)$ with respect to the norm $\|\cdot\|_{k,p,\Omega}$. Our definition enables us to extend functions from $W_0^{k,p}(\Omega)$ by zero outside Ω even when $k \notin \mathbb{N}^N$ and to avoid the non-trivial problem of boundedness of the extension operator as a mapping from $W_0^{k,p}(\Omega)$ in $W^{k,p}(\mathbb{R}^N)$ (cf. [14; 3.4.3] where a corresponding assertion is proved for isotropic spaces under restrictive assumptions on Ω and k).

2.8. The Lebesgue measure of a set $A \subset \mathbb{R}^N$ will be denoted by $|A|$. If $\Omega = \mathbb{R}^N$, then the sign of domain will be usually omitted in notations of spaces, norms and integrals. We shall make use of the convention $1/q = 0$ if $q = \infty$.

3. ENTROPY NUMBERS

3.1. Throughout the whole paper we shall suppose that

$$1 \leqq p < \infty, \quad p \leqq q \leqq \infty \quad \text{and } \frac{1}{\omega} > \frac{1}{p} - \frac{1}{q}. \tag{3.1}$$

Our aim is to prove the existence of continuous embedding

$$I : W_0^{k,p}(\Omega) \to L^q(\Omega) \tag{3.2}$$

and to investigate the entropy numbers $e_n(I)$ of (3.2).

First of all we shall prove an auxiliary embedding result.

3.2. LEMMA: Let G be an open set in $\mathbb{R}^N$ and V a λ-horn of radius $h > 0$. Let $1 \leqq r \leqq \infty$ be such that

$$\frac{1}{q} - \frac{1}{p} \leqq \frac{1}{r} \leqq \frac{1}{q} + \min(0, \frac{1}{\omega} - \frac{1}{p}). \tag{3.3}$$

If $k_i \in \mathbb{N}$ for some $i = 1,\dots,N$, suppose that the second inequality (3.3) is strict whenever $\omega = p > 1$ or $\omega > p = 1$. If $k \notin \mathbb{N}^N$, suppose that the first inequality (3.3) is strict and that the second one is strict whenever $\omega \leqq p$.

Then there exists a constant $c > 0$ independent of h and G such that

$$\|f\|_{q,G} \leqq c\|\phi\|_{r,G}[h^{-N(1/p-1/q+1/r)}\|f\|_{p,G+V} + \sum_{i=1}^{N} M_i(f;k_i,p,G+V)] \tag{3.4}$$

for every $\phi \in L^r(G)$ and $f \in W^{k,p}(G+V)$.

PROOF: Let $f \in W^{k,p}(G+V)$. Assume the functions f and $D_i^{[k_i]} f$ to be extended by zero outside $G+V = \{x+y;\ x \in G,\ y \in V\}$. According to [3; II.7] there exist functions $K_i, L_j \in C_0(R^N)$ $(i = 0,\dots,N,\ j$ such that $k_j \in \mathbb{N})$ and $L_\ell \in C_0^\infty(R^1)$ $(k_\ell \notin \mathbb{N})$ such that

$$f(x) = h^{-N} \int f(x+y) K_0(y:h^{1/\lambda})dy + \int_0^h \sum_{i=1}^{N} v^{-1-N+N/\omega} J_i(x,v)dv, \tag{3.5}$$

where $y:h^{1/\lambda} = (y_1 h^{-1/\lambda_1},\dots,y_N h^{-1/\lambda_N})$ and

$$J_i(x,v) = D_i^{k_i} f(x+y) K_i(y:h^{1/\lambda})dy \text{ if } k_i \in \mathbb{N}, \tag{3.6}$$

$$J_i(x,v) = v^{-\frac{N}{\omega}(1+1/k_i)} \tag{3.7}$$

$$\times \int_{-\infty}^{\infty}\int \Delta_i^{[k_i]+1}(R^1) f(x_1+y_1,\dots,x_i+y_i+t,\dots,x_N+y_N) K_i(y:v^{1/\lambda})$$

$$\times L_i(tv^{-1/\lambda_i})dydt \text{ if } k_i \notin \mathbb{N}.$$

The inequalities (3.3) yield $0 \leqq \frac{1}{q} - \frac{1}{r} \leqq \frac{1}{p} \leqq 1$. Hence, putting $\frac{1}{s} = \frac{1}{q} - \frac{1}{r}$ we have $p \leqq q \leqq s \leqq \infty$ and

$$\frac{1}{\omega} \geqq \frac{1}{p} + \min(0, \frac{1}{\omega} - \frac{1}{p}) \geqq \frac{1}{p} - \frac{1}{s}. \tag{3.8}$$

Let $i = 1,\dots,N$ be such that $k_i \in \mathbb{N}$. If $\omega = p = 1$, then

$$\frac{1}{\omega} = 1 \geqq 1 - \frac{1}{s} = \frac{1}{p} - \frac{1}{s}$$

with the equality if and only if $s = \infty$. If $\omega > p > 1$, then (3.8) yields

$$\frac{1}{s} \geqq \frac{1}{p} - \frac{1}{\omega} > 0,$$

i.e. $s < \infty$. If $\omega < p$, then the first inequality (3.8) is strict. Finally, if $\omega = p > 1$ or $\omega > p = 1$, then according to the assumptions the second inequality (3.8) is strict. Thus, we have $1 \leqq p \leqq s \leqq \infty$ and $\frac{1}{\omega} \geqq \frac{1}{p} - \frac{1}{s}$ with $1 < p < s < \infty$ or $p = 1$, $s = \infty$ in the case when $\frac{1}{\omega} = \frac{1}{p} - \frac{1}{s}$. These are assumptions of Lemma 10.1 in [3] and using the technique of the proof of that lemma we obtain the estimate

$$\left\|\int_0^h v^{-1-N+N/\omega} J_i(x,v)dv\right\|_{s,G} \leqq c_1 \|D_i^{k_i} f\|_{p,G+V} \tag{3.9}$$

with a constant $c_1 > 0$ independent of f, G and h.

Now, let $i = 1,\dots,N$ be such that $k_i \notin \mathbb{N}$. The inequalities (3.8) yield $\frac{1}{\omega} \geqq \frac{1}{p} - \frac{1}{s}$ and according to the assumptions $1 \leqq p < s < \infty$. Hence, we can use the techniques of [3; IV.16] to obtain the estimate

$$\left\|\int_0^h v^{-1/N+N/\omega} J_i(x,v)dv\right\|_{s,G} \tag{3.10}$$

$$\leqq c_2 \int_0^1 \int_{G+V} |\Delta_i(G+V)D_i^{[k_i]} f(x,x_i+h)|^p \, h^{-1-\kappa_i p} \, dxdh$$

$$\leqq c_2 \, M_i(f;k_i,p,G+V),$$

where $c_2 > 0$ is a constant which does not depend on f, G and h.

The first summand in (3.5) can be estimated by means of the Young inequality:

$$\left\|h^{-N} \int f(x+y)K(y:h^{1/\lambda})dy\right\|_{s,G} \leqq h^{-N} \|K(\cdot:y^{1/\lambda})\|_{s_1} \|f\|_{p,G+V} \tag{3.11}$$

$$= \|K\|_{s_1} h^{-N(1/p-1/s)} \|f\|_{p,G+V},$$

where $1/s_1 = 1 - 1/p + 1/s$.

The Hölder inequality yields

$$\|f\phi\|_{q,G} \leqq \|\phi\|_{r,G} \|f\|_{s,G} \tag{3.12}$$

and the inequality (3.4) follows from (3.5), (3.9)-(3.12).

3.3. <u>COROLLARY</u>: Let the parameters ω, p, q, r satisfy the assumptions of Lemma 3.2 and let $h > 0$. Then there exists $c > 0$ such that the inequality

$$\|f\phi\|_{q,\Omega} \leq c \sup_{x\in\mathbb{R}^N} \|\phi\|_{r,B_h(x)\cap\Omega} \|f\|_{k,p} \tag{3.13}$$

holds for every $f \in W_0^{k,p}(\Omega)$ and for every measurable function ϕ on $\mathbb{R}^N$.

<u>PROOF</u>: Let $h > 0$, $f \in W_0^{k,p}(\Omega)$ and ϕ be given, let f be extended by zero outside Ω. There exist 2^N intervals G_j and λ-horns V_j of radius h/2 such that

$$B_h(0) = \bigcup_{j=1}^{2^N} G_j = \bigcup_{j=1}^{2^N} (G_j+V_j).$$

Lemma 3.2 yields

$$\|g\phi\|_{q,B_h(0)} \leq c_1 \sum_{j=1}^{2^N} \|\phi\|_{r,G_j} \|g\|_{k,p,G_j+V_j} \tag{3.14}$$

$$\leq c_2 \|\phi\|_{r,B_h(0)} \|g\|_{k,p,B_h(0)}$$

for every $g \in W^{k,p}(B_h(0))$, where c_1, c_2 do not depend on g and Φ. Hence, tessellating $\mathbb{R}^N$ by intervals $B_n = B_h(x_n)$ and using the fact $q \geqq p$ we have

$$\|f\phi\|_{q,\Omega} \leq \left(\sum_{n=1}^{\infty} \|f\phi\|^q_{q,B_n}\right)^{1/q} \leq c_2\left(\sum_{n=1}^{\infty} \|\phi\|^q_{r,B_n} \|f\|^q_{k,p,B_n}\right)^{1/q}$$

$$\leq c_2 \sup_{x\in\mathbb{R}^N} \|\phi\|_{r,B_h(x)}\left(\sum_{n=1}^{\infty} \|f\|^p_{k,p,B_n}\right)^{1/p}$$

and (3.13) follows since $\{B_n \times B_n\}_n$ is a system of pair-wise disjoint sets in $\mathbb{R}^N \times \mathbb{R}^N$ and the function ϕ can be assumed to vanish outside Ω.

3.4. <u>REMARK</u>: Of course, the constant c in (3.13) depends on h.

<u>3.5</u>. We shall need the following easy consequence of V.V. Borzov's estimate for ε-entropy of the unit ball of $W^{k,p}(B_1)$ in $L^q(B_1)$ (cf. [10;2.3]):

LEMMA: There exists a constant $c > 0$ such that for every $R > 0$ the entropy numbers of the embedding $I_R : W^{k,p}(B_R(0)) \to L^q(B_R(0))$ satisfy

$$e_n(I_R) \leqq c\, R^{N(\frac{1}{\omega} - \frac{1}{p} + \frac{1}{q})}\, n^{-1/\omega}, \quad n \in \mathbb{N}. \tag{3.15}$$

PROOF: Let U be the unit ball in $W^{k,p}(B_1(0))$ and let $\varepsilon > 0$. The ε-entropy $H_\varepsilon(U;L^q(B_1(0)))$ is defined as the dyadic logarithm of the minimal cardinality of ε-net of the set $I_1(\bar{U})$ in $L^q(B_1(0))$. V.V. Borzov [7] proved the estimate

$$H_\varepsilon(U;L^q(B_1(0))) \leqq c_1 \varepsilon^{-\omega}$$

with $c_1 > 0$ independent of ε. It can easily be reformulated in terms of entropy numbers by

$$e_n(I_1) \leqq c_2\, n^{-1/\omega}. \tag{3.16}$$

The estimate (3.15) follows from (3.16) by a simple use of the definition of $e_n(I_R)$ and of linear transform of coordinates which maps $B_1(0)$ onto $B_R(0)$ (cf. (2.4) and (2.5)).

3.6. THEOREM: Let p, q, k satisfy the assumptions (3.1). Suppose that for some numbers h, $t > 0$

$$\sup_{x \in \Omega \setminus B_R(0)} |\Omega \cap B_h(x)| = O(R^{-Nt}) \text{ as } R \to \infty. \tag{3.17}$$

Then there exists compact embedding (3.2) and

$$e_n(I) = O(n^{-\mu/\omega}) \text{ as } n \to \infty, \tag{3.18}$$

where

$$\mu \leqq \min\left(\frac{t}{t+1}, \frac{t}{t + q(1/\omega - 1/p + 1/q)}\right),$$

the inequality being strict in any of the following cases:

(i) $\omega = p = 1$ or $\omega < p$, $k \notin \mathbb{N}^N$,

(ii) $\omega > p = 1$, $k_i \in \mathbb{N}$ for some $i = 1,\ldots,N$,

(iii) $\omega = p > 1$, k arbitrary.

PROOF: We shall follow the ideas of [9] (see also [10, 11]). Let us define for every $R > 0$ the mappings $E_R : W_0^{k,p}(\Omega) \to W^{k,p}(B_R(0))$, $F_R : L^q(B_R(0)) \to L^q(\Omega)$ and $G_R : W_0^{k,p}(\Omega) \to L^q(\Omega)$ by $E_R f(x) = G_R f(x) = f(x)$ and $F_R f(x) = f(x)$ if $x \in \Omega \cap B_R(0)$, otherwise $E_R f(x) = G_R f(x) = 0$ and $F_R f(x) = 0$. Then $G_R = F_R \circ I_R \circ E_R$, $\|F_R\| = \|E_R\| = 1$ and according to the properties of entropy numbers (see [13])

$$e_n(I) \leqq e_n(G_R) + \|I - G_R\| \leqq \|F_R\|\, e_n(I_R)\, \|E_R\| + \|I - G_R\| . \tag{3.19}$$

By Corollary 3.3, taking the characteristic function of $\Omega \setminus B_R(0)$ for ϕ we have

$$\|f\|_{q,\Omega\setminus B_R(C)} \leqq c_1 \sup_{x\in\mathbb{R}^N} |(\Omega\setminus B_R(0)) \cap B_h(x)|^{1/r} \|f\|_{k,p} \tag{3.20}$$

for any $f \in W_0^{k,p}(\Omega)$ and $R > 0$, where $0 < \frac{1}{r} \leqq \frac{1}{q} + \min\,(0, \frac{1}{\omega} - \frac{1}{p})$ with strict inequalities in the cases (i)-(iii). Hence, by (3.17) and (3.20) there exist $c_2 > 0$ and $R_1 > 0$ such that

$$\|I - G_R\| \leqq c_2\, R^{-Nt},\ R \geqq R_1.$$

Using this inequality and (3.15), we get from (3.19)

$$e_n(I) \leqq c_3(R^{N(1/\omega-1/p+1/q)}\, n^{-1/\omega} + R^{-Nt/r}),\ R \geqq R_1.$$

If we put here $R = n^{\nu/\omega}$ with $\nu = [\frac{Nt}{r} + N(\frac{1}{\omega} - \frac{1}{p} + \frac{1}{q})]^{-1}$, we obtain the desired estimate (3.18).

3.7. REMARK: Corollary 3.3 and Theorem 3.6 extend Theorems 3.1 and 3.3 of [10] in various directions. Firstly, we allow r and μ to reach the limit values in cases (iii) and (iv) of Theorem 3.6 (in [10] there are open intervals for these parameters). The proof of Theorem 3.1 in [10] contains in fact the proof of boundedness of the embedding operator (3.2), but the technique is not subtle enough. Secondly, 'isotropic' techniques are used in [10]; it yields an appearance of a worse parameter $\max(k_1,\dots,k_N)$ in place of N/ω in the formula for μ. Finally, our approach allowed to extend both assertions for Sobolev-Slobodeckij spaces.

3.8. Now we shall make another approach to the problem which is based on the use of a Poincaré-type inequality. The idea goes back to R.A. Adams [1] and H. König [8] (see also [10,11]). Recall, that Ω, p, q and k are assumed to satisfy assumptions of Sections 2.3 and 3.1.

3.9. DEFINITION: Let $i = 1,\dots,N$ and let A, B be measurable sets in $\mathbb{R}^N$. We shall say that A is i-full with respect to B if

$$|P_iB \setminus P_iA|_{N-1} = 0,$$

where P_i is the orthogonal projection onto the hyperplane $x_i = 0$ and $|\cdot|_{N-1}$ stands for the (N-1)-dimensional Lebesgue measure.

3.10. LEMMA. Let $i = 1,\dots,N$, $h > 0$ and let $A \subset \mathbb{R}^N$ be i-full with respect to $B_h(0)$. Let $k_ip > 1$ and suppose, moreover, that $p < q$ if $k \notin \mathbb{N}^N$ and that $q < \infty$ if $k \notin \mathbb{N}^N$, $\omega \leq p$ or if $k_j \in \mathbb{N}$ for some $i = 1,\dots,N$ and $\omega = p > 1$.

Then there exists a constant $c > 0$ which depends only on p,q and k and such that the inequality

$$\|f\|_{q,B_h(0)} \leq c\, h^{N(1/\omega-1/p+1/q)} \sum_{i=1}^{N} M_i(f;k_ip,B_h(0)) \tag{3.21}$$

holds for every $f \in C^\infty(B_h(0))$ such that $\operatorname{supp} f \subset B_h(0) \setminus A$.

PROOF: Let $1 \leq p < \infty$ and $\ell > 0$ be such that $\ell p > 1$. Denote $M(\cdot;\ell,p,(-1,1))$ the corresponding seminorm in $W^{\ell,p}(-1,1)$ (cf. (2.2) and (2.3)). Then there exists a constant $c_1 > 0$ such that

$$\|g\|_{p,(-1,1)} \leq c_1\, M(g;\ell,p,(-1,1)) \tag{3.22}$$

holds for every function $g \in C^\infty(-1,1))$ which vanishes in a neighbourhood of some point $t \in (-1,1)$. In fact, if such a constant does not exist, then we can find a sequence of functions $g_n \in C^\infty((-1,1))$ with $t \in (-1,1) \setminus \operatorname{supp} g_n$ such that

$$1 \quad \|g_n\|_{p,(-1,1)} > n\, M(g_n;\ell,p,(-1,1)). \tag{3.23}$$

Since $W^{\ell,p}(-1,1)$ is compactly embedded in $L^p(-1,1)$ (cf. [3; VI.26]), there is a subsequence of $\{g_n\}$ which converges to g in $L^p(-1,1)$ and, by (3.23), also in $W^{\ell,p}(-1,1)$. It follows from (3.23) that $M(g;\ell,p,(-1,1)) = 0$, i.e. g is a polynomial of degree not greater than $[\ell]-1$. If $\ell \leq 1$ or $\ell > 1$ then $W^{\ell,p}(-1,1)$ is continuously embedded in $C([-1,1])$ or $C^{[\ell]-1}([-1,1])$ and therefore $g^{(m)}(t) = 0$ for $m = 0$ or $m = 0,\ldots,[\ell]-1$, respectively. Hence $g = 0$, which contradicts the normalization condition in (3.23).

Now let the assumptions of the lemma be fulfilled and let $f \in C^\infty(B_h(0))$ with $\operatorname{supp} f \subset B_h(0)\setminus A$. We define the function $\tilde{f} \in C^\infty(B_1(0))$ and the set $\tilde{A}$ by $\tilde{f}(x) = f(x{:}h^{-1/\lambda})$, $\tilde{A} = \{x;\ x{:}h^{-1/\lambda} \in A\}$. Then $\tilde{A}$ is i-full with respect to $B_1(0)$ and $\operatorname{supp} \tilde{f} \subset B_1(0)\setminus\tilde{A}$. Without loss of generality we can suppose that $P_iA = P_iB_1(0)$. Since the assumptions of Lemma 3.2 are satisfied with $r = \infty$ and $\phi \equiv 1$, we obtain from (2.4), (2.5), (3.14) and (3.22)

$$\begin{aligned}
\|f\|_{q,B_h} &= h^{N/q}\,\|\tilde{f}\|_{q,B_1} \leq c_2 h^{N/q}\,\|\tilde{f}\|_{k,p,B_1} \\
&= c_2 h^{N/q}\Big[\Big(\int_{P_iB_1}\int_{-1}^{1}|\tilde{f}(x',x_i)|^p dx_i dx'\Big)^{1/p} + \sum_{i=1}^{N} M_i(\tilde{f};k_i,p,B_1)\Big) \\
&\leq c_2 h^{N/q}\Big[c_1\Big(\int_{P_iB_1} M(\tilde{f}(x',\cdot);k_i,p,(-1,1))^p dx'\Big)^{1/p} \\
&\qquad + \sum_{i=1}^{N} M_i(\tilde{f};k_i,p,B_1)\Big] \\
&\leq c_3 h^{N/q}\sum_{i=1}^{N} M_i(\tilde{f};k_i,p,B_1) \\
&= c_3 h^{N/\omega-N/p+N/q}\sum_{i=1}^{N} M_i(f;k_ip,B_h)
\end{aligned}$$

(we used the fact that for every $x' \in P_iB_1(0)$ there exists $x_i \in (-1,1)$ such that $(x',x_i) \in \tilde{A} \subset B_1(0) \smallsetminus \operatorname{supp} \tilde{f})$.

3.11. THEOREM. Let p, q, k satisfy the assumptions (3.1) and let i=1,...,N. Moreover, let $p < q$ if $k \notin \mathbb{N}^N$, let $q < \infty$ if $k \notin \mathbb{N}^N$ and $\omega \leqq p$ or if $k_j \in \mathbb{N}$ for some j and $\omega = p > 1$ and let $p > 1/k_i$ whenever $k_i \in (0,1)$. Suppose that there exist numbers R_0, $t > 0$ and a positive function $h = h(R)$, $R > 0$ such that

$$h(R) = O(R^{-t}) \text{ as } R \to \infty \tag{3.24}$$

and the set $\mathbb{R}^N \smallsetminus \Omega$ is i-full with respect to any $B_{h(R)}(x), x \in \mathbb{R}^N \smallsetminus B_R(0)$, $R \geqq R_0$.

Then there exists compact embedding (3.2) and the estimate (3.18) holds with

$$\mu = \frac{t}{t+1}. \tag{3.25}$$

PROOF: It can be made by the same way as the proof of Theorem 3.6. The only difference consists in an estimate alternating (3.20). Let $R > R_0 + 1$, $h = h(R-1)$ and let $f \in W_0^{k,p}(\Omega)$ be extended by zero outside Ω. We tessellate $\mathbb{R}^N$ by intervals $B_j = B_h(x_j)$, $j \in \mathbb{N}$, and put $J = \{j;\ B_j \cap (\Omega \smallsetminus B_R(0)) \neq \emptyset\}$. Then according to the assumptions and Lemma 3.10 (we use also the fact that $p \leqq q$) we obtain

$$\begin{aligned}
\|f\|_{q,\Omega\smallsetminus B_R(0)} &\leqq \Big(\sum_{j\in J} \|f\|_{q,B_j}^q\Big)^{1/q} \\
&\leqq c_1 h^{N(1/\omega-1/p+1/q)}\Big(\sum_{i=1}^N \sum_{j\in J} M_i(f;k_i,p,B_j)^q\Big)^{1/q} \\
&\leqq c_1 h^{N(1/\omega-1/p+1/q)}\Big(\sum_{i=1}^N \sum_{j\in J} M_i(f;k_i,p,B_j)^p\Big)^{1/p} \\
&\leqq c_2 h^{N(1/\omega-1/p+1/q)} \sum_{i=1}^N M_i(f;k_i,p,\Omega),
\end{aligned}$$

and so

$$\begin{aligned}
\|f\|_{q,\Omega\smallsetminus B_R(0)} &\leqq c_3 (R-1)^{-Nt(1/\omega-1/p+1/q)} \|f\|_{k,p,\Omega} \\
&\leqq c_4 R^{-Nt(1/\omega-1/p+1/q)} \|f\|_{k,p,\Omega}
\end{aligned}$$

for R large enough.

3.12. COROLLARY: Let $(\min_{1\leq i\leq n} k_i)^{-1} < p < q < \infty$.

Let there exist numbers R_0, $t > 0$ and a positive function h satisfying (3.24) and with the property that for every $x \in \mathbb{R}^N \setminus B_R(0)$, $R \geq R_0$ there exists $i = 1,\dots,N$ such that $\mathbb{R}^N \setminus \Omega$ is i-full with respect to $B_{h(R)}(x)$. Then there exists compact embedding (3.2) and the estimate (3.18) holds with μ given by (3.25).

3.13 REMARK: Theorem 3.11 is a counterpart to Theorem 3.7 of M. Namasivayam [10]. Our extension consists in considering spaces of fractional order and in the improvement of the formula for μ. In [10] the 'isotropic' techniques are used together with the chain of embeddings $W_0^{k,p}(\Omega) \to W_0^{\nu-1,p}(\Omega) \to L^q(\Omega)$, where $\nu = \min k_i$ and $(\nu-1)/N > 1/p - 1/q$. (Let us mention, that the parameter $(\nu-1)$ can be directly replaced by ν since in Theorem 3.7 of [10] it is assumed $p > 1$.) On the other hand, in [10] a larger class of domains is considered.

4. EXAMPLES OF QUASIBOUNDED DOMAINS

4.1. We shall outline several examples of domains which satisfy the assumptions of Theorems 3.6 and/or 3.11. The detailed calculations are left to the reader. The parameters k, λ, ω are as in Section 2.1.

4.2. EXAMPLE: Let $N \in \mathbb{N}$ and $t > 0$. Let b_j, $j = 1,\dots,N$, be positive continuous non-increasing functions on $[0,\infty)$. The trompet-shaped domain

$$\Omega_j = \{x = (x',x_j) \in \mathbb{R}^N;\ |x'| < b_j(|x_j|),\ x_j \in \mathbb{R}^1\}$$

satisfies the condition (3.17) if and only if $b_j(z) = O(z^{-t\omega k_j/(N-1)})$ as $z \to \infty$. It fulfils the assumptions of Theorem 3.11 if and only if $j \neq i$ and $b_j(z) = O(z^{-tk_j/k_i})$ as $z \to \infty$.

4.3. EXAMPLE: Let $t, c > 0$ be given. For $m \in \mathbb{Z}^N \setminus \{0\}$ define $h_m = 1 - c|m|_k^{-Nt}$, $B_m = B_{h_m}(m)$ and $\Omega = \mathbb{R}^N \setminus \bigcup_m \bar{B}_m$. Then Ω satisfies the assumptions of Theorem

3.6 with h = 1. Obviously, Ω does not satisfy the assumptions of Theorem 3.11.

4.4. EXAMPLE: Let $N = 2$ and $t > 0$. Define the systems of curvilinear rays

$$S_n^{(1)} = \{x = (x_1,x_2);\ |x_2|^{\lambda_2} = |2^j x_1|^{\lambda_1},\ j = 1,\ldots,2^{n-1}, |x|_k \geq 2^{n\lambda_1/(1+t)}\},$$

$$S_n^{(2)} = \{x = (x_1,x_2);\ |x_1|^{\lambda_1} = |2^j x_2|^{\lambda_2},\ j = 1,\ldots,2^{n-1}, |x|_k \geq 2^{n\lambda_2/(1+t)}\},$$

$n \in \mathbb{N}$. The domain $\Omega = \mathbb{R}^N \setminus \bigcup_{n=1}^{\infty} (S_n^{(1)} \cup S_n^{(2)})$ is an anisotropic analogue of the well-known 'spiny urchin' (see e.g. [2]), which satisfies the assumptions of Corollary 3.12. Clearly, for any t, h > 0 the condition (3.17) is not fulfilled.

4.5. EXAMPLE: Let $N = 2$, $t > 0$ and $h > 0$. We introduce the generalized polar coordinates (r,θ) in $\mathbb{R}^2$ by $|x_1|^{\lambda_1} \operatorname{sgn} x_1 = r\cos\theta$, $|x_2|^{\lambda_2} \operatorname{sgn} x_2 = r\sin\theta$. The spiral-shaped domain

$$\Omega = \{(r,\theta);\ 0 < \theta < \infty,\ h\theta + 1 < r < h\theta + 1 + c(h\theta + 1)^{-2t}\}$$

satisfies condition (3.17).

References

[1] R.A. Adams, Capacity and compact imbeddings, J. Math. Mech. 19 (1970), 923-929.

[2] R.A. Adams, Sobolev spaces. Academic Press, New York - San Francisco - London 1975.

[3] O.V. Besov, V.P. Il'in and S.M. Nikol'skij, Integral representations of functions and imbedding theorems (Russian), Izdat. Nauka, Moscow, 1975.

[4] M.S. Birman, and M.Z. Solomjak, Piecewise polynomial approximations of functions of the classes W_p^α (Russian), Mat. Sb. 73 (1967), 331-355.

[5] M.S. Birman and M.Z. Solomjak, Quantitative analysis in Sobolev imbedding theorems and applications to spectral theory, 10th Math. School Kiev 1974, 5-189 (Russian); Amer. Math. Soc. Translations (2) 114 (1980), 1-132.

[6] V.V. Borzov, On some applications of piecewise polynomial approximations of functions of anisotropic classes W_p^r (Russian), Dokl. Akad. Nauk SSSR 198 (1971), 499-501.

[7] V.V. Borzov, Some applications of theorems on piecewise polynomial approximations of functions of anisotropic classes W_p^r. In: Problems of mathematical physics 6 (Russian), Izdat. Leningrad. Univ., Leningrad, 1972, 53-67.

[8] H. König, Operator properties of Sobolev imbeddings over unbounded domains, J. Funct. Anal. 24 (1977), 32-51.

[9] H. König, Approximation numbers of Sobolev imbeddings over unbounded domains, J. Funct. Anal. 29 (1978), 74-87.

[10] M. Namasivayam, Embeddings of anisotropic Sobolev spaces on unbounded domains, Ann. Mat. Pura Appl. 144 (1986), 157-172.

[11] M. Namasivayam, Thesis. Univ. of Sussex, Brighton 1985.

[12] S.M. Nikol'skij, Approximation of functions of several variables and embedding theorems. Springer-Verlag, Berlin - Heidelberg - New York, 1975.

[13] A. Pietsch, Operator ideals. VEB Deutscher-Verlag der Wissenschaften, Berlin, 1978.

[14] H. Triebel, Theory of function spaces. Akademische Verlagsgesellschaft Geest & Portig K.-G., Leipzig, Birkhäuser Verlag, Basel - Boston - Stuttgart, 1983.

J. Rakosnik
Math. Inst. Acad. Sci.
Zitna 25
11567 Prague 1
Czeckoslovakia

H. TRIEBEL

Local means, oscillations, atoms and exotic pseudodifferential operators in function spaces

1. INTRODUCTION

Let R^n be the Euclidean n-space and let $f \in D'(R^n)$, where the latter stands for the space of all complex-valued distributions on R^n. There exist many devices to measure smoothness as well as local and global behaviour of f, in particular if f is a regular distribution, i.e. a locally integrable function. We remind the reader of derivatives and differences together with the related spaces of Hölder, Sobolev and Besov type, but also of more sophisticated techniques connected with Fourier analytical methods, approximations and representations via smooth building blocks. The aim of this paper is to outline some recent developments of this type connected with the spaces B^s_{pq} and F^s_{pq} on R^n and on domains.

2. SOME DEFINITIONS

2.1. Let $B = \{y|\ |y| < 1\}$ be the unit ball in R^n and let $k_0(y)$ and $k(y)$ be two C^∞ functions in R^n with

$$\text{supp } k_0 \subset B, \quad \text{supp } k \subset B, \tag{1}$$

$$\int_{R^n} k_0(y)\, dy \neq 0 \quad \text{and} \quad \int_{R^n} k(y)\, dy \neq 0. \tag{2}$$

Let $M \in N$ (the collection of all natural numbers) and $k_M(y) = \Delta^M k(y)$, where

$$\Delta^M = \Big(\sum_{j=1}^n \frac{\partial^2}{\partial x_j^2}\Big)^M$$

is the Mth power of the Laplacian. We introduce the local means

$$k_N(t,f)(x) = \int_{R^n} k_N(y)\, f(x+ty)\, dy, \quad x \in R^n,\ t > 0, \tag{3}$$

$N \in N_0 = N \cup \{0\}$ and $f \in D'(R^n)$. Let $L_p(R^n)$ with $0 < p \leq \infty$ be the usual L_p

spaces on R^n quasi-normed in the usual way.

2.2. DEFINITION: Let $s \in \mathbb{R}$, $0 < q \leq \infty$ and $0 < p \leq \infty$. Let $N \in \mathbb{N}$ with $2N > \max(s, n(1/p - 1)_+)$. Let k_o and k_N be the above functions.

(i) Then $B^s_{pq}(R^n)$ is the collection of all $f \in D'(R^n)$ such that

$$\|k_o(1,f)|L_p(R^n)\| + \left(\int_0^1 t^{-sq} \|k_N(t,f)(\cdot)|L_p(R^n)\|^q \frac{dt}{t}\right)^{1/q} \qquad (4)$$

(usual modification if $q = \infty$) is finite.

(ii) Let in addition $p < \infty$, then $F^s_{pq}(R^n)$ is the collection of all $f \in D'(R^n)$ such that

$$\|k_o(1,f)|L_p(R^n)\| + \left\|\left(\int_0^1 t^{-sq}|k_N(t,f)(\cdot)|^q \frac{dt}{t}\right)^{1/q}|L_p(R^n)\right\| \qquad (5)$$

(usual modification if $q = \infty$) is finite.

2.3. REMARK: Both $B^s_{pq}(\mathbb{R}^n)$ and $F^s_{pq}(\mathbb{R}^n)$ are quasi-Banach spaces. They are independent of $k_o(y)$, $k(y)$ and N. The above definition is based on [20], see also [19]. A systematic study of these spaces may be found in [18], where we used a somewhat different definition. We only mention that these two scales of spaces cover many well-known function spaces, such as

(i) the fractional Sobolev spaces $H^s_p = F^s_{p2}$ with $s \in \mathbb{R}$ and $1 < p < \infty$ (with the classical Sobolev spaces $W^m_p = H^m_p$ with $m \in \mathbb{N}_o$ and $1 < p < \infty$ as a subcase),

(ii) the Hölder-Zygmund spaces $C^s = B^s_{\infty\infty}$ with $s > 0$,

(iii) the classical Nikol'skij-Besov spaces $\Lambda^s_{pq} = B^s_{pq}$ with $s > 0$, $1 < p < \infty$, $1 \leq q \leq \infty$,

(iv) the (inhomogeneous) Hardy spaces $H_p = F^0_{p2}$ with $0 < p < \infty$.

2.4. DEFINITION: Let Ω be a bounded C^∞ domain in $\mathbb{R}^n$. Let s,p,q be the same numbers as in section 2.2 (in particular $p < \infty$ in the case of the F^s_{pq}-spaces). Then $B^s_{pq}(\Omega)$ is the restriction of $B^s_{pq}(\mathbb{R}^n)$ on Ω and $F^s_{pq}(\Omega)$ is the restriction of $F^s_{pq}(\mathbb{R}^n)$ on Ω.

2.5. REMARK: Of course, $B^s_{pq}(\Omega)$ and $F^s_{pq}(\Omega)$ are considered as subspaces of $D'(\Omega)$. They are quasi-normed in the usual way via

$$\|f|B^s_{pq}(\Omega)\| = \inf \|g|B^s_{pq}(\mathbb{R}^n)\| , \tag{6}$$

where the infimum is taken over all $g \in B^s_{pq}(\mathbb{R}^n)$ with $f = g|\Omega$ (restriction of g on Ω in the sense of $D'(\Omega)$). Similarly for $\|f|F^s_{pq}(\Omega)\|$. We have again the special cases mentioned in section 2.3.

3. LOCAL MEANS

3.1. We defined the spaces $B^s_{pq}(\mathbb{R}^n)$ and $F^s_{pq}(\mathbb{R}^n)$ via the local means (3) which have a high degree of flexibility. On the other hand we mentioned in section 2.3 some special cases which are originally defined on the basis of different principles: derivatives, differences, Fourier analytical decompositions, traces of harmonic functions and temperatures in $\mathbb{R}^{n+1}_+$ on $\mathbb{R}^n$ (identified with $x_{n+1} = 0$), etc. The question arises whether all these quite different devices can be subordinated to means of type (3) now with more general kernels $k_N(y)$. Then one needs more information about f, for instance $f \in F^s_{pq}(\mathbb{R}^n)$. It is quite clear that the class of admissible kernels depends on such an *a priori* knowledge of f. A systematic study of this problem has been given in [20]. It comes out that all these quite different techniques can be treated via means of type (3) based on more general classes of admissible kernels. To give an impression we assume that f is a continuous function on $\mathbb{R}^n$. Then the usual differences

$$\Delta^N_h f(x) = \sum_{j=0}^{N} (-1)^{N-j} f(x+jh),\ N \in \mathbb{N},\ h \in \mathbb{R}^n, \tag{7}$$

can be represented as local means,

$$\Delta^N_{th} f(x) = \int_{\mathbb{R}^n} \Big(\sum_{j=0}^{N} (-1)^{N-j} \delta_{jh}\Big)(y)\, f(x+ty)\, dy,\ t > 0, \tag{8}$$

where δ_{jh} is the δ-distribution with jh as its off-point (appropriate interpretation). Let n = 1, then the problem is to clarify under what circumstances, i.e. for which s,p,q and $f \in F^s_{pq}(\mathbb{R}^n)$, the kernel in (8) is admissible, i.e. the means $k_N(t,f)$ in (5) with n = 1 can be replaced by $\Delta^N_t f$. If n > 1, then it seems to be appropriate in our context to replace Δ^N_{th} in (8) by means of differences. We shall not discuss all these problems in generality, but we restrict ourselves to a recent result connected with means of differences of type (8) in domains.

3.2. Let Ω be a bounded C^∞ domain in $\mathbb{R}^n$. Let $N \in \mathbb{N}$, $t > 0$ and $x \in \Omega$, then

$$V^N(x,t) = \{h \mid h \in \mathbb{R}^n,\ x + \tau h \in \Omega,\ |h| < t,\ 0 \leqq \tau \leqq N\} \tag{9}$$

is the maximal truncated cone of radius t and with its vertex in the origin such that $x + NV^N(x,t) \subset \Omega$. Let $\Delta^N_h f(x)$ be given by (7). Let $f \in L_1(\Omega)$, then

$$d^N_t f(x) = \frac{1}{|V^N(x,t)|} \int_{V^N(x,t)} |\Delta^N_h f(x)|\,dh,\ x \in \Omega,\ t > 0,\ N \in \mathbb{N} \tag{10}$$

are local means of differences.

3.3. THEOREM: Let Ω be a bounded C^∞ domain in $\mathbb{R}^n$.

(i) Let $0 < p \leqq \infty$, $0 < q \leqq \infty$ and $s > n(1/p-1)_+$. Let $N \in \mathbb{N}$ with $N > s$. Then

$$B^s_{pq}(\Omega) = \{f \mid f \in L_1(\Omega),$$
$$\|f|L_p(\Omega)\| + \left(\int_0^1 t^{-sq}\, \|d^N_t f|L_p(\Omega)\|^q \frac{dt}{t}\right)^{1/q} < \infty\} \tag{11}$$

(usual modification if $q = \infty$).

(ii) Let $0 < p < \infty$, $0 < q \leqq \infty$ and $s > n\max((1/p-1)_+, (1/q-1)_+)$. Let $N \in \mathbb{N}$ with $N > s$. Then

$$F^s_{pq}(\Omega) = \{f \mid f \in L_1(\Omega),$$
$$\|f|L_p(\Omega)\| + \left\|\left(\int_0^1 t^{-sq}\, d^N_t f(\cdot)^q \frac{dt}{t}\right)^{1/q}\Big|L_p(\Omega)\right\| < \infty\}$$

(usual modification if $q = \infty$).

3.4. REMARK: Of course, (11) and (12) must be understood in the sense of equivalent quasi-norms. The theorem remains valid if one replaces Ω by $\mathbb{R}^n$ (or $\mathbb{R}^n_+$) and $L_1(\Omega)$ by $L_1^{loc}(\mathbb{R}^n)$ (or $L_1^{loc}(\mathbb{R}^n_+)$).

3.5. REMARK: Characterizations of function spaces via differences (and derivatives) have a long history. We recall the Hölder-Zygmund spaces and the classical Nikol'skij-Besov spaces. Characterizations of this type for fractional Sobolev spaces have been given in the late 1960s and the early 1970s by Strichartz and Lizorkin; see [18: 2.5.10, Remark 3] for detailed references. Extensions of these results to the spaces $F^s_{pq}(\mathbb{R}^n)$ are due to Kaljabin and the author: see [18: 2.5.9-2.5.12], where we gave many references, and also [19] for a more recent systematic study. The counterpart of (12) for $F^s_{pq}(\mathbb{R}^n)$ coincides with the corollary in [18: 2.5.11]. However the above intrinsic characterizations of $F^s_{pq}(\Omega)$ caused a lot of trouble (the corresponding assertions for $B^s_{pq}(\Omega)$ are much easier). The first decisive step in this direction is due to Kaljabin [13, 14] who proved (12) under the restrictions $s > 0$, $1 < p < \infty$, $1 < q < \infty$. A full proof of the above theorem may be found in [23]. We refer also to Seeger [16].

4. ATOMS

4.1. The idea is to represent functions or distributions belonging to certain function spaces as sums of simpler functions, building blocks or atoms. The development we have in mind began in 1974 with Coifman's atomic representations of elements of Hardy spaces; see [3,15]. This branch of the theory of function spaces has been studied in great style in the St Louis school around Rochberg, Taibleson and Weiss. Recently Frazier and Jawerth obtained in [10-12] atomic representations for both B^s_{pq} and F^s_{pq} spaces; we refer also to [2]. As far as atomic representations for B^s_{pq} spaces are concerned, we have nothing new to say compared with [10]. On the other hand we describe a new atomic representation for F^s_{pq} spaces which is influenced by the work of Frazier and Jawerth (we used their representations for B^s_{pq} spaces) and by the atomic representations for tent spaces due to Coifman, Meyer and Stein; see [4]. This representation has been announced in [21], and full proofs may be found in [22]. There is a large variety of applications of atomic representations to diverse problems in the theory of function spaces, among them mapping properties of (exotic) pseudodifferential and

Fourier integral operators. We refer to the papers by Frazier and Jawerth and [21, 22].

4.2. Next we follow closely [21]. We introduce the collection $\mathbb{Q} = \{Q_{\nu k}\}$ of all cubes $Q_{\nu k}$ in $\mathbb{R}^n$ with sides parallel to the axes, centred at $2^{-\nu}k$ and with side length $2^{-\nu}$, where $\nu \in \mathbb{N}_0$ and $k \in \mathbb{R}^n$ stands for the lattice points with integer-valued components. If Q is a cube in $\mathbb{R}^n$ and if $r > 0$ then rQ is the cube concentric with Q and with side length r times the side length of Q. We write $(\nu,k) < (\nu',k')$ if $\nu \geq \nu'$ and $Q_{\nu k} \subset 2\, Q_{\nu' k'}$.

4.3. DEFINITION: Let $-\infty < s < \infty$ and $0 < p \leq 1 < q \leq \infty$.

(i) A function $a(x)$ is called an s-atom if it is supported by a cube with side length 1 and $|D^\gamma a(x)| \leq 1$ for all γ with $|\gamma| \leq (1 + [s])_+$.

(ii) Let Q be a cube in $\mathbb{R}^n$ (with sides parallel to the axes).

Then a function $a(x)$ is called a (Q,s,q)-atom if

$$\operatorname{supp} a \subset Q, \tag{13}$$

$$\int_{\mathbb{R}^n} x^\alpha a(x)\, dx = 0 \quad \text{where} \quad |\alpha| \leq n(1/p-1) - s \tag{14}$$

and

$$|D^\gamma a(x)| \leq |Q|^{-1/q+s/n-|\gamma|/n} \quad \text{for all } \gamma \text{ with } |\gamma| \leq ([s] + 1)_+. \tag{15}$$

(iii) The distribution g is called an (s,p,q)-atom if there exists a cube $Q_{\mu 1} \in \mathbb{Q}$ such that

$$g = \sum_{(\nu,k)<(\mu,\ell)} d_{\nu k}\, a_{\nu k}(x), \tag{16}$$

where $a_{\nu k}$ are $(Q_{\nu k},s,q)$-atoms and $d_{\nu k}$ are complex numbers with

$$\Big(\sum_{(\nu,k)<(\mu,\ell)} |d_{\nu k}|^q \Big)^{1/q} \leq |Q|^{1/q-1/p} \tag{17}$$

(usual modification if $q = \infty$).

4.4. REMARK: If $s > n(1/p-1)$ then (14) means that no moment conditions are required. Furthermore we always assume that the functions $a(x)$ in (i) and (ii) have classical derivatives up to the required order. The atoms in (ii) coincide with the corresponding atoms in [10]; see also [2]. The atoms in (iii) belong both to $F^s_{pq}(R^n)$ and $F^s_{qq}(R^n)$. The convergence in (16) must be understood in this sense.

4.5. THEOREM: Let $-\infty < s < \infty$ and $0 < p \leqq 1 < q \leqq \infty$. Then $f \in D'(R^n)$ is an element of $F^s_{pq}(R^n)$ if and only if it can be represented as

$$f = \sum_{j=1}^{\infty} (\mu_j a_j + \lambda_j g_j) \tag{18}$$

where a_j is an s-atom, g_j is an (s,p,q)-atom, μ_j and λ_j are complex numbers with

$$(\sum_{j=1}^{\infty} |\mu_j|^p + |\lambda_j|^p)^{1/p} < \infty . \tag{19}$$

The infimum of (19) with respect to all representations (18) is an equivalent quasi-norm in $F^s_{pq}(R^n)$.

4.6. REMARK: The following observation is of crucial importance for applications: Let $L \in \mathbb{N}$ and $M \in \mathbb{N}$ with

$$L \geqq n(1/p-1) - s \quad \text{and} \quad M \geqq (1 + [s])_+ .$$

Then the theorem remains valid if one replaces $n(1/p-1) - s$ in (14) by L and $(1 + [s])_+$ in section 4.3(i) and in (15) by M. In other words one may assume from the very beginning that all the atoms in section 4.3(i,ii) are as smooth as one wants including a control about moment conditions and the growth of the derivatives. Somewhat different atomic representations of $F^s_{pq}(R^n)$ spaces may be found in [11,12].

5. APPLICATIONS TO EXOTIC PSEUDODIFFERENTIAL OPERATORS

5.1. As we said atomic representations for F^s_{pq} spaces allow many applications (see [11,12,21,22]), among them applications to pseudodifferential and

Fourier integral operators. We restrict ourselves to a very rough description of how atoms can be used in order to prove mapping properties for pseudodifferential operators. Let $\nu \in \mathbb{R}$; then $S_1^{\nu\ \mathrm{loc}}$ is the collection of all functions $\Phi(x,\xi)$ in $\mathbb{R}^{2n}$ such that for any $R > 0$

$$\sup_{|x|\leq R} |D_x^\alpha D_\xi^\beta \Phi(x,\xi)| \leq c_{R,\alpha,\beta} \quad |\xi|^{\nu+|\alpha|-|\beta|}, \quad |\xi| \geq 1, \tag{20}$$

complemented by some mild (and unimportant) conditions about the behaviour of $\Phi(x,\xi)$ if $|\xi| \leq 1$. The exotic pseudodifferential (or better Fourier integral) operators which can be treated on the basis of atomic representations look like

$$Af(x) = \int_{\mathbb{R}^n} e^{ix\xi+i\ \Phi(x,\xi)} \ a(x,\xi)\ \hat{f}(\xi)\ d\xi, \tag{21}$$

where $\Phi(x,\xi) \in S_1^{\nu\ \mathrm{loc}}$ is a real phase function whereas $a(x,\xi) \in S_1^{-\mu\ \mathrm{loc}}$ may be complex and $\hat{f}$ stands for the Fourier transform of f.

5.2. MAPPING PROPERTIES: Let $0 < p \leq 1 < q \leq \infty$, $s > n(1/p-1)$, $N \in \mathbb{N}$ with $N > 1 + 1/p$, $\nu \geq 0$, $\mu \geq 0$ and

$$\mu \geq \nu(1 + [s] + N), \tag{22}$$

complemented by some mild conditions about the behaviour of $\Phi(x,\xi)$ and $a(x,\xi)$ for small $|\xi|$, then

$$A : F_{pq}^{s\ \mathrm{com}}(\mathbb{R}^n) \to F_{pq}^{s\ \mathrm{loc}}(\mathbb{R}^n) \tag{23}$$

where it is quite clear what is indicated by the superscripts 'com' ($\sim$ compact) and 'loc' ($\sim$ local).

5.3. BASIC PRINCIPLE: Let g be an (s,p,q)-atom in the sense of section 4.3(iii) based on the unit cube $Q = Q_{00} \in \mathbb{Q}$. Then

$$Ag = CG + \tilde{G}(x), \tag{24}$$

where G is again an (s,p,q)-atom based on, say, 2Q, the complex number C depends only on the coefficients $c_{R,\alpha,\beta}$ from (20) and their counterparts for $a(x,\xi)$ with $|\alpha| \leqq 1+[s]$ and $|\beta| \leqq N$, and

$$\tilde{G} \in C^K(\mathbb{R}^n), \quad |D^\gamma \tilde{G}(x)| (1 + |x|)^\rho \leqq c, \quad |\gamma| \leqq K, \tag{25}$$

where both K and ρ can be chosen as large as one wants (in dependence on N). Very roughly, A preserves an (s,p,q)-atom beside constants and modulo smooth functions where the decay of the derivatives can be prescribed. This observation and the theorem in section 4.5 are the basis to prove assertions of type 5.2. Details and sharp formulations may be found in [22] (or the announcement [21]).

5.4. IMPROVEMENTS: Assertions of type 5.2 can be improved essentially if one combines the results from section 5.2 with known mapping properties of type $A : L_2^{com} \to L_2^{loc}$, complex interpolation and duality. For example, if $\Phi(x,\xi)$ and $a(x,\xi)$ are independent of x, $0 < p < \infty$, $s \in \mathbb{R}$, $\nu > 0$ and $\mu > |\nu n(1/p - 1/2|$ then A map $F^{s\ com}_{p2}(\mathbb{R}^n)$ into $F^{s\ loc}_{p2}(\mathbb{R}^n)$ (Hardy-Sobolev spaces). This is a sharp result for this class of operators, usually called strongly singular; see [9, 17].

6. OSCILLATIONS

6.1. The idea to use oscillations and local approximations of functions by polynomials in order to express smoothness properties has a long history. It starts with the space BMO of all functions of bounded mean oscillations by John and Nirenberg in 1961. It was systematically used by Campanato and other mathematicians in the mid-1960s and led to the Morrey-Campanato spaces and their use in connection with partial differential equations. The Russian school used local approximation spaces in connection with spaces of Besov type: we mention in particular Il'in and Brudnyj. In the 1970s Calderón and Scott used sharp maximal functions connected with local approximations by polynomials in order to study Sobolev spaces. More precise descritpions and in particular detailed references of this, nowadays older, part of the history may be found in [23]. More recently, Dorronsoro [6-8] gave characterizations of Besov spaces and fractional Sobolev spaces in terms of oscillations. As far as Besov spaces are concerned we refer also to the survey by Wallin [24]. An extension of these results to F^s_{pq} spaces is due to Seeger [16]. In this context we mention also the spaces C^s_p introduced by DeVore and Sharpley in [5]; see also [1]. It comes out that these spaces C^s_p coincide with $F^s_{p\infty}$. We follow the treatment given in [23] and refer to this paper for a more detailed comparison of known and new results and a more careful discussion of the above and further references.

6.2. OSCILLATIONS: Let Ω be a bounded C^∞ domain in $\mathbb{R}^n$. Let $x \in \Omega$ and $t > 0$, then

$$B(x,t) = \{y | y \in \Omega,\ |x-y| < t\}$$

is the intersection of Ω with the ball centred at x and of radius t. Let $0 < u \leq \infty$, $M \in \mathbb{N}_0$ (i.e. a natural number or 0) and $f \in L_u(\Omega)$. Then

$$\operatorname{osc}^M_u f(x,t) = \inf \left(\frac{1}{|B(x,t)|} \int_{B(x,t)} |f(y)-P(y)|^u\, dy\right)^{1/u} \tag{26}$$

(usual modification if $u = \infty$) where the infimum is taken over all polynomials P(y) of degree less than or equal to M. (Of course, osc^M_u depends on Ω, but this will not be indicated.)

6.3. THEOREM: (i) Let $0 < p < \infty$, $0 < q \leq \infty$, $1 \leq r \leq \infty$ and

$$s > n(1/p-1/r)_+ \quad \text{and} \quad s > n(1/q-1/r)_+.$$

Let $0 < u \leq r$ and $M \in \mathbb{N}_0$ with $M \geq [s]$, then

$$F^s_{pq}(\Omega) = \{f | f \in L_{\max(p,r)}(\Omega),$$
$$\|f|L_p(\Omega)\| + \|\left(\int_0^1 t^{-sq} \operatorname{osc}^M_u f(\cdot,t)^q \frac{dt}{t}\right)^{1/q} |L_p(\Omega)\| < \infty\} \tag{27}$$

(modification if $q = \infty$) in the sense of equivalent quasi-norms.

(ii) Let $0 < p \leq \infty$, $0 < q \leq \infty$, $1 \leq r \leq \infty$ and $s > n(1/p-1/r)_+$. Let $0 < u \leq r$ and $M \in \mathbb{N}_0$ with $M \geq [s]$, then

$$B^s_{pq}(\Omega) = \{f | f \in L_{\max(p,r)}(\Omega),$$
$$\|f|L_p(\Omega)\| + (\int_0^1 t^{-sq} \|osc_u^M f(\cdot,t)| L_p(\Omega)\|^q \frac{dt}{t})^{1/q} < \infty\}$$

(modification if $q = \infty$) in the sense of equivalent quasi-norms.

<u>6.4.</u> <u>REMARK</u>: The theorem remains valid if one replaces Ω by R^n or R^n_+. We shall not discuss special cases of this theorem connected with the above-mentioned papers. But we wish to emphasize that in the case of $q = \infty$ the integral over t in (27) must be interpreted as

$$f^M_{s,u}(x) = \sup_{0<t<1} t^{-s} osc_u^M f(x,t),$$

which is a sharp maximal function of the type used by DeVore, Sharpley Bojarski and their predecessors. This makes clear that function spaces defined via sharp maximal functions are closely connected with spaces of type $F^s_{p\infty}$.

References

[1] B. Bojarski, Sharp maximal operators of fractional order and Sobolev imbedding inequalities, Bull. Polish Acad. Sci., Ser. Math., 33 (1985), 7-16.

[2] H.-Q. Bui, Representation theorems and atomic decomposition of Besov spaces, Math. Nachr., 132 (1987), 301-311.

[3] R.R. Coifman, A real variable characterization of H^p, Studia Math., 51 (1974), 269-274.

[4] R.R. Coifman, Y. Meyer and E.M. Stein, Some new function spaces and their applications to harmonic analysis, J. Funct. Anal., 62 (1985), 304-335.

[5] R.A. DeVore and R.C. Sharpley, Maximal functions measuring smoothness, Mem. Amer. Math. Soc., 47:293 (1984), 1-115.

[6] J.R. Dorronsoro, A characterization of potential spaces, Proc. Amer. Math. Soc., 95 (1985), 21-31.

[7] J.R. Dorronsoro, Poisson integrals of regular functions, Trans. Amer. Math. Soc., 297 (1986), 669-685.

[8] J.R. Dorronsoro, On the differentiability of Lipschitz-Besov functions, Trans. Amer. Math. Soc., 303 (1987), 229-240.

[9] C. Fefferman and E.M. Stein, H^p spaces of several variables, Acta Math., 129 (1972), 137-193.

[10] M. Frazier and B. Jawerth, Decomposition of Besov spaces, Indiana Univ. Math. J., 34 (1985), 777-799.

[11] M. Frazier and B. Jawerth, The ϕ-transform and applications to distribution spaces, In Function Spaces and Applications, Proc. Lund 1986, Lect. Notes Math. 1302, Springer (1988), pp.223-246.

[12] M. Frazier and B. Jawerth, A discrete transform and decompositions of distribution spaces (to appear).

[13] G.A. Kaljabin, Function classes of Lizorkin-Triebel type in domains with Lipschitz boundary (Russian), Dokl. Akad. Nauk SSSR, 271 (1983), 795-798; English translation in Sov. Math. Dokl., 28 (1983), 156-159.

[14] G.A. Kaljabin, Theorems on extension, multipliers and diffeomorphisms for generalized Sobolev-Liouville classes in domains with a Lipschitz boundary (Russian), Trudy Mat. Inst. Steklov, 172 (1985), 173-185; English translation in Proc. Steklov Inst. Math., 3 (1987), 191-205.

[15] R. Latter, A characterization of $H^p(R^n)$ in terms of atoms, Studia Math., 62 (1978), 93-101.

[16] A. Seeger, A note on Triebel-Lizorkin spaces, In Approximation Theory and Function Spaces, Proc. Banach Center Sem., Warsaw, Spring 1986 (to appear).

[17] P. Sjölin, An H^p inequality for strongly singular integrals, Math. Z., 165 (1979), 231-238.

[18] H. Triebel, Theory of function spaces, Birkhäuser (1983); Akad. Verlagsgesellschaft (1983).

[19] H. Triebel, Einige neuere Entwicklungen in der Theorie der Funktionenräume, Jahresber. Deutsch. Math.-Verein, 89 (1987), 149-178.

[20] H. Triebel, Characterizations of Besov-Hardy-Sobolev spaces: a unified approach, J. Approx. Theory, 52 (1988), 162-203.

[21] H. Triebel, Atomic representations of F^s_{pq} spaces and Fourier integral operators, In Seminar Analysis of the Karl-Weierstrass-Institute 1986/87, Teubner-Texte Math. 106, Teubner (1988), pp. 297-305.

[22] H. Triebel, Atomic decompositions of F^s_{pq} spaces. Applications to exotic pseudo-differential and Fourier integral operators, Math. Nachr. (to appear).

[23] H. Triebel, Local approximation spaces. Zeitschr. Analysis Anwendungen (to appear).

[24] H. Wallin, New and old function spaces, In Function Spaces and Applications, Proc. Lund 1986, Lect. Notes Math. 1302, Springer (1988), pp. 97-114.

Hans Triebel
Sektion Mathematik
Universität Jena
DDR-6900 Jena
DDR

3. BOUNDARY VALUE PROBLEMS FOR SINGULAR DOMAINS

P. GRISVARD

Singular solutions of elliptic problems and applications to the exact controllability of hyperbolic problems

1. WHAT ARE THE SINGULAR SOLUTIONS OF ELLIPTIC BOUNDARY VALUE PROBLEMS AND WHAT IS THEIR USE?

1.1. INTRODUCTION: We say that the solution of a second-order elliptic boundary value problem is singular when its gradient is unbounded and thus may become infinite at some point. This may happen when:

(i) the operators have discontinuous coefficients,

(ii) the data are discontinuous,

(iii) the boundary of the domain has corner edges or vertices.

I shall focus on this third case.

1.2 SOME PRACTICAL EXAMPLES:

(i) Lightning conductor

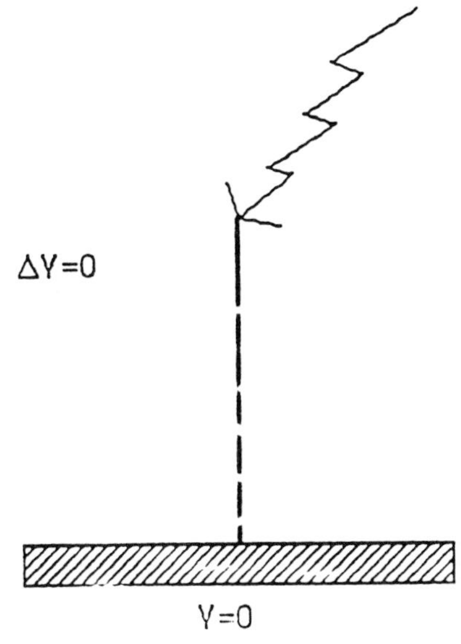

Lightning is attracted by the tip of an iron pole connected to the ground and thus kept at zero potential. The lightning is evidence of a very large electric field E in the vicinity of the pole tip. One has

$$E = - \text{grad } V$$

where V is the electric potential. In other words the gradient of V tends to infinity at the pole tip. Here V is a harmonic function, outside the ground and the pole, with the Dirichlet boundary condition V = 0 at any point connected to the ground.

(ii) Heating

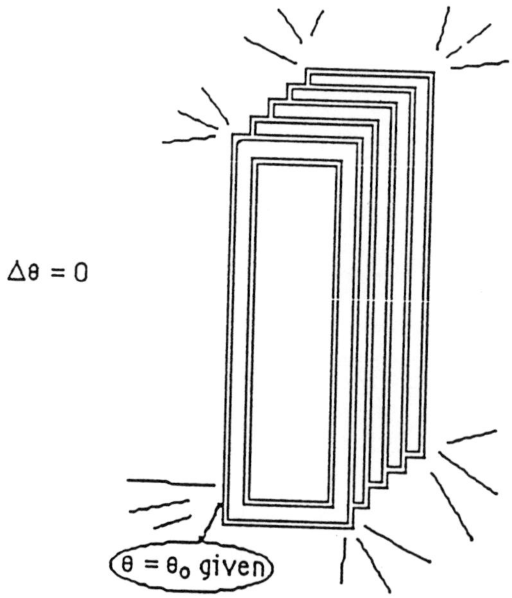

Heaters are more efficient when they have many edges. The heat flow q is large near vertices and edges. One has

$$q = - k \text{ grad } \theta$$

where θ is the temperature distribution. The gradient of θ tends to infinity at the edges. Here θ is a harmonic function outside the heater with the Dirichlet condition θ = a given θ_0 (the temperature of the water inside) on the surface of the heater. Here singular solutions are welcomed!

(iii) <u>Wrapping</u>

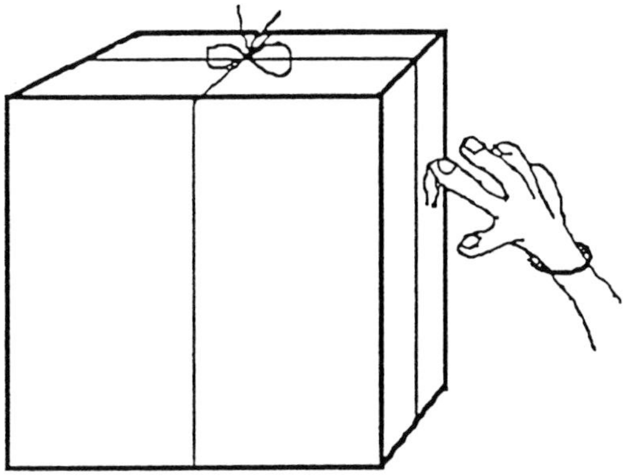

Any small cut in the plastic wrapping of a parcel makes it much easier to open it. Indeed the stresses, i.e. the body forces, tend to become very large at the tip of the cut. A little additional extra force may cause rupture.

Here large stresses are welcomed. They are not so in water-dams or in nuclear reactors.

The constitutive law of an elastic material reads

$$\sigma = A(Du + {}^tDu)/2$$

where σ denotes the stress tensor, u the displacement vector-field and therefore $(Du + {}^tDu)/2$ is the strain tensor. Here u is the solution of a second-order elliptic system and the gradient of some component of the displacement vector field tends to infinity at the tip of the cut or crack in the elastic material.

1.3. <u>A MODEL PROBLEM, NOTATION</u>: I shall describe singular solutions of the Neumann problem for the Laplace equation in 2d or 3d. The 2d domain Ω_2 is as follows.

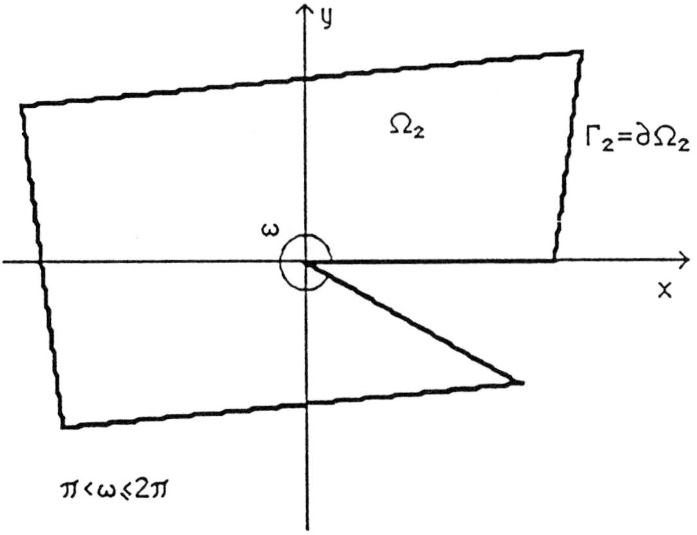

For simplicity, as we shall see later, it is convenient to assume that Ω_2 has only one reentrant corner, at 0, with measure ω defined by $0 < \theta < \omega$ in the usual polar coordinates for which

$$x + iy = re^{i\theta}.$$

We allow the possibility that $\omega = 2\pi$ for cuts.

The 3d domain will be

$$\Omega_3 = \Omega_2 \times]0,\pi[$$

the third variable will be denoted by z.

The classical formulation of the Neumann problem is as follows (Ω being either Ω_2 or Ω_3). Given $f: \Omega \to \mathbf{R}$ and $g: \Omega \to \mathbf{R}$, one looks for $u: \Omega \to \mathbf{R}$ such that

$$\Delta u = f \quad \text{in } \Omega$$
$$\partial u/\partial n = g \text{ on } \Gamma \qquad (1)$$

where $\partial/\partial n$ denotes the differentiation in the direction of the outward unitary normal vector n on Γ wherever it exists (i.e. away from corners and edges). The necessary condition on the data for the existence of a solution is

$$\int_\Omega f \, dx = \int_\Gamma g \, ds.$$

One actually shows existence and uniqueness of a weak solution u, up to the addition of a constant, meaning $u \in H^1(\Omega)$ such that

$$-\int_\Omega \text{grad } u \text{ grad } v \, dx = \int_\Omega fv \, dx - \int_\Gamma gv \, ds \qquad (2)$$

for every $v \in H^1(\Omega)$ the Sobolev space defined by

$$v, \|\text{grad } v\| \in L^2(\Omega).$$

This is achieved by applying the variational method (whether or not Γ has corners, edges or vertices) as in [10] and [12].

Our main problem is to decide whether the weak solution is classical (or strong). Clearly one has

(1) ⇒ (2) by integration by parts

(2) ⇒ (1) when u is regular

In other words given f and g regular can we expect u to be regular? The answer is no when singular solutions occur.

1.4. <u>TWO-DIMENSIONAL MODEL RESULTS</u>: Let $u : \Omega_2 \to \mathbb{R}$ be a solution of (2) with f square integrable and g regular enough then

$$u = u_R + cr^\alpha \cos \alpha\theta$$

where $\alpha = \pi/\omega < 1$, grad u_R is bounded, and

$$s = r^{\alpha} \cos \alpha\theta$$

has unbounded gradient:

$$\|\text{grad } s\|^2 = |\partial s/\partial r|^2 + r^{-2}|\partial s/\partial\theta|^2 = \alpha^2 r^{2(\alpha-1)} \to +\infty, \text{ as } r \to +\infty.$$

Here the factor c is a global function of the data f and g, meaning that c depends also on the values of f and g far from the corner under consideration. In other words c may differ from zero even when f and g vanish near the corner. This is typical in the case of the lightning conductor.

The above breakdown of the solution u is particularly valuable for the numerical computation of u. Actually the usual finite element codes are reliable for an approximate computation of the regular part u_R of u. A sufficient accuracy in the computation of the singular part cs, and in particular of the factor c which is physically the most relevant, is usually achieved by:

(i) either a mesh refinement near 0

(ii) or preferably by augmenting with s the space of trial functions in the finite element method.

A more precise statement relies on the use of Sobolev spaces: Given $f \in L^2(\Omega_2)$ and $g \in H^{1/2}(\Gamma_2)$, one has $u_R \in H^2(\Omega_2)$, i.e. u_R has square integrable second derivatives. This is not enough to ensure that grad u_R is bounded but one has the following (see [3]):

THEOREM 1: Let $u \in H_0^1(\Omega_2)$ be a solution of (2) with $f \in L^p(\Omega_2)$ and $g \in W^{1-1/p,p}(\Gamma_2)$ then

$$u = u_R + cs$$

where $u_R \in W^{2,p}(\Omega_2)$.

This means that the pth powers of the second derivatives of u_R are integrable. When $p > 2$ this implies the boundedness of grad u_R by Sobolev's theorem. Note that this statement allows a piecewise constant f and this is of practical interest.

Similar results hold for the Dirichlet problem if one replaces the cosines by sines. Also the corresponding functions for mixed boundary conditions and for the elasticity system with various boundary conditions are known (see [4] and [6]).

1.5. <u>THREE-DIMENSIONAL MODEL RESULTS:</u> The domain is now $\Omega_3 = \Omega_2 \times]0,\pi[$:

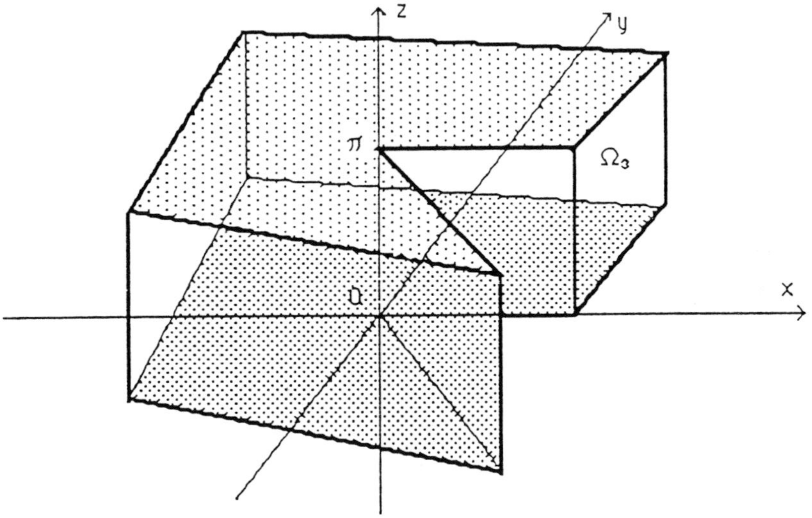

Observe that this polygon is a very special one in the sense that only three faces meet at any vertex and the edges parallel to the z'Oz axis all cut the other faces at right angles. In other words the vertices have this special feature that all the corresponding angles but one are right angles.

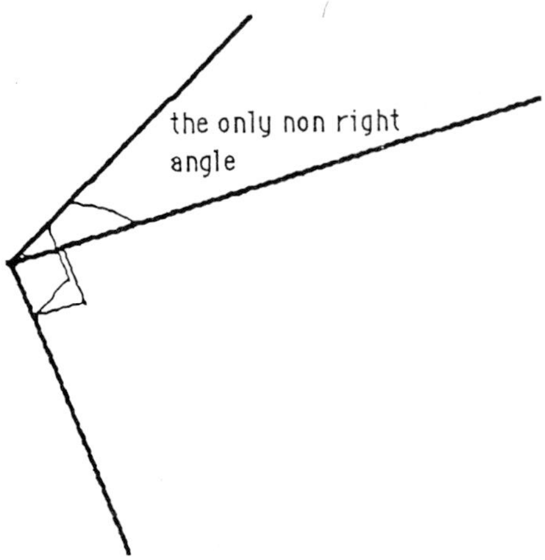

Let $u:\Omega_3 \to \mathbb{R}$ be a solution of (2) then we can give two types of statements:

STATEMENT OF THE FIRST KIND: If f and g are regular then

$$u = u_R + c(z)s$$

where grad u_R is bounded, and c' is bounded.

STATEMENT OF THE SECOND KIND (see [5]):

THEOREM 2: Let $u \in H^1_o(\Omega_o)$ be a solution of (2) with $f \in L^2(\Omega_3)$ and $g \in H^{1/2}(\Gamma_3)$ then

$$u = u_R + \{\Sigma_{k \geq 0}\, c_k e^{-kr} \cos kz\}s$$

where $u_R \in H^2(\Omega_3)$ and

$$\Sigma_{k \geq 0}\, k^{2(1-\alpha)} c_k^{\;2} < +\infty.$$

Here in 3d the condition that $u_R \in H^2(\Omega_3)$ is far from being enough for the boundedness of grad u_R. In addition the convergence of the series which represents the singular part of the solution is very poor due to the mild decay at infinity of the sequence c_k.

Before proceeding it will be useful to comment on the structure of the singular part of the solution

$$u_S = \{\Sigma_{k\geq 0} c_k e^{-kr}\cos kz\}\mathcal{s}.$$

It can be rewritten as

$$u_S = \{\int_0^{\pi} K(r,z,\zeta)\phi(\zeta)d\zeta\}\mathcal{s}$$

where

$$\phi(z) = \Sigma_{k\geq 0} c_k \cos kz \in H^{1-\alpha}(]0,\pi[)$$

and K is a Poisson kernel that approaches the Dirac distribution as $r \to 0$ and consequently one has

$$u_S \approx \phi(z)\,\mathcal{s}\,, \text{ as } r \to 0.$$

More simply if Ω_3 were $\Omega_2 \times \mathbb{R}$ instead of $\Omega_2 \times]0,\pi[$ then u_S would be

$$u_S = \{K\star\phi\}\mathcal{s}$$

with $\phi \in H^{1-\alpha}(\mathbb{R})$ and $\hat{K}(r,\zeta) = e^{-r|\zeta|}$ hence $K(r,z) = r/\pi(r^2 + z^2)$.

Further results are needed to obtain a bounded grad u_R.

IMPROVED STATEMENT OF THE SECOND KIND:

We shall have to add more singular terms and to use either Sobolev spaces of fractional order or Sobolev spaces built over L^p with $p \neq 2$.

An easy consequence of Theorem 2 is

THEOREM 3: Let $u \in H^1(\Omega_3)$ be a solution of (2) with $f \in H^{1/2+\varepsilon}(\Omega_3)$ and $g \in H^{1+\varepsilon}(\Gamma_3)$ with $\varepsilon > 0$ then

$$u = u_R + \{\Sigma_{k\geq 0} c_k e^{-kr}\cos kz\}r^{\alpha}\cos\alpha\theta + \{\Sigma_{k\geq 0} d_k e^{-kr}(1+rk)\cos kz\}r^{2\alpha}\cos 2\alpha\theta$$

where $u_R \in H^{5/2+\varepsilon}(\Omega_3)$ and

$$\Sigma_{k\geq 0}\ k^{2(1-\alpha)} c_k^2 < +\infty,\quad \Sigma_{k\geq 0}\ k^{2(1-2\alpha)} d_k^2 < +\infty$$

and the second sum occurs only when $\omega > \pi/(3/4 + \varepsilon/2)$.

Under such assumptions the gradient of u_R is bounded but a piecewise constant f is not allowed by $H^{1/2+\varepsilon}(\Omega_3)$.

An alternative statement is

THEOREM 4: Let $u \in H^1(\Omega_3)$ be a solution of (2) with $f \in W^{\varepsilon,p}(\Omega_3)$ with an $\varepsilon > 0$ and $g \in W^{1+\varepsilon-1/p,p}(\Gamma_3)$ then

$$u = u_R + \{\Sigma_{k\geq 0}\ c_k e^{-kr} \cos kz\} r^{\alpha} \cos \alpha\theta + \{\Sigma_{k\geq 0}\ d_k e^{-kr}(1+rk)\cos kz\} r^{2\alpha}\cos 2\alpha\theta$$

where $u_R \in W^{2,p}(\Omega_3)$ and

$$\phi(z) = \Sigma_{k\geq 0}\ c_k \cos kz \in W^{2-2/p-\alpha}(]0,\pi])$$

$$\psi(z) = \Sigma_{k\geq 0}\ d_k \cos kz \in W^{2-2/p-2\alpha}(]0,\pi])$$

and the second sum occurs only when $\omega > \pi/(1/2-1/p)$.

This statement is presumably true also when $\varepsilon = 0$. However it is enough to allow a piecewise constant f and the corresponding u_R has bounded gradient by Sobolev's theorem applied with $p > 3$. More 3d results are to be found in Kadlec [8], Kondratiev [9] and Dauge [1].

2. MATHEMATICAL ANALYSIS, AN OUTLINE

2.1. THE SIMPLEST PROBLEM: Let us consider a weak solution $u \in H_0^1(\Omega_2)$ of the Dirichlet problem $\Delta u = f$ with a given square integrable f. We follow the derivation in [2]. We ask under which conditions will u belong to $H^2(\Omega_2)$? For this purpose we consider the operator

$$\Delta : H^2(\Omega_2) \cap H_0^1(\Omega_2) \to L^2(\Omega_2)$$

and try to describe its range. We exclude the case when Ω_2 has cracks for simplicity. Here we denote Ω_2 by Ω.

2.2. THE CLOSED RANGE PROPERTY: It follows from the inequality

$$\|u\|_{H^2(\Omega)} = \{\Sigma_{|\alpha|\leq 2} \int_\Omega |D^\alpha u|^2 \, dydy\}^{1/2} \leqq C \|\Delta u\|_{L^2(\Omega)}$$

$$= C \{ \int_\Omega |\Delta u|^2 \, dxdy\}^{1/2},$$

for every $u \in H^2(\Omega) \cap H_0^1(\Omega)$, that we shall derive below.

Bounds for u and its first derivatives are well known and obtained by integration by parts.

Bounds for the second derivatives are first derived for $u \in C^3(\bar{\Omega}) \cap H_0^1(\Omega)$, a dense subspace of $H^2(\Omega) \cap H_0^1(\Omega)$ for the $H^2(\Omega)$ norm:

$$\|\Delta u\|^2 = \|D_1^2 u + D_2^2 u\|^2$$

$$= \|D_1^2 u\|^2 + \|D_2^2 u\|^2 + 2 \int_\Omega D_1^2 u D_2^2 u \, dxdy.$$

This last integral is calculated integrating by parts:

$$\int_\Omega D_1^2 u D_2^2 u \, dx\, dy = - \int_\Omega D_1 u \, D_1 D_2^2 u \, dxdy + \int_\Gamma D_1 u \, D_2^2 u \, dy$$

$$= \int_\Omega |D_1 D_2 u|^2 \, dxdy + \int_\Gamma D_1 u \, D_1 D_2 u \, dx + \int_\Gamma D_1 u D_2^2 u \, dy$$

$$= \|D_1 D_2 u\|^2 + \int_\Gamma D_1 u \, d(D_2 u).$$

We shall show that the last integral over Γ vanishes. Denote by Γ_j, $j = 1,2,\ldots,N$, the sides of Γ, and by S_{j-1} and S_j the end points of Γ_j following the direct orientation. Differentiating the boundary condition along Γ_j leads to

$$\lambda_j D_1 u + \mu_j D_2 u = 0$$

for some real λ_j and μ_j. Then two cases are possible:

(i) If $\mu_j = 0$ one has $D_1 u = 0$ and hence

$$\int_{\Gamma_j} D_1 u \, d(D_2 u) = 0.$$

(ii) If $\mu_j \neq 0$ one has $D_2u = -(\lambda_j/\mu_j)D_1u$ and hence

$$\int_{\Gamma_j} D_1u \, d(D_2u) = -(\lambda_j/2\mu_j)[(D_1u)^2(S_j) - (D_1u)^2(S_{j-1})]$$

On the other hand one has at S_j:

$$\lambda_j D_1u + \mu_j D_2u = 0$$

$$\lambda_{j+1} D_1u + \mu_{j+1} D_2u = 0$$

and consequently $D_1u(S_j) = 0$ for every j.

Summing up we conclude that

$$\|\Delta u\|^2 = \|D_1^2u\|^2 + \|D_2^2u\|^2 + 2\|D_1D_2u\|^2.$$

2.3. THE ADJOINT PROBLEM: Let $v \in L^2(\Omega)$ be orthogonal to

$$\Delta(H^2(\Omega) \cap H_0^1(\Omega))$$

i.e.

$$\int_\Omega \Delta u \, v \, dxdy = 0$$

for every $u \in H^2(\Omega) \cap H_0^1(\Omega)$ (which contains $\mathcal{D}(\Omega)$). Obviously v is harmonic

$$\Delta v = 0 \text{ in } \Omega.$$

Taking the restriction of such a v to Γ_j is meaningful:

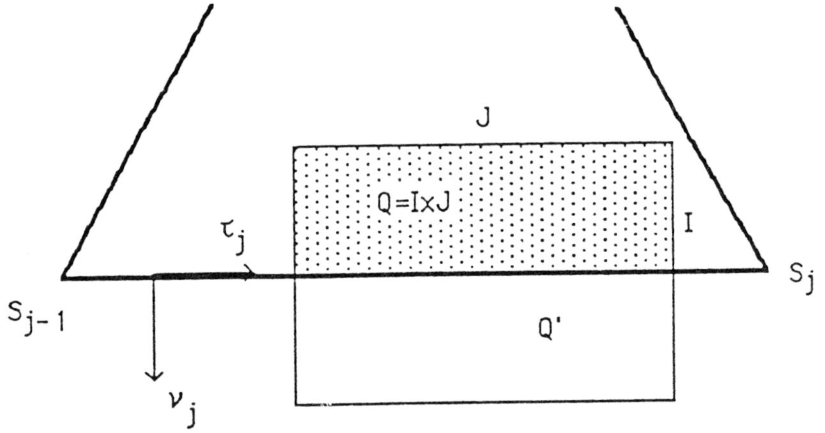

Using the notation which is self-explantory on the figure one has:

$$v \in L^2(I\,;L^2(J)) \subset L^2(I\,;H^{-2}(J))$$

and

$$\partial^2 v/\partial\nu_j^2 \in L^2(I\,;H^{-2}(J)) \text{ since } \partial^2 v/\partial\nu_j^2 = -\,\partial^2 v/\partial\tau_j^2$$

Thus v is continuously differentiable with values in $H^{-2}(J)$ and $v|\Gamma_j$ and $\partial v/\partial\nu_j|\Gamma_j$ are well-defined in $H^{-2}(J)$.

This allows one to claim that v vanishes on Γ_j for every j:

$$\int_\Omega \Delta u \; v \; dxdy = \Sigma_j \; \langle \partial u/\partial\nu_j ; v\rangle = 0,$$

where the brackets on the Γ_j's are meaningful for smooth enough u.

Reflexion through J produces a continuation V of v which is harmonic in QUQ' and this shows that v is regular up to interior points of Γ_j, i.e. varying I and J,

$$v \in C^\infty(\bar{\Omega} \setminus \{S_1, S_2, \ldots\}).$$

This regularity of v makes it expandible in sines near each corner

$$v = \Sigma_{k\in\mathbf{Z},k\neq 0} \; c_k^j \; r_j^{k\pi/\omega_j} \sin(k\pi\theta_j/\omega_j)$$

in a neighbourhood V_j of S_j. Here r_j, θ_j denote the polar coordinates related to S_j, meaning that r_j is the distance to S_j and $\theta_j = 0$ on Γ_{j+1}. There are two cases:

(i) The convex case $\omega_j < \pi (\Rightarrow \pi/\omega_j > 1)$. The condition that $v \in L^2(\Omega)$ excludes

$$r_j^{k\pi/\omega_j} \sin(k\pi\theta_j/\omega_j)$$

for $k < 0$ and therefore

$$v = \Sigma_{k\geq 1} \; c_k^j \; r_j^{k\pi/\omega_j} \sin(k\pi\theta_j/\omega_j) \in H^1(V_j).$$

(ii) The non-convex case $\omega_j > \pi$ ($\Rightarrow \pi/\omega_j \in]1/2,1[$). The condition that $v \in L^2(\Omega)$ excludes

$$r_j^{k\pi/\omega_j} \sin(k\pi\theta_j/\omega_j)$$

for $k < -1$ and therefore

$$v + c_{-1}^j \; r_j^{-\pi/\omega_j} \sin(\pi\theta_j/\omega_j) = \Sigma_{k\geq 1} \; c_k^j \; r_j^{k\pi/\omega_j} \sin(k\pi\theta_j/\omega_j) \in H^1(V_j).$$

Globally this implies that

$$v + \Sigma_{\omega j>\pi} \; c_{-1}^j \; r_j^{-\pi/\omega_j} \sin(\pi\theta_j/\omega_j) \; \chi\,(r_j) \in H^1(\Omega),$$

where χ is a cut-off function

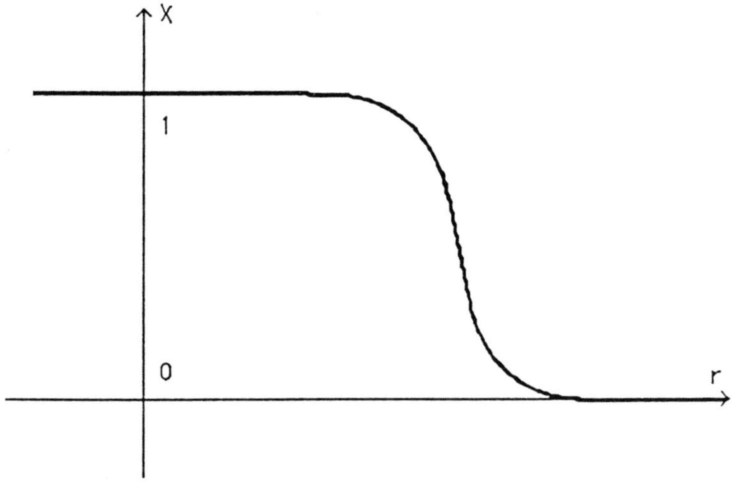

Since uniqueness holds in $H^1(\Omega)$ and v is a solution of the homogeneous problem this shows that $\Delta(H^2(\Omega) \cap H_0^1(\Omega))$ has codimension M, the number of non-convex corners of Ω, in $L^2(\Omega)$.

2.4. <u>BACK TO SINGULAR SOLUTIONS</u>: We look for the appropriate M-dimensional space S such that

$$\Delta(\{H^2(\Omega) \oplus S\} \cap H_0^1(\Omega)) = L^2(\Omega).$$

Define

$$S_j = r_j^{\pi/\omega j} \sin(\pi\theta/\omega_j) \ \chi(r_j)$$

if $\omega_j > \pi$. These are M linearly independent functions belonging to $H_0^1(\Omega)$ and not to $H^2(\Omega)$ and such that $\Delta S_j \in L^2(\Omega)$. S is generated by S_j for $\omega_j > \pi$ and, therefore, for every $f \in L^2(\Omega)$ there exists a unique $u \in H_0^1(\Omega)$ such that $\Delta u = f$ in Ω and furthermore

$$u = u_R + \Sigma_{\omega j > \pi} \, c_j S_j$$

where $u_R \in H^2(\Omega)$.

This is Theorem 1 of section 1 in the particular case when $p = 2$. When $p \neq 2$ the inequality that implies the closed range property is derived, locally, by Fourier transform techniques relying basically on Mikhlin's multiplier theorem.

3d results like Theorem 2 of section 1 are derived from the above 2d results by Fourier series techniques with respect to z.

3. APPLICATION TO THE EXACT CONTROLLABILITY OF HYPERBOLIC PROBLEMS

3.1. THE PROBLEM, NOTATION: Given u_0 and $u_1 : \Omega \to \mathbb{R}$, we look for a time T and a 'control' $v : \Sigma =]0,T] \times \Gamma \to \mathbb{R}$ such that u is the solution of

$$u'' - \Delta u = 0 \text{ in } Q = \Omega \times]0,T[$$

$$u(0) = u_0, u'(0) = u_1$$

$$u = v \text{ on } \Sigma$$

and is such that

$$u(T) = u'(T) = 0.$$

Here we consider the functions as depending on the time t only with vector values in spaces of functions of $x \in \Omega$. Thus $u(0)$ and $u(T)$ mean respectively u at time 0 and at time T. Also u' and u'' denote the first and second derivatives in time.

In other words we consider a solution u of the homogeneous wave equation with Cauchy data u_0 and u_1. We want to force u to rest at time T, also to be determined and perhaps long enough, by acting on u through the Dirichlet data v. This is called exact Dirichlet boundary controllability for the wave equation.

3.2. THE HILBERT UNIQUENESS METHOD OF LIONS: This is an approach by duality introduced in [11]. We integrate the wave equation in u against a 'test' function ϕ solution of

$$\phi'' - \Delta\phi = 0 \text{ in } Q$$

$$\phi(0) = \phi_0, \phi'(0) = \phi_1$$

$$\phi = 0 \text{ on } \Sigma.$$

Assuming

$$u(T) = u'(T) = 0$$

one gets after integration by parts

$$\int_\Sigma v\ \partial\phi/\partial\nu\ dtds = \int_\Omega (u_1\phi_0 - u_o\phi_1)dx\ .$$

Assuming $u_o \in L^2(\Omega)$ and $u_1 \in H^{-1}(\Omega)$ the right-hand side of this identity is given linear form in $\phi := \{\phi_o,\phi_1\} \in V = H_0^1(\Omega) \times L^2(\Omega)$ while the left-hand side is a bilinear form in v and ϕ. To make a long story short we shall accept the idea that the whole problem of finding v reduces to proving some sort of coerciveness. Substituting

$$v = \partial\phi/\partial\nu$$

can we show that

(i) $\phi \to \{\int_\Sigma (\partial\phi/\partial\nu)^2\ dtds\}^{1/2}$

is a norm on V

(ii) equivalent to that of

$$H_0^1(\Omega) \times L^2(\Omega)?$$

3.3. MULTIPLIERS: The central problem is to understand the behaviour of $\partial\phi/\partial\nu$. We recall a technique that goes back to Rellich: one can show that $\phi \in H_0^1(\Omega)$ and $\Delta\phi \in L^2(\Omega)$ imply $\partial\phi/\partial\nu \in L^2(\Gamma)$ by evaluating

$$\int_\Omega \Delta\phi\ m.\nabla\phi\ dx$$

where $m \in C^1(\bar{\Omega};\mathbb{R}^n)$ is a suitable vector field such that

$$m.\nu > 0 \text{ on } \Gamma.$$

Ho has considered the particular vector field

$$m(x) = x-x_0$$

where x_0 is arbitrary for the time being. One gets

$$\int_\Omega \Delta\phi\; m.\nabla\phi\; dx = -\int_\Omega \nabla\phi.\nabla(m.\nabla\phi)dx + \int_\Gamma m.\nu(\partial\phi/\partial\nu)^2\; ds \tag{3}$$

provided the integrals are meaningful and consequently

$$\int_\Omega \Delta\phi\; m.\nabla\phi\; dx = (1/2)\int_\Gamma m.\nu(\partial\phi/\partial\nu)^2\; ds - (1-n/2)\int_\Omega |\nabla\phi|^2\; dx$$

and

$$0 = \int_Q (\phi'' - \Delta\phi)\; m.\nabla\phi\;\; dtdx$$

$$= (n/2)\int_Q (\phi')^2 dx + (1-n/2)\int_Q |\nabla\phi|^2 dtdx - (1/2)\int_\Sigma m.\nu(\partial\phi/\partial\nu)^2\; dtds$$
$$+ \int_\Omega \phi' m.\nabla\phi\; dx\Big|_0^T.$$

Denoting by E the total energy

$$E = (1/2)\int_\Omega (|\nabla\phi_0|^2 + |\phi_1|^2)\; dx$$

we obtain for large enough T_0 (which can be calculated accurately)

$$(T-T_0)E \leq (1/2)\int_\Sigma m.\nu(\partial\phi/\partial\nu)^2\; dtds$$

and therefore there exists a constant C such that

$$E \leq C(T - T_0)^{-1}\int_{\Sigma(+)} m.\nu(\partial\phi/\partial\nu)^2\; dtds \tag{4}$$

for $T > T_o$, where

$$\Sigma(+) = \{x \in \Sigma;\ m.\nu > 0\}.$$

This gives a positive answer to the questions asked at the end of section 3.2. Indeed

$$\phi \to \{\int_\Sigma m.\nu(\partial\phi/\partial\nu)^2 dtds\}^{1/2}$$

defines a norm on V that is stronger than the norm of

$$H_0^1(\Omega) \times L^2(\Omega).$$

Proving the converse inequality by applying the above-mentioned Rellich technique is quite easy and the two norms are actually equivalent. However the above calculations are not always possible since in (3) it has been assumed that

$$\phi \in H_0^1(\Omega) \text{ and } \Delta\phi \in L^2(\Omega)$$

imply that

$$\phi \in H^2(\Omega) \text{ and } (\partial\phi/\partial\nu) \in L^2(\Gamma).$$

3.4. <u>THE CASE OF POLYGONS</u>: As we have seen in section 1 the assumptions that

$$\phi \in H_0^1(\Omega) \text{ and } \Delta\phi \in L^2(\Omega)$$

imply

$$\phi = \phi_R + c\,S$$

where $\phi_R \in H^2(\Omega)$ and $S = r^\alpha \sin\alpha\theta$, $\alpha = \pi/\omega$. It is clear therefore that

$$\phi \notin H^2(\Omega) \text{ when } \omega > \pi$$

and that

$$(\partial\phi/\partial\nu) \notin L^2(\Gamma) \text{ when } \omega = 2\pi.$$

Then the case $\omega = 2\pi$, i.e. when Ω has a cut, is clearly hopeless since we can not even write the quantity

$$\int_\Sigma (\partial\phi/\partial\nu)^2 \, dtds.$$

From now on let us assume that Ω has no cut and therefore $\omega < 2\pi$. We have $\alpha > 1/2$ and consequently $S \in H^s(\Omega)$ for $s < \alpha + 1$. This allows $s > 3/2$: there exists $\varepsilon > 0$ such that

$$\phi \in H^{\varepsilon+3/2}(\Omega).$$

This implies that

$$\Delta\phi \text{ and } \nabla(m.\nabla\phi) \in H^{\varepsilon-1/2}(\Omega) = H^{-\varepsilon+1/2}(\Omega)^*$$

and that

$$m.\nabla\phi \text{ and } \nabla\phi \in H^{\varepsilon+1/2}(\Omega) \subset H^{-\varepsilon+1/2}(\Omega).$$

Since the spaces $H^{-\varepsilon+1/2}(\Omega)$ and $H^{-\varepsilon+1/2}(\Omega)^*$ are in duality we can give a meaning to the integrals in (3) as follows:

$$\int_\Omega \Delta\phi \, m.\nabla\phi \, dx = -\langle\nabla\phi,\nabla(m.\nabla\phi)\rangle + (1/2)\int_\Gamma m.\nu(\partial\phi/\partial\nu)^2 \, ds. \qquad (3')$$

There is no need of modification in the subsequent calculations and the conclusion is that there is exact Dirichlet boundary controllability for the wave equation in a polygon with no cut after a long enough time T_0.

3.5. <u>THE CASE OF POLYHEDRA</u>: All the same we can consider a 3d open Ω with polyhedral boundary. If we assume that Ω has no crack we can show the existence of $\varepsilon > 0$ such that $\phi \in H_0^1(\Omega)$ and $\Delta\phi \in L^2(\Omega)$ imply $\phi \in H^{\varepsilon+3/2}(\Omega)$. Again (3') holds then and the subsequent calculations are valid. The conclusion is

<u>THEOREM 5</u>: Assume that Ω is a bounded open subset of R^3 with a polyhedral boundary and with no crack, then there exists a time T_0 such that for every pair

$$u_0 \in L^2(\Omega),\ u_1 \in H^{-1}(\Omega)$$

and every time $T > T_0$ there exists $v \in L^2(\Sigma)$ such that u the (weak) solution of

$$u'' - \Delta u = 0 \text{ in } Q = \Omega \times]o,T[$$

$$u(0) = u_0, u'(0) = u_1$$

$$u = v \text{ on } \Sigma$$

fulfils

$$u(T) = u'(T) = 0.$$

Details are to be found in [7].

Actually we have proven a little more since we can impose that v should have its support in

$$\Sigma(+) = \{x \in \Sigma; (x-x_0), \nu > 0\}$$

where x_0 is any point in R^3. The part of Σ on which one needs to impose a control v may be small enough. However this leaves aside the case when Ω has cracks, which is an open problem and is obviously very important in practice. Some particular cases are solved in [7]. Similar difficulties arise when one considers the wave equation with mixed boundary conditions on a regular domain Ω.

References

[1] M. Dauge, Régularités et singularités des solutions des problèmes aux limites elliptiques sur des domaines singuliers de type à coins, Thèse d'état, Nantes (1986).

[2] P. Grisvard, Alternative de Fredholm relative au problème de Dirichlet dans un polygone ou un polyèdre 1, Boll. Unione Mat. Ital. (4), 5 (1972), 132-164.

[3] P. Grisvard, Elliptic problems in non-smooth domains, Monogr. Surv. Math. no. 24, Pitman (1985).

[4] P. Grisvard, Problèmes aux limites dans les polygones, mode d'emploi, Bulletin de la Direction des études et recherches de l'EDF, Serie C: Mathématiques et Informatique, no. 1 (1985), 21-59.

[5] P. Grisvard, Edge behavior of the solution of an elliptic problem, Math. Nach., 132 (1987), 281-299.

[6] P. Grisvard, Singularities in elasticity theory, in Applications of Multiple Scaling in Mechanics, ed. P.G. Ciarlet and E. Sanchez-Palencia, Collection 'Recherches en Mathématiques Appliquées' no. 4, Masson (1987), pp.134-150.

[7] P. Grisvard, Controlabilité exacte des solutions de l'équation des ondes en présence de singularités, Preprint Nice n.153 (1989); to appear in J. Math. Pures Appl., Paris.

[8] Kadlec , On the regularity of the solution of the Poisson problem on a domain with boundary locally similar to the boundary of a convex open set, Czech. Math. J., 14:89 (1964), 386-393.

[9] V.A. Kondratiev, Boundary value problems for elliptic equations in domains with conical or angular points, Trans. Moscow Math. Soc. (1967) 227-313.

[10] Lions , Equations opérationnelles et Problèmes aux Limites, Springer (1961).

[11] Lions , The J. Von Neumann Lecture, SIAM National Meeting, Boston (1986).

[12] Lions and Magenes , Problèmes aux Limites Non Homogénes et Applications, Dunod (1968).

P. Grisvard
Laboratoire de Mathématiques
I.M.S.P.
Université de Nice
Parc Valrose
F. 06034 Nice-Cedex
France

L. PÄIVÄRINTA

The behaviour of fields near a corner of a flat plate

0. INTRODUCTION

Thin objects have often crucial influence on physical fields near to them. We have in mind situations like an electromagnetic field near a thin superconductor, acoustic potential on a sound-soft or sound-hard object or a crack in an isotropic homogeneous elastic body. The fields turn out to get singular near the boundary of the body. Especially interesting is the case of corners on the boundary. What is the singular behaviour near these points? The problem often reduces to integral equations on the object Ω:

$$Tu = f \qquad (0.1)$$

where f is some known smooth function on Ω (in scattering problems the incident field) and T the corresponding integral operator. In several cases T is a pseudodifferential operator with principal symbol either $|\xi|$ or $|\xi|^{-1}$. What makes these operators interesting is the fact that they do not possess the so-called transmission property (cf. [2]). Roughly speaking this means that the solution u of (0.1) can be singular near the boundary $\partial\Omega$ even if f is smooth.

The case of corner points was first studied for differential boundary value problems in the pioneering work of V. Kontratjev [3]. Since then a growing theory for these questions with several physical applications has been developed. We refer here to the monograph of P. Grisward [2]. Until recent years the corresponding theory for pseudodifferential operators has been rather incomplete although the basic contribution of B. Plamenewskij dates back to late sixties [8]. However, very recently the field has been most active and several essential features characteristic for the singular behaviour of the solutions of pseudodifferential boundary value problems have been cleared up (cf. [12], [5], [6], [11], [7] and [10]).

1. PHYSICAL EXAMPLE

The simplext physical example we have in mind is the problem to determine the charge distribution on a thin perfect conductor in the electrostatic case. Hence we study a bounded thin condenser Ω loaded with a charge density ρ. It produces a harmonic electric potential V taking a constant value q on Ω and vanishing at the infinity. For simplicity we suppose that $\Omega \subset \mathbb{R}^2 = \{x_3 = 0\} \subset \mathbb{R}^3$, that is the condenser is assumed to be flat. According to the boundary conditions for Maxwell's equations ρ is equal to the jump of the normal component of the electric field through Ω:

$$\rho(x) = - \left.\frac{\partial V}{\partial x_3}\right|_{x_3=0+} + \left.\frac{\partial V}{\partial x_3}\right|_{x_3=0} .$$

Since V is harmonic in $\mathbb{R}^3 \setminus \Omega$ and vanishes at infinity we may use Poissons integral formula to get

$$V(x) = F^{-1}(e^{-|x_3||\xi|}\,\hat{V}_0(\xi))(x_1,x_2) \tag{1.2}$$

where $x = (x_1,x_2,x_3)$, $V_0 = V|_{\mathbb{R}^2}$ and $F\hat{V}_0 = V_0$ stands for the two-dimensional Fourier-transform of V_0. Plugging (1.2) into the equation (1.1) we get

$$\rho(x) = 2F^{-1}(|\xi|\hat{V}_0(\xi))(x) \tag{1.3}$$

for $x \in \Omega$. According to (1.1) the density ρ vanishes outside Ω and we may invert (1.3) in the whole plane $\mathbb{R}^2$ to obtain

$$2V_0(x) = F^{-1}(\frac{1}{|\xi|}\,\hat{\rho}(\xi))(x), \ x \in \mathbb{R}^2. \tag{1.4}$$

Our physical problem is to determine the charge distribution ρ on Ω if the potential difference g between Ω and infinity is measured. Since $V_0|_\Omega = g$ we have from (1.4)

$$T\rho(x) = 2g \tag{1.5}$$

where T is a pseudodifferential operator with symbol $|\xi|^{-1}$ i.e.

$$Tu(x) = F^{-1}\left(\frac{\hat{u}(\xi)}{|\xi|}\right). \tag{1.6}$$

Note that (1.5) is exactly of the form (0.1) with a constant right hand side in this case.

We do not go into other physical problems but only mention in passing that in the scattering problem of electromagnetic waves the normal derivative of the scattered electric field satisfies (in principal symbol level) the equation (1.5) with the incident field in the right hand side. Moreover, if we have a flat crack Ω in a homogeneous isotropic solid then the displacement u and the pressure p are connected through the equation

$$p(x) = \frac{\mu}{1 - \nu} F^{-1}(|\xi|\hat{u}(\xi)), \; x \in \Omega$$

where μ is the shear modulus and ν is the Poisson ratio (cf. [4]).

2. CORNER SYMBOL

We are interested in the solutions u of

$$Tu(x) = f(x) \tag{2.1}$$

for

$$Tu(x) = \int_{\Omega} \frac{1}{|x - y|} u(y)dy \tag{2.2}$$

and especially in the singularities of u near the possible corners of $\partial\Omega$. Hence we assume that $\Omega \subset \mathbf{R}^3$ is compact surface having a boundary $\partial\Omega$ that is smooth outside a finite set of singular points.

Since on the plane

$$F\left(\frac{1}{|x|}\right)(\xi) = c \frac{1}{|\xi|}$$

the operator T defined by (2.2) is a pseudodifferential operator with principal symbol $c \frac{1}{|\xi|}$. Hence it is clear that T extends to a bounded linear operator between Sobolev-spaces

$$T : \dot{H}^s(\Omega) \to H^{s+1}(\Omega).$$

Here $\dot{H}^s(\Omega)$ means the subspace of the distributions in $H^s(\mathbf{R}^2)$ having their support in $\bar{\Omega}$. If Ω is smooth it is possible to show by using the boundary symbolic calculus developed by S. Rempel and B.-W. Schulze the following

2.1. THEOREM: The operator

$$T : \dot{H}^s(\Omega) \to H^{s+1}(\Omega)$$

is an isomorphism if and only if $-1 < s < 0$.

PROOF Cf. [5] Th. 2.6: Let us now turn to the case of a corner point $p \in \partial\Omega$. We pick up a neighbourhood of p in $\mathbf{R}^3$ so that $U \cap \Omega$ is mapped onto a part U' of a cone C^α in $\mathbf{R}^2$.

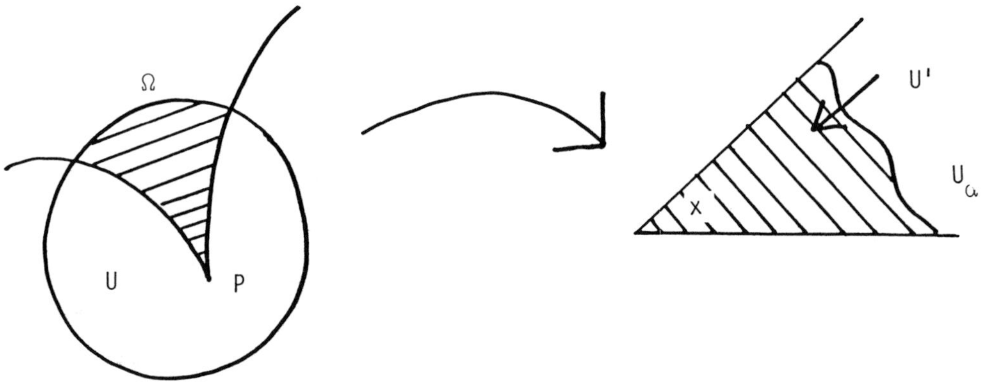

Figure 1

The cone C_α is supposed to have its vertex at the origin and the opening angle α. It can be shown (cf. [6]) that only the principal part of the operator affects to the strength of the singularity of the solution. By the coordinate chart around p the operator T is transformed to an operator T' on C_α. Since the principal symbol of T' is again $c \frac{1}{|\xi|}$ we may study the operator

$$T\, u(x) = \int_{C_\alpha} \frac{1}{|x - y|} u(y)dy, \; x \in C_\alpha. \tag{2.3}$$

As is common for differential boundary value problems we pass onto polar coordinates (r,ϕ) where C_α is represented as $r > 0$ and $0 < \phi < \alpha$. After doing this the equation $T_\alpha u = v$ looks like

$$T_\alpha u(r,\phi) = \int_0^\infty \frac{d\rho}{\rho} \int_0^{2\pi} d\psi \frac{v(\rho,\psi)}{((r/\rho)^2 - 2(r/\rho)\cos(\phi - \psi) + 1)^{1/2}} . \tag{2.4}$$

Since the kernel of T_α depends only on r/ρ it is convenient to take the Mellin transform with respect to r of both sides of (2.4). By definition the Mellin transform of a function $f \in C_0^\infty(R_+)$ is

$$Mf(z) := \hat{f}(z) := \int_0^\infty t^{z-1} f(t)dt, \; z \in \mathbb{C}. \tag{2.5}$$

We make use of the convolution theorem

$$M(\int_0^\infty f(r/\rho)g(\rho)d\rho/\rho)(z) = Mf(z)Mg(z) \tag{2.6}$$

to get

$$\widetilde{T_\alpha u}(z,\phi) = \int_0^\alpha m(z,\phi - \psi)\tilde{u}(z,\psi)d\psi = \tilde{v}(z,\phi) \tag{2.7}$$

where

$$m(z,\phi) = \int_0^\infty \frac{t^{z-2}dt}{(t^2 - 2t\cos\phi + 1)^{1/2}} . \tag{2.8}$$

The main idea of this approach is by studying the equation $T_\alpha u = v$ or equivalently the equation (2.7) to determine the poles of the solution $\tilde{u}$. This is important since, roughly speaking, the behaviour r^γ near the origin for a function $f \in C_0^\infty(\bar{R}_+)$ implies a simple pole at $-\gamma$ in the Mellin side and vice versa. Accordingly, it follows that it is enough to find out the zeros of the operator $M_\alpha(z)$

$$M_\alpha(z)u(\phi) = \int_0^\alpha m(z,\phi - \psi)u(\psi)d\psi \tag{2.9}$$

in suitable function spaces. Note that by the zero of an operator function it is meant the value of z where it is not invertible. The operator function $M_\alpha(z)$ is called the corner symbol of T_α.

3. BASIC PROPERTIES OF THE CORNER SYMBOL

The usual tool in situations where one is tempted to determine the zeros of an operator function is the meromorphic Fredholm theory (cf. [9]). It says the following: If an operator-valued meromorphic function f taking values as a Fredholm operator is invertible at least at one point then it is invertible everywhere except a discrete set of exceptional values (the zeroes of f). Hence we need to study the Fredholm properties of $M_\alpha(z)$. We are able to prove the following

3.1. THEOREM: The operator

$$M_\alpha(z) : \dot{H}^s(0,\alpha) \to H^{s+1}(0,\alpha)$$

is a Fredholm operator with index zero for each $z \in \mathbb{C}$. Moreover, $M_\alpha(z)$ is invertible if $\operatorname{Im} z \neq 0$ or $1 < z < 2$.

PROOF: We will only give an outline here. The complete proof will appear in [6], in Theorem 3.11 and Remark 3.2. Our starting point is the observation that $(t^2 - 2tx + 1)^{-1/2}$ is the generating function of the Legendre polynomials i.e.

$$\frac{1}{\sqrt{t^2 - 2tx + 1}} = \sum_{n=0}^{\infty} P_n(x)t^n. \tag{3.1}$$

Plugging this to the definition (2.9) yields us after some calculations the following representation of $M_\alpha(z)$:

$$M_\alpha(z)u(\phi) = \sum_{m=-\infty}^{\infty} b_m(z)\hat{u}(m)e^{im\phi}$$

where $\hat{u}(m)$ are the Fourier-coefficients of u and

$$b_m(z) = \frac{1}{2}\;\frac{\Gamma(\frac{m+z-1}{2})\Gamma(\frac{m-z+2}{2})}{\Gamma(\frac{m+z}{2})\;\Gamma(\frac{m-z+3}{2})}\,. \tag{3.3}$$

From (3.3) and from Stirling's formula we see that $b_m(z) = \frac{\pi}{|m|} + O(|m|^{-2})$. Hence it is clear that $\mathbf{M}_\alpha(z)$ maps $\dot{H}^s(0,\alpha)$ continuously to $H^{s+1}(0,\alpha)$ for every $s \in \mathbb{R}$ and $z \in \mathbb{C}$ (except for $z \in \mathbb{Z}$ which are the poles of $\mathbf{M}_\alpha(z)$). Moreover we see that $\mathbf{M}_\alpha(z)$ coincides modulo a compact operator with pseudo-differential operators having symbols $|\eta|^{-1}$ where η is the dual variable to $\phi \in (0,\alpha)$. In fact, we get for distribution kernels

$$K_{M_\alpha(z)}(\phi,\psi) = \sum_{m=-\infty}^{\infty} b_m(z)e^{im(\phi-\psi)} \sim \sum_{m=-\infty}^{\infty} (1 + |m|\;)^{-1}e^{im(\phi-\psi)}$$

$$\sim \int_{-\infty}^{\infty} (1 + |\eta|)^{-1}e^{i\eta(\phi-\psi)}d\eta \sim K_{Op(|\eta|^{-1})}(\phi,\psi)$$

where ψ means the equivalence of the corresponding operators modulo compact operators. But for the operator $S = Op(|\eta|^{-1}) : \dot{H}^s(0,\alpha) \to H^{s+1}(0,\alpha)$ defined by

$$Su(x) = F^{-1}(\frac{\hat{u}(\eta)}{|\eta|})(x)$$

it is well-known that it is an isomorphism for $-1 < s < 0$ (cf. [5]). Hence $M_\alpha(z)$ is a Fredholm operator with index zero for $-1 < s < 0$. A more careful study of the coefficients $b_m(z)$ show that $\mathbf{M}_\alpha(z)$ is additionally injective for $\operatorname{Im} z \neq 0$ (cf. [6]). From (3.3) we can also see that $b_m(z) > 0$ for $1 < z < 2$. Hence $M\,(z)$ is also injective for $1 < z < 2$. The injectivity together with the Fredholm-property proves that $M_\alpha(z)$ is invertible for all z with $\operatorname{Im} z \neq 0$ or $1 < z < 2$. □

According to the previous theorem the zeros of $\mathbf{M}_\alpha(a)$ lie in the real axis. Since the zeros that are greater than 2 correspond to solutions with infinite energy we are mainly interested in the largest zero of $M_\alpha(z)$ smaller than 1. If this value of z is called z_α then the solution u of (2.1) behaves like r^{-z_α} for $r \to 0$.

We have

3.2. <u>THEOREM</u>: For each α, $0 \leqq \alpha < 2\pi$, the operator $\mathbf{M}_\alpha(z)$ has exactly one zero z_α in the interval $0 < z < 1$.

PROOF: Define the operator $\tilde{M}_\alpha(z) : \dot{H}^s(0,\alpha) \to H^{s+1}(0,\alpha)$

$$\tilde{M}_\alpha(z)u(\phi) = \sum_{|m|>0} b_m(z)\hat{u}(m)e^{im\phi}. \tag{3.4}$$

According to (3.3) $b_m(z) > 0$ for $|m| > 0$ if $0 < z < 3$ and hence the operator $\tilde{M}_\alpha(z)$ is an isomorphism for these values of z.

Since

$$M_\alpha(z)u(\phi) = b_0(z)\hat{u}(0) + \tilde{M}_\alpha(z)u(\phi) \tag{3.5}$$

we may conclude as follows: if z is a zero for $M_\alpha(z)$ there exist a function u_z such that

$$\tilde{M}_\alpha u_z(\phi) = 1$$

for every $\phi \in (0,\alpha)$. Hence by (3.5) the problem of finding z_α is reduced to find the zero of the function

$$\int_0^\alpha \tilde{M}_\alpha^{-1}(z)(1)(\phi)d\phi + \frac{1}{b_0(z)} . \tag{3.6}$$

Since $b_0(0) = 0$, $\lim_{z\to -1+0} b_0(z) = \infty$ and $\lim_{z\to 1-0} b_0(z) = -\infty$ the claim follows from the fact that $\int_0^\alpha \tilde{M}_\alpha^{-1}(z)(1)(\phi)d\phi$ is increasing and positive in z. □

At the end of this section we mention that the values z_α can be obtained analytically for some geometrically simple cases: It is easy to check that $z_{2\pi} = 0$. This means that for $\alpha = 2\pi$ the singularity vanishes which is, of course, obvious from the physical point of view. For $\alpha = \pi$ the function $v(\phi) = \sin^{1/2}(\phi)$ can be proved to be in the kernel of $M_\alpha(1/2)$ and consequently $z_\pi = 1/2$ and the solution u of (2.4) has a behaviour

$$u(r,\phi) \sim r^{-1/2} \sin^{-1/2}(\phi)$$

for small r. The interesting limiting case $\alpha \to 0$ can also be handled analytically. We obtain that $z_\alpha \to 1$ as $\alpha \to 0$. We can even get the asymptotic speed of the convergence. The result is described in terms of a Sobolev norm of a constant:

$$1 - z^\alpha \sim \|1\|_{H^{1/2}(0,\alpha)} \sim \log \left(\frac{1}{\alpha}\right)^{-1/2}. \tag{3.7}$$

We found it surprising that analysis of function spaces such as the calculation of the above Sobolev norm of a constant can yield very concrete information of physical objects as here the singularity of the magnetic field near a small corner of a thin superconductor. All the claims represented here are proved in [6]. The second equivalence in (3.7) is due to E.Saksman.

4. NUMERICAL SOLUTIONS

Except the cases $\alpha = 2\pi$, $\alpha = \pi$ and $\alpha \to 0$ the values z_α must be computed numerically from the equation

$$\int_0^\alpha \tilde{M}_\alpha^{-1}(z)(1)(\phi)d\phi = -\frac{1}{\overline{b_0(z)}} . \tag{4.1}$$

The key to the numerical solution is the fact that $\tilde{M}_\alpha$ is a pseudodifferential operator with principal symbol $c|\xi|^{-1}$. And hence by [5] the solution u_α of

$$\tilde{M}_\alpha u_\alpha(\phi) = 1 \tag{4.2}$$

has an inverse squareroot singularity near the endpoints of $(0,\alpha)$. Since $\tilde{M}_\alpha$ is elliptic in the interior it is also clear that $u_\alpha(\phi)$ is smooth away from the endpoints 0 and α. Consequently if we define a weight function

$$w(\phi) = \left(\left(\frac{\alpha}{2}\right)^2 - \left(\phi - \frac{\alpha}{2}\right)^2\right)^{1/2}$$

the function $v_\alpha(\phi) = w(\phi)u_\alpha(\phi)$ is smooth and bounded in $(0,\alpha)$. This means that the Fourier series of v converges rapidly. The use of this and the Fourier series discretization of (4.2) yield a reasonably stable solution of v_α and hence u_α. By connecting this to any reasonable root-finding method we obtain an effective numerical solution to z_α. The details can be found again in [6].

References

[1] P. Grisvard, Elliptic problems in non-smooth domains. Monographs and Studies in Math. 24. Pitman, London 1985.

[2] L. Hörmander, The analysis of linear partial differential operators III. Springer-Verlag, Grundl. Math. Wiss. Band 274, Berlin 1985.

[3] V.A. Kontratjev, Boundary problems for elliptic equations in domains with conical points. Trudy Mosk. Mat. Ob. 16 p. 209-292 (1967).

[4] P.A. Martin, Orthogonal polynomial solutions for pressurized elliptiocal cracks. Q. Tl. Mech. appl. Math. 39 p. 269-287 (1986).

[5] L. Päivärinta and S. Rempel, A deconvolution problem with the kernel $\frac{1}{|x|}$ on the plane. Appl. Anal. 26 p. 105-128 (1987).

[6] L. Päivärinta and S. Rempel, Corner singularities of solutions to $\Delta^{\pm 1/2}u = f$ in two dimensions. To appear.

[7] T.v. Petersdorf and E. Stephan, Decompositions in edge and corner singularities for the solution of the Laplacian in a polyhedron. TH Darmstadt, Preprint 1150, 1988.

[8] B. Plamenewsky, Boundedness of singular integrals in spaces with a weight. Mat. sb. 5 p. 539-558 (1968).

[9] M. Reed and B. Simon, Methods of modern mathematical physics, IV. Analysis of Operators. Academic Press 1978.

[10] S. Rempel, Elliptic pseudo-differential operators on manifolds with corners and edges. This issue.

[11] S. Rempel and B.-W. Schulze, Asymptotics for elliptic mixed boundary problems. Pseudodifferential and Mellin operators in spaces with conormal singularity. Akademie Verlag, Mathematical research 50, Berlin 1988.

[12] E. Stephan, Boundary integral equations for screen problems in $\mathbb{R}^3$. J. Integral Equations and Operator Theory 10 (1987).

L. Päivärinta
Department of Mathematics
University of Helsinki
Hullituskatu 15
00100 Helsinki,
Finland

S. REMPEL

Elliptic pseudodifferential operators on manifolds with corners and edges

A theory of pseudo differential boundary value problems on manifolds with edges was developed in [10]. Here we perform the next, more complex, generalization to the case of geometric corners arising at intersection points of edges. Despite the numerous papers on this topic (see [2], [3],[4], [5],[6],[8],[10],[11], even for differential boundary value problems some important points were not treated in a satisfactory way. Near the corners the singular terms from the edge interplay with the proper corner singularity. Here we give only the results and examples. Full proofs will be given elsewhere.

1. GEOMETRIC ASSUMPTIONS

Let X be a closed, bounded subset of $\mathbb{R}^n$ with boundary X. Let $\partial_0 X = X \setminus \partial X$ be the interior of X and assume that ∂X consists of three disjoint parts $\partial_1 X, \partial_2 X$ and $\partial_3 X$. Here $\partial_1 X$ is defined as the maximal open subset of ∂X where ∂X is locally represented as $\{x \in \mathbb{R}^n : f(x) = 0\}$ with suitable real-valued function $f \in C^\infty(\mathbb{R}^n)$, $df \neq 0$ if $f = 0$. $\partial_2 X$ is the maximal open subset of $\partial X \setminus \partial_1 X$ where ∂X is locally represented as $\{x \in \mathbb{R}^n : f(x) = g(x) = 0\}$ with suitable $f, g \in C^\infty(\mathbb{R}^n)$, df and dg linearly independent at $f = g = 0$. Assume that $\partial_3 X = \partial X \setminus (\partial_1 X \cup \partial_2 X)$ is locally represented as $\{x \in \mathbb{R}^n : f(x) = g(x) = h(x) = 0\}$ with suitable $f, g, h \in C^\infty(\mathbb{R}^n)$, df, dg, dh linearly independent at $f = g = h = 0$. Clearly, $\partial_1 X$ is the smooth part of the boundary, $\partial_2 X$ consists of edges locally formed by intersection of two transversal hypersurfaces and at $\partial_3 X$ we have an intersection of three transversal hypersurfaces. Below we shall restrict ourselves to the important special case of dimension 3, where $\partial_1 X$ consists of parts of two-dimensional hypersurfaces, $\partial_2 X$ consists of smooth arcs and $\partial_3 X$ consists of a finite number of isolated points. Then each point in $\partial_1 X$ possesses a neighborhood diffeomorphic to $\mathbb{R}^3_+ = \{x = (x_1, x_2, x_3) \in \mathbb{R}^3 : x_3 > 0\}$. For neighborhoods of points in $\partial_2 X$ the local model is the diedre $\mathbb{R}^3_{++} = \{x \in \mathbb{R}^3 : x_2 > 0,\ x_3 > 0\}$ and for points in $\partial_3 X$ the octant $\mathbb{R}^3_{+++} = \{x \in \mathbb{R}^3 : x_1 > 0,\ x_2 > 0,\ x_3 > 0\}$. Sometimes it is more transparent to take as model the tangent cone to the boundary points; for instance, if we want to reflect the metric structure of $X \subset \mathbb{R}^3$. Then the model diedre is $\mathbb{R} \times C(0,\alpha) \subset \mathbb{R}^3$ where $C(0,\alpha) \subset \mathbb{R}^2$ denotes the open cone with vertex in the origin and angle α. Representing points x in $\mathbb{R}^3$ by $(r,\omega) \in \mathbb{R}_+ \times S^2$, $r = |x|$, $\omega = x/|x|$, the tangent cone to a point in $\partial_3 X$ appears as $\mathbb{R}_+ \times N$, $N \subset S^2$ is defined as the interior or the exterior of a geodesic triangle on S^2 formed by the big circles of intersection of the three tangent planes with S^2. We denote the cone in $\mathbb{R}^3$ defined by a subset $M \subset S^2$ by $C(M)$. Let us remark that the whole discussion below covers also the situation when more than three edges meet in a corner point. Then the base of the tangent cone is a geodesic polygon in S^2. Moreover, it is not hard to generalize the results below to the case of arbitrary dimensions and a three-step stratification of the manifold X of conical type.

2. FUNCTION SPACES

The function spaces on X are defined by means of a partition of unity and coordinates referring to the local models. We work in L^2-based Sobolev spaces H^s_{loc}, $s \in \mathbb{R}$, and suitable specifications of the behavior of the functions on the different parts of the boundary which are encoded in weights and asymptotics (singular terms).

Recall that in [7],[8] and [10] the definitions and basic properties of the spaces near $\partial_1 X$ and $\partial_2 X$ are given. Near $\partial_1 X$ besides the smoothness degree s there appears a weight $\gamma_1 \in \mathbb{R}$ and a finite positive divisor $p_1 = \{(p^1_j, m^1_j)\}$, $p^1_j \in \mathbb{C}$, $m^1_j \in \mathbb{N}$, $\mathrm{Re}(p^1_j) \geq 1/2-\gamma_1$ giving the Hilbert space $K^{s;P_1}(\mathbb{R}^3_+)$, $P_1 = (\gamma_1, p_1)$. To guarantee the invariance of this space with respect to coordinate changes respecting the boundary we must assume that for each $\ell \in \mathbb{N}$ we have $\{(p^1_j-\ell, m^1_j)\} \cap \{\mathrm{Re}(z) \geq 1/2-\gamma_1\} \leq p_1$. An analogous condition is assumed below also for all other singularity types. Near the edge $\partial_2 X$ besides P_1 determining the behavior near the plane surfaces of the diedre $\mathbb{R} \times C(0,\alpha)$ there appears $P_2 = (\gamma_2, p_2)$, where $\gamma_2 \in \mathbb{R}$ is another weight and $p_2 = \{(p^2_j, m^2_j, L^2_j)\}$, $p^2_j \in \mathbb{C}$, $\mathrm{Re}(p^2_j) \geq 1-\gamma_2$, $m^2_j \in \mathbb{N}$, L^2_j are finite-dimensional subspaces of $K^{\infty;P_1}(0,\alpha)$.

Finally let us consider a neighborhood of a corner point, i.e. $C(N) \subset \mathbb{R}^3$, $N \subset S^2$ a domain bounded by a geodesic polygon. Let $\gamma_3 \in \mathbb{R}$ be another weight and $p_3 = \{(p^3_j, m^3_j, L^3_j)\}$, $p^3_j \in \mathbb{C}$, $\mathrm{Re}(p^3_j) \geq 3/2-\gamma_3$, $m^3_j \in \mathbb{N}$, L^3_j are finite-dimensional subspaces of $K^{\infty;P_1,P_2}(N)$. Then the singular functions

$$\omega(\rho)\rho^{-p_j^3}\log^k(\rho)v(\tilde{x}),\ 0 \leqq k \leqq m_j^3,\ v \in L_j^3 \tag{2.1}$$

$\omega \in C_0^\infty(\bar{R}_+)$, $\omega = 1$ near 0, $\rho = |x|$, $\tilde{x} = x/|x|$, $x \in R^3$, belong to $K_{loc}^{\infty;P_1P_2}(C(N)\setminus 0)$ and span a space denoted by $E(p_3)$.

Now we define

$$K^{s:P_1,P_2,P_3}(C(N)) = K^{s;P_1,P_2,(\gamma_3,0)}(C(N)) + E(p_3)$$

where the weighted space $K^{s:P_1,P_2,(\gamma_3,0)}(C(N))$ is defined as follows. Denote by $\varepsilon:R_+ \to R$ a diffeomorphism given by $\tau = \log(t)$ for $t < 1$ and $\tau = t$ for $t > 2$. Choose an arbitrary smooth positive function $g(t)$ on R_+ which is constant for large t and is equal to t for small t. Let π be the change to polar coordinates

$$\pi: R^3\setminus 0 \to R_+ \times S^2.$$

Then

$$C(N) \overset{\pi}{\to} R_+ \times N \overset{\varepsilon}{\to} R \times N$$

transforms the model corner $C(N)$ into the manifold with edges $R \times N$. On $R \times N$ we have the *global* space $K^{s;P_1,P_2}(R \times N)$ defined by means of a partition of unity on N only (not on R) and tangential Fourier transform with respect to R. Then we set

$$K^{s;P_1,P_2,(\gamma_3,0)}(C(N)) = \pi^*(g(\rho)^{\gamma_3}\varepsilon^*(K^{s;P_1,P_2}(R \times N))).$$

It is clear that the arbitrariness of the choice of ε and g does not affect the space. Moreover, when localized in the interior we get simply H_{loc}^s, near each point of $\partial_1 X$ $K_{loc}^{s;P_1}$ and near each point of $\partial_2 X$ $K_{loc}^{s;P_1,P_2}$. Therefore it makes sense to define $K^{s;P_1,P_2,P_3}(X)$ by means of a certain atlas of local coordinates. Obviously, we have continuous embeddings

$$K^{s;P_1,P_2,P_3}(X) \to K^{s';P_1',P_2',P_3'}(X)$$

for $s \geqq s'$, $\gamma_j \geqq \gamma_j'$, $p_j \leqq p_j'$, $P_j = (\gamma_j,p_j)$, $P_j' = (\gamma_j',p_j')$, $j = 1,2,3$.

PROPOSITION 1: The above embedding is compact iff $s_j > s_j'$, $\gamma_j > \gamma_j'$ for $j = 1,2,3$.

3. OPERATORS ON X

We describe here the most transparent version of an operator calculus with principal symbols. The hierarchy of symbols is enlarged by the corner symbols $\partial_3\sigma$ attached to each point of $\partial_3 X$. They are quite similar to the symbols near a conical point (see [7],[8],[9]). The difference lies, of course, in the fact that now the base of the cone itself has conical corner points. Therefore in the Mellin image it takes values in operators in the algebra of boundary value problems on the base. It is clear that the classes which contain classical boundary value problems, their parametrices, etc., will also contain boundary, coboundary operators, etc. Compared with the case of manifolds with edges, where in general 3×3 matrices of operators appear here one has in general 4×4 matrices of operators acting from function spaces on $(\partial_0 X,\partial_1 X,\partial_2 X,\partial_3 X)$ into function spaces on $(\partial_0 X,\partial_1 X,\partial_3 X,\partial_3 X)$. In order to simplify the notation we restrict ourselves mainly to operators acting from X to X.

Among all linear continuous operators between the mentioned function spaces we distinguish a class with a structure which is described by symbol classes and classes of negligible operators. Then an operator convention is described associating a linear continuous mapping in the given function spaces with each given symbol. It involves some auxiliary choices; however the result is unique modulo negligible operators.

For given function spaces we call negligible all operators which are continuous even into the space with smoothness order and weights better by one. More precisely, let $P = (P_1,P_2,P_3)$ be given, $P_j = (\gamma_j,p_j)$ and fix an order $m \in \mathbb{R}$. Let $Q = (Q_1,Q_2,Q_3)$, $Q_j = (\lambda_j,q_j)$, be such that $\lambda_j = \gamma_j - m$. We denote by N the subspace of all operators in $L(K^{s;P},K^{s-m;Q})$ which belong to $L(K^{s;P},K^{s-m+1;R^+})$ where $R = ((\gamma_j-m+1,q_j))$ and

$$K^{s;R^+} = \lim_{\overleftarrow{\varepsilon\to 0}} K^{s;R^\varepsilon}$$

where $R^\varepsilon = ((\gamma_j + \varepsilon, r_j))$ if $R = ((\gamma_j, r_j))$.

Remark that we do not pose assumptions about the adjoints (as in [10]), which cannot be formed for all boundary value problems in the class. It is clear that the negligible operators form a left ideal. Among all operators with the property that their restrictions to spaces with better weight map continuously into the corresponding space with better weight they are even a two-sided ideal. Note that for a class with $\ell \in \mathbb{N}$ leading symbol components one must choose as the negligible operators those which improve the smoothness order and all weights by ℓ.

Now we give the definition of the symbols. The symbol levels attached to $\partial_1 X$ and $\partial_2 X$ were described in [10]. As the only modification we drop here the assumptions about the adjoints. The new corner symbols are constructed in two steps. First we have the class $S_{3,M}$ of proper corner symbols without interference with the other symbolic levels. In a second step we take into account the effect of the interior, the boundary and the edge symbols.

We have the following commutative diagram of topological vector spaces and linear continuous mappings

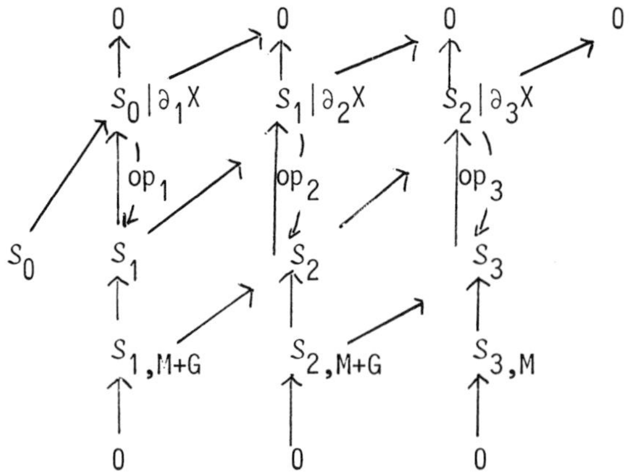

where the fixed order m is not indicated, S_0 denotes the principal interior symbols, S_1 the principal boundary symbols and S_2 the principal edge symbols of order m which are realized in the function spaces under consideration. The classes $S_{j,M+G}$ denote the proper boundary, edge and corner symbols. The mappings $S_j \to S_j|\partial_{j+1}X$ are simply the restrictions. The mappings $S_j \to S_{j-1}|\partial_j X$ recover the higher symbolic level form the lower one. By op_j we denote operator conventions. By construction the second and third columns are exact. We define S_3 and $S_{3,M}$ such that also the fourth column is exact.

S_3 consists of operators on the model cone C(N) at the corner. First we define the class of proper corner symbols $S_{3,M}$. They themselves obey another symbolic level, consisting of meromorphic families with values in $\mathcal{W}^{-\infty}(N)$ which are rapidly decreasing at parallels to the imaginary axis and poles with finite range operators in the principal part of the Laurent expansion at each pole. Let a(z) be such a family. Assume in addition that there is no pole at $3/2 - \gamma_3$. Then we have an associate Mellin operator on C(N) given by

$$op_M(a)u(t) = \frac{1}{2\pi i} \underset{\mathrm{Re}\ z=3/2-\gamma_3}{} t^{-z}a(z)Mu(z)\ dz$$

where u is considered as a function of t (the polar radius) with values in some function space on $N \subset S^2$, for instance $K^{\infty;P_1,P_2}(N)$. If we set

$$H^{s;P_1,P_2,(\gamma_3,0)}(C(N)) = \pi^*(t^{\gamma_3}\varepsilon^*(K^{s;P_1,P_2}(\mathbb{R} \times N)))$$

and $H^{s;P_1,P_2,P_3}(C(N))$ the sum with $E(p_3)$ then $op_M(a)$ acts continuously as operator

$$t^{-m}op_M(a): H^{s;P_1,P_2,P_3}(C(N)) \to H^{s';Q_1,Q_2,(\gamma_3-m,0)}(C(N))$$

for each $s' \in \mathbb{R}$. Moreover its restriction to spaces with better weights acts continuously into the corresponding space with improved weight. The corner symbols in $S_{3,M}$ are in one-to-one correspondence with their Mellin symbols, being families of operators on Re $z = 3/2 - \gamma_3$. They are transplanted to the manifold by localization near the origin and push forward with the coordinate mappings.

It remains to define the operator convention op_3. We proceed as in [10],

4.2.3. With the principal interior, boundary and edge symbols taken at the corner we associate an operation on C(N) in the given function spaces as follows. The base N of the cone is covered by a finite atlas. Then the construction is carried out in local coordinates. The case of an interior coordinate patch or one with smooth boundary is treated as in the case of a cone with smooth boundary. Here the interior and the boundary symbols take part. Consider now a coordinate patch which contains a corner of N. Then we get from the homogeneous interior and boundary symbols an operator

$$K^{s;P_1,P_2}(N) \to K^{s-m;Q_1,Q_2}(N)$$

which can differ only by an operator in $S_{2,M+G}$ from that arising from the edge principal symbol. Thus we obtain a Mellin symbol with values in $\mathfrak{W}^m(N)$ which gives rise to an operator on the cone C(N). The arbitrary choices of the atlas, the partition of unity, etc., do not affect the resulting operator modulo $S_{3,M}$. Therefore the construction of the fourth column in the diagram is finished. At the same time the exactness is clear.

Concerning operations with symbols it is most important to be sure that compositions do not lead beyond the class. This is easily checked for the proper corner symbols in $S_{3,M}$. Then it is a simple algebraic argument that it remains true for the extension by $\mathrm{op}_3(S_2|\partial_3 X)$.

Now the class $\mathfrak{W}(X)$ is defined as the subset of operators

$$K^{s;P}(X) \to K^{s-m;Q}(X)$$

which belong to the class $\mathfrak{W}$ (see [10]) in all points outside the corners and admit near each corner the representation as S_3 operator modulo negligible operators in N.

Note that the diagram also contains the compatibility condition between the different symbol components. In fact, those and only those tuples of symbols occur as a symbol tuple of operators, which are mapped to each other in the diagram.

Only for brevity of notation have we fixed the order of the operators. It is clear that also the case of systems is covered, where the different components may have different orders.

4. COMPOSITIONS AND ELLIPTICITY

The main result about the class $\mathcal{W}$ says that it is stable under compositions.

THEOREM 2: Let $A \in \mathcal{W}$ and $B \in \mathcal{W}$ be such that the composition $A \circ B$ is defined. Then $A \circ B$ belongs to $\mathcal{W}$ and has the composition of the symbols as its symbol.

The only point to prove is the local result near the corner. It follows from the stability under compositions of the corner symbols and the fact that the negligible operators form a two-sided ideal. □

Next we want to detect the Fredholm property in terms of the symbols. It is clear that the conditions which are known to be necessary for the Fredholm property on compact manifolds with edges must be necessarily assumed. In addition at each corner point a condition appears.

THEOREM 3: An operator $A \in \mathcal{W}$ defines a Fredholm operator iff the principal interior, the principal boundary, the principal edge symbol are families of bijections and the Mellin symbol of the corner symbol takes values in the bijective operators on $(N, \partial_1 N, \partial_2 N)$.

The obvious candidate for a parametrix in our class is the operator associated with the inverse interior, boundary and edge symbol, and the corner symbol with the inverse Mellin symbol. By an algebraic argument it can be shown that this symbol tuple satisfies the compatibility conditions. For the corresponding operator $A^{(-1)}$ we get that

$$1 - A \circ A^{(-1)} \text{ and } 1 - A^{(-1)} \circ A$$

are compact operators. In fact, they increase the smoothness and all weights by 1 and the compactness is a consequence of Proposition 1.

The necessity of the conditions in Theorem 3 follows as usual from the reconstruction procedure of the symbols out of the operator.

5. EXAMPLES

Let us illustrate our results in the case of different boundary value problems for the Laplace operator, namely Dirichlet, Neumann and coupling conditions. Since the results depend only on local information we may restrict ourselves to a single corner point. Classically the boundary problems for the Laplace operator are reduced to integral equations on the boundary by Green's formula. In the presence of edges and/or corners the standard compactness results break down (see [1]). However the pseudodifferential nature of the equations on the boundary is preserved. Thus the Fredholm alternative for identity plus compact operator must be replaced by the ellipticity condition from section 4.

Consider the problem of determining a solution u to the Dirichlet or Neumann problem for the Laplace operator in the function spaces of section 1. We have to study the mappings

$$\binom{\Delta}{B} : K^{s;P}(X) \to \begin{array}{c} K^{s-2;Q_1}(X) \\ \oplus \\ K^{s';Q_2}(\partial_1 X) \end{array}$$

where B denotes either Dirichlet or Neumann conditions and s' equals correspondingly s-1/2 or s-3/2. We assume that the singularity types P, Q_1, Q_2 are such that the ellipticity conditions at $\partial_1 X$ and $\partial_2 X$ are satisfied. Then for almost all weights γ_3 the ellipticity condition is satisfied, because the set of poles of the inverse of the Mellin symbol of the corner symbol is discrete in the complex plane. It is very interesting to determine these poles explicitly. This gives an eigenvalue problem for the Laplace-Beltrami operator on the spherical triangle on S^2. It is treated numerically, for instance, in [12]. Our approach also covers pseudo differential operators. It is therefore reasonable to use the integral equations from potential theory. For the Dirichlet problem we may treat the model problem at a corner by means of an Ansatz as potential of a double layer. Then the Fredholm property and the leading singularity of the solution can be found by solving a generalized eigenvalue problem on the one-dimensional curve on S^2 composed of parts of three big circles. In a similar way the Neumann and the transmission problems can be treated. Details will be given elsewhere.

References

[1] M. Costabel, Boundary integral operators on curved polygons, Ann. Mat. Pura Appl., 33 (1983), 305-326.

[2] M. Dauge, Régularités et singularités des solutions de problèmes aux limites elliptiques sur les domains singuliers de type à coins, Thesis, Nantes (1986).

[3] P. Grisvard, Elliptic problems in non-smooth domains, Monogr. Stud. Math. no. 24, Pitman (1985).

[4] V.A. Kondrat'ev, Boundary problems for elliptic equations in domains with conical points, Trudy Mosk. Mat. Ob., 16 (1967), 209-292.

[5] T. v. Petersdorf and E. Stephan, Decompositions in edge and corner singularities for the solution of the Laplacian in a polyhedron, Preprint 1150, TH Darmstadt (1988).

[6] V.G. Maz'ja and B.A. Plamenevskij, L^p estimates of solutions of elliptic boundary problems in domains with edges, Trudy Mosk. Mat. Ob., 37 (1978), 49-93.

[7] L. Päivärinta and S. Rempel, A deconvolution problem with the kernel $1/|x|$ on the plane, Appl. Anal., 26 (1987), 105-128.

[8] L. Päivärinta and S. Rempel, The corner behavior of solutions to the equation $\Delta^{\pm 1/2}u = f$ in two dimensions, to appear.

[9] S. Rempel and R.-W. Schulze, Complete Mellin and Green symbolic calculus in spaces with conormal asymptotics, Ann. Global Anal. Geom., 4:2 (1986), 137-224.

[10] S. Rempel and B.-W. Schulze, Asymptotics for elliptic mixed boundary problems. Pseudo-differnetial and Mellin operators in spaces with conormal singularity, Math. Res., no. 50, Akademie-Verlag (198).

[11] B.-W. Schulze, Corner Mellin operators and reduction of orders with parameters, Preprint, IMath (1988).

[12] H. Walden and R.B. Kellogg, Numerical determination of the fundamental eigenvalue for the Laplace operator on a spherical domain, J. Eng. Math., 11 (1977), 299-318.

Stephan Rempel
Karl-Weierstraß-Institut für Mathematik
der Akademie der Wissenschaften der DDR
Mohrenstraße 39
DDR-1086 Berlin
DDR

4. NONLINEAR PROBLEMS

J. BERKOVITS

A degree theoretic approach to semilinear wave equation

1. INTRODUCTION

Let H be a real separable Hilbert space. In this paper we study semilinear equations in H of the form

$$Au + Bu = h, \tag{1}$$

where A is a densely defined, closed linear operator having infinite-dimensional kernel and $\operatorname{Im} A = (\operatorname{Ker} A)^{\perp}$ so that Im A is closed. Moreover, we assume that the inverse $A^{-1}:\operatorname{Im} A \to \operatorname{Im} A$ is compact. The nonlinear part $B : H \to H$ is usually assumed to be monotone. The assumptions made on A are motivated by the *semilinear wave equation*

$$\begin{aligned}
&u_{tt} - u_{xx} + g(t,x,u) = h(t,x),\ (t,x) \in \mathbb{R} \times (0,\pi),\\
&u(t,0) = u(t,\pi) = 0 \text{ for all } t \in \mathbb{R}, \qquad (2)\\
&u \text{ is } 2\pi\text{-periodic in } t.
\end{aligned}$$

Denote by C^2 the space of twice continuously differentiable functions $u : \mathbb{R} \times [0,\pi] \to \mathbb{R}$ such that $u(t,0) = u(t,\pi) = 0$ for all $t \in \mathbb{R}$ and u is 2π-periodic in t. Let $\Omega = (0,2\pi) \times (0,\pi)$. The 2π-periodic extension in t of any $u \in C_0^\infty(\Omega)$ belongs to C^2 and $C_0^\infty(\Omega)$ is dense in $L^2(\Omega)$, the Hilbert space of measurable square integrable real functions with the usual inner product $(\cdot,\cdot)$ and norm $\|\cdot\|$. Hence it is natural to consider (2) in $L^2(\Omega)$. Let $g : \Omega \times \mathbb{R} \to \mathbb{R}$ be a function satisfying Caratheodory conditions, i.e. $g(\cdot,\cdot,u)$ is measurable for each $u \in \mathbb{R}$ and $g(t,x,\cdot)$ is continuous on $\mathbb{R}$ for almost every $(t,x) \in \Omega$. Moreover, we assume that there exists $c > 0$ and $h_1 \in L^2(\Omega)$ such that

$$|g(t,x,u)| \leqq c|u| + h_1(t,x)$$

for almost every $(t,x) \in \Omega$ and all $u \in \mathbf{R}$. These assumptions guarantee that the Nemytskii operator T generated by g, $T(u)(t,x) \equiv g(t,x,u(t,x))$, is bounded and continuous from $L^2(\Omega)$ into itself. For any $h \in L^2(\Omega)$ the *generalized solution* of (2) is any $u \in L^2(\Omega)$ with

$$(u, v_{tt} - v_{xx}) + (T(u), v) = (h, v) \tag{3}$$

for all $v \in C^2$. Obviously for any classical solution of (2), $(u_{tt} - u_{xx}, v) = (u, v_{tt} - v_{xx})$. Let $\phi_{mn}(t,x) = \pi^{-1} \exp(imt) \sin nx$, $m \in \mathbf{Z}$, $n \in \mathbf{Z}_+$. Each $u \in L^2(\Omega)$ can be presented as a Fourier series

$$u = \sum_{m \in \mathbf{Z}, n \in \mathbf{Z}_+} u_{mn} \phi_{mn}$$

where $u_{mn} = (u, \phi_{mn})$ and $\bar{u}_{mn} = u_{-m,n}$ since u is real. The abstract realization of the wave operator $\partial^2/\partial t^2 - \partial^2/\partial x^2$ in $L^2(\Omega)$ is linear operator $L : D(L) \to L^2(\Omega)$ defined by

$$Lu = \sum_{m \in \mathbf{Z}, n \in \mathbf{Z}_+} (n^2 - m^2) u_{mn} \phi_{mn},$$

where

$$D(L) = \{u \in L^2(\Omega) \mid \sum_{m \in \mathbf{Z}, n \in \mathbf{Z}_+} |n^2 - m^2|^2 |u_{mn}|^2 < \infty\}.$$

One can verify that $u \in L^2(\Omega)$ is a generalized solution of the semilinear wave equation if and only if

$$u \in D(L), \; Lu + T(u) = h \text{ in } L^2(\Omega). \tag{4}$$

The operator $L : D(L) \to L^2(\Omega)$ is densely defined, self-adjoint, closed and $\operatorname{Im} L = (\operatorname{Ker} L)^{\perp}$. L has a pure point spectrum of eigenvalues

$$\sigma(L) = \{n^2 - m^2 \mid n \in \mathbf{Z}_+, m \in \mathbf{Z}\}$$

with corresponding eigenvectors ϕ_{mn}. Clearly $\sigma(L)$ is unbounded from below and from above. Any eigenvalue $\lambda \neq 0$ has finite multiplicity but

Ker $L = \overline{\text{sp}}\{\phi_{nn}, \phi_{-n,n}\}$ is infinite dimensional. Let L_0 be the restriction of L to $D(L) \cap \text{Im } L$. Then the right inverse of L_0,

$$L_0^{-1} : \text{Im } L \to \text{Im } L \cap D(L)$$

is compact. Hence equation (4) is an example of equations of the form (1).

The semilinear wave equation and other corresponding hyperbolic problems have been studied by many authors, see for instance [3], [4], [6], [7], [11] and [12]. This paper is closely related to the works of Mawhin and Willem, see [8], [9] and [10]. Their approach is based on a combination of monotone operator theory and Leray-Schauder degree theory. Using the same methods, Browder [5] gives some ideas on how to construct a degree theory, which can be applied to the semilinear wave equation. The construction by Browder is included here as a special case.

2. MAPPINGS OF MONOTONE TYPE

Let H be a real, separable Hilbert space with inner product $(\cdot,\cdot)$ and norm $\|\cdot\|$. We recall that a map $f : H \to H$ is *monotone*, denoted $f \in (\text{MON})$, if $f(u) - f(v), u - v) \geqq 0$ for all u and v in H, and it is *strongly monotone*, denoted $f \in (\text{MON})_S$, if there exists a continuous strictly increasing function $k : [0,\infty) \to [0,\infty)$ with $k(0) = 0$ such that $(f(u) - f(v), u - v) \geqq k(\|u - v\|)\|u - v\|$ for all u and v in H. The map f is said to be of *class* (S_+), if for any sequence $\{u_n\}$ converging weakly to u in H, $u_n \rightharpoonup u$, for which $\limsup(f(u_n), u_n - u) \leqq 0$ we have $\{u_n\}$ converging strongly to u in H, $u_n \to u$. The map f is *pseudomonotone*, denoted $f \in (\text{PM})$, if for any sequence $\{u_n\}$, $u_n \rightharpoonup u$, for which $\limsup(f(u_n), u_n - u) \leqq 0$, we have $(f(u_n), u_n - u) \to 0$ and $f(u_n) \rightharpoonup f(u)$. We say that f is *quasimonotone*, denoted $f \in (\text{QM})$, if for any sequence $\{u_n\}$, $u_n \rightharpoonup u$, we have $\limsup(f(u_n) - f(u), u_n - u) \geqq 0$. We assume that all mappings considered are *bounded*, i.e. they take bounded sets into bounded sets, and *demicontinuous*, i.e. $u_n \to u$ implies $f(u_n) \rightharpoonup f(u)$. Note that any map of the *Leray-Schauder type* $I + C$, C compact, is contained in the class (S_+). One can prove that the following inclusions hold:

$$\begin{array}{ccccccc} (\text{LS}) & \longrightarrow & (S_+) & \longrightarrow & (\text{PM}) & \longrightarrow & (\text{QM}) \\ (\text{MON})_S & \longrightarrow & (\text{MON}) & \nearrow & (\text{COMP}) & \nearrow & \end{array}$$

We close this section by the following perturbation result which is needed in the sequel.

PROPOSITION 1: $f \in (QM)$ if and only if $g + f \in (S_+)$ for all $g \in (S_+)$

For classification and properties of mappings of monotone type we refer to [1].

3. MAPPINGS OF THE FORM Qg + Pf

Let H be a real, separable Hilbert space and M a closed subspace of H with the corresponding orthogonal projection $Q : H \to M$. Denote $N = M^{\perp}$ and $P = I - Q$. Let $G \subset H$ be an open bounded set. We consider mappings $F : \bar{G} \to H$ of the form

$$F = Qg + Pf$$

where $g = I + C$, C compact, and f is a mapping of monotone type. The connection to the semilinear wave equation $Lu + T(u) = h$ is the following. Let us choose $H = L^2(\Omega)$, $M = \text{Im } L$, $N = \text{Ker } L$, $g = I + L_0^{-1}QT$ and $f = T$. Then

$$F = Q(I + L_0^{-1}QT) + PT$$

and we have

LEMMA 1: For any $h \in L^2(\Omega)$

(i) $u \in D(L) \cap \bar{G}$ and $Lu + T(u) = h$

iff

(ii) $u \in \bar{G}$ and $F(u) = \hat{h}$, where $\hat{h} = (L_0^{-1}Q + P)h$.

PROOF: Assume $u \in \bar{G}$ and $F(u) = \hat{h}$, i.e.

$$Q(u + L_0^{-1}QT(u)) + PT(u) = (L_0^{-1}Q + P)h.$$

Since $\text{Ker } L = P(H) \subset D(L)$ and $L_0^{-1}Q(H) \subset \text{Im } L \cap D(L)$ we necessarily have $u \in D(L)$. Using the fact that $L_0^{-1}Q + P$ is the right inverse of $L + P$ we get

$$Lu + T(u) = h.$$

Obviously the converse is also true. □

Note that if we take $f = g = I + C \in (LS) \subset (S_+)$, then

$$F = Qg + Pf = I + C,$$

and thus the class of Leray-Schauder type mappings is contained to our class as a special case.

4. CONSTRUCTION OF A DEGREE FUNCTION

Let $G \subset H$ be an open bounded set and $f : \bar{G} \to H$ a bounded, demicontinuous map of class (S_+) and $g = I + C : \bar{G} \to H$ a map of Leray-Schauder type. Let

$$F = Q(I + C) + Pf : \bar{G} \to H$$

where $Q : H \to M$ and $P : H \to M^\perp = N$ are orthogonal projections. Let $\{N_n\}$ be a sequence of finite-dimensional subspaces of N such that $N_n \subset N_{n+1}$ for all n and $\cup_{n=1}^\infty N_n$ is dense in N. Denote by P_n the orthogonal projection from H onto N_n. Let

$$F_n = I + QC + P_n f - P_n : \bar{G} \to H,$$

which is clearly a Leray-Schauder type mapping. For any $y \in H$, let $y_n = Qy + P_n y$. We shall use the following obvious facts:

(i) $PP_n = P_n P = P_\ell P_n = P_n P_\ell = P_n$ for all $\ell \geq n$,

(ii) if $u_k \rightharpoonup u$ (resp. $u_k \to u$) and $n_k \to \infty$, then $P_{n_k} u_k \rightharpoonup Pu$ (resp. $P_{n_k} u_k \to Pu$).

Our next result shows how the family $\{F_n\}$ is related to F.

<u>LEMMA 2</u>: Assume $y \notin F(\partial G)$. Then there exists n_0 such that

$$y_n \notin F_n(\partial G) \text{ for all } n \geq n_0.$$

PROOF: Since $(F - y)_n = F_n - y_n$ we may assume without loss of generality that $y = 0$. If the assertion does not hold, there exist sequences $\{n_k\}$, $n_k \to \infty$, and $\{u_k\} \subset \partial G$ with $F_{n_k}(u_k) = 0$, i.e.

$$Qu_k + QC(u_k) = 0 \tag{5}$$

$$Pu_k + P_{n_k} f(u_k) - P_{n_k} u_k = 0 \tag{6}$$

At least for a subsequence we can write $u_k \rightharpoonup u$, $C(u_k) \to c$ and $f(u_k) \rightharpoonup w$. It follows from (5) that $Qu_k \to Qu$. By (6), we first conclude $Pu_k = P_{n_k} u_k$ and consequently $P_{n_k} f(u_k) = 0$ and $Pw = 0$. Hence,

$$\begin{aligned}\limsup(f(u_k),u_k - u) &= \limsup(f(u_k),Pu_k - Pu)\\ &= \limsup(f(u_k),Pu_k) = \limsup(f(u_k),P_{n_k} u_k)\\ &= \limsup(P_{n_k} f(u_k),u_k) = 0,\end{aligned}$$

which implies $u_k \to u \in \partial G$. Thus $F(u) = 0$, where $u \in \partial G$; a contradication, completing the proof. □

By the previous lemma the Leray-Schauder degree $d_{LS}(F_n,G,y_n)$ is well-defined for all $n \geq n_0$. In fact, its value remains stable as n goes to infinity:

LEMMA 3: Assume $y \notin F(\partial G)$. Then there exists $n_1 \geq n_0$ such that

$$d_{LS}(F_n,G,y_n) = \text{constant} \quad \text{for all } n \geq n_1.$$

PROOF: Again we can take $y = 0$. Suppose that for some sequence $\{n_k\}$, $n_k \geq n_0$, $n_k \to \infty$, one has

$$d_{LS}(F_{n_k},G,0) \neq d_{LS}(F_{n_{k+1}},G,0).$$

By the well-known properties of the Leray-Schauder degree there necessarily exist sequences $\{u_k\} \subset \partial G$ and $\{t_k\} \subset (0,1)$ with

$$(1 - t_k)F_{n_k}(u_k) + t_kF_{n_{k+1}}(u_k) = 0$$

which can be equivalently stated as

$$Qu_k + QC(u_k) = 0 \tag{7}$$

$$Pu_k + \{(1 - t_k)P_{n_k} + t_kP_{n_{k+1}}\}(f(u_i) - u_k) = 0 \tag{8}$$

where we can assume $u_k \rightharpoonup u$, $C(u_k) \to c$ and $f(u_k) \rightharpoonup w$. As in Lemma 2 we get from (7) and (8) $Qu_k \to Qu$, $Pu_k = P_{n_{k+1}}u_k$, $P_{n_k}f(u_k) = 0$ and $Pw = 0$. Equation (8) now implies

$$P_{n_{k+1}}f(u_k) = -(1 - t_k)t_k^{-1}[P_{n_{k+1}}u_k - P_{n_k}u_k]$$

and we can write

$$\begin{aligned}\lim\sup(f(u_k),u_k - u) &= \lim\sup(f(u_k),P_{n_{k+1}}u_k)\\ &= \lim\sup\{-(1 - t_k)t_k^{-1}\|P_{n_{k+1}}u_k - P_{n_k}u_k\|^2\}\\ &\leq 0\end{aligned}$$

implying $u_k \to u \in \partial G$. Consequently, $F(u) = 0$ with $u \in \partial G$, a contradiction which completes the proof. □

It is now relevant to define

$$d(F,G,y) = \lim_{n\to\infty} d_{LS}(F_n,G,y_n)$$

whenever $y \notin F(\partial G)$. Degree function d satisfies the following properties:

(a) If $d(F,G,y) \neq 0$, then $y \in F(G)$.

(b) (Addivity of domains) If G_1 and G_2 are open disjoint subsets of G and $y \notin F(\bar{G}\setminus(G_1 \cup G_2))$, then

$$d(F,G,y) = d(F,G_1,y) + d(F,G_2,y).$$

(c) (Invariance under homotopy) Let $F_0 : \bar{G} \to H$ and $F_1 : \bar{G} \to H$ be any two mappings for which the degree d is defined. Then the *affine homotopy* between F_0 and F_1 is the family $F_t = (1 - t)F_0 + tF_1$, $0 \leqq t \leqq 1$, of mappings from $\bar{G}$ into H. If y_t, $0 \leqq t \leqq 1$, is a continuous curve in H with $y_t \notin F_\ell(\partial G)$ for all t in [0,1], then

$$d(F_t,G,y_t) = \text{constant in } t \text{ on } [0,1].$$

REMARK: The homotopy invariance property can be extended to a broader class of homotopies of form

$$F_t = Qg_t + Pf_t, \quad 0 \leqq t \leqq 1,$$

where g_t, $0 \leqq t \leqq 1$, is a Leray-Schauder type homotopy and f_t, $0 \leqq t \leqq 1$, a bounded homotopy of class (S_+); see [2].

(d) For the identity I we have $d(I,G,y) = +1$ whenever $y \in G$.

One can prove that the degree function d constructed above is unique [2], i.e. any degree function defined for the same class of mappings satisfying (a), (b), (c) and (d) coincides with d. Moreover, the degree theory can be uniquely extended (in a slightly modified form) to the case $F = Q(I+C) + Pf$, where f is quasimonotone. By Proposition 1, $f + \varepsilon I \in (S_+)$ for all $\varepsilon > 0$ and one can easily prove that if $y \notin \overline{F(\partial G)}$, then there exists $\varepsilon_0 > 0$ such that

$$y \notin F_\varepsilon(\partial G) = [Q(I + C) + P(f + \varepsilon I)](\partial G)$$

for all $0 < \varepsilon < \varepsilon_0$. Hence we can define a generalized degree function d_{QM} by setting

$$d_{QM}(F,G,y) = \lim_{\varepsilon \to 0+} d(F_\varepsilon,G,y)$$

whenever $y \notin \overline{F(\partial G)}$. The condition (a) now has the form

(a') If $d_{QM}(F,G,y) \neq 0$, then $y \in \overline{F(G)}$.

For further details we refer to [2].

5. APPLICATIONS

We consider here only the simplest possible affine homotopy and show how it can be applied to the semilinear wave equation. Let $F = Q(I + C) + Pf:\bar{G} \to H$, where C is compact and f is a bounded, demicontinuous mapping. Then we have the following

THEOREM 1: Assume $0 \in G$ and

$$((1 - t)I + tF)(u) \neq 0$$

for all $u \in \partial G$ and $0 \leqq t \leqq 1$.

(i) If $f \in (S_+)$, then $d(F,G,0) = +1$ and the equation $F(u) = 0$ has at least one solution in G.

(ii) If $f \in (QM)$, then the equation $F(u) = 0$ is *almost solvable* in the sense that $0 \in \overline{F(\bar{G})}$.

PROOF: If $f \in (S_+)$, then

$$d(F,G,0) = d(I,G,0) = +1$$

and the proof of (i) is complete. Let now $f \in (QM)$. If $0 \in \overline{F(\partial G)} \subset \overline{F(\bar{G})}$ then the assertion (ii) is true. Hence we can assume $0 \notin \overline{F(\partial G)}$. Then there exists $\varepsilon_1 > 0$ such that

$$[(1 - t)I + tF_\varepsilon](u) \neq 0$$

for all $u \in \partial G$, $0 \leqq t \leqq 1$, $0 < \varepsilon < \varepsilon_1$. Indeed, otherwise we can find sequences $\{\varepsilon_n\}$, $\varepsilon_n \to 0+$, $\{t_n\} \subset [0,1]$, $t_n \to t$, and $\{u_n\} \subset \partial G$, $u_n \rightharpoonup u$, such that $(1 - t_n)u_n + t_n F_{\varepsilon_n}(u_n) = 0$, i.e.

$$Qu_n + t_n QC(u_n) = 0$$

$$(1 - t_n + t_n\varepsilon_n)Pu_n + t_n Pf(u_n) = 0.$$

If $t_n \to 1$, then $F(u_n) \to 0$ implying $0 \in \overline{F(\partial G)}$. Thus we can assume $t_n \to t \neq 1$. At least for a subsequence $Qu_n \to Qu$ and hence

$$0 \leq t \limsup(f(u_n),u_n - u) = \limsup(t_n Pf(u_n),u_n - u)$$

$$= \limsup \{-(1 - t_n + t_n\varepsilon_n)(Pu_n,u_n - u)\}$$

$$= (1 - t)\limsup\{- \|Pu_n - Pu\|^2\} \leq 0$$

which implies $Pu_n \to Pu$ and consequently $u_n \to u \in \partial G$. Hence $(1-t)u+tF(u) = 0$, $u \in \partial G$, a contradiction. The assertion (ii) now follows, since

$$d_{QM}(F,G,0) = d(F_\varepsilon,G,0) = d(I,G,0) = +1$$

which implies $0 \in \overline{F(G)} \subset \overline{F(\bar{G})}$. □

Theorem 1 can be used to prove directly the existence result of Brezis and Nirenberg [4] for generalized L^2-solutions of the semilinear wave equation $Lu + T(u) = 0$ (see also [10]). The nonlinear part T is generated by the Caratheodory function $g:\Omega \times \mathbf{R} \to \mathbf{R}$. Assume that $g(t,x,\cdot)$ is non-decreasing and for every $u \in \mathbf{R}$ and a.e. $(t,x) \in \Omega$

$$\eta|u| - h_1(t,x) \leq |g(t,x,u)| \leq \gamma|u| + h_2(t,x),$$

where $\eta > 0$, $0 < \gamma < 3$ and $h_1,h_2 \in L^2(\Omega)$. Then

(i) $(Tu,u) \geq (1/\gamma')\|Tu\|^2 - C$ for all $u \in H$, where $C > 0$ and $\gamma < \gamma' < 3$,

(ii) $\|T(u)\| \to \infty$ as $\|u\| \to \infty$.

These conditions imply the existence of $R > 0$ such that

$$Lu + (1 - t)Pu + tT(u) \neq 0$$

for all $\|u\| = R$, $0 \leq t \leq 1$. Indeed, suppose $Lu_n + (1-t)Pu_n + t_nT(u_n) = 0$, where $\{t_n\} \subset (0,1]$ and $\|u_n\| \to \infty$. From the definition of L one can easily see that $\|Lu_n\| \geq \|Qu_n\|$ and $(Lu_n,u_n) \geq -(1/3)\|Lu_n\|^2$. By (i) and the fact that $\|Lu_n\| = t_n \|QT(u_n)\| \leq t_n \|T(u_n)\|$ we get

$$0 = (Lu_n, u_n) + (1 - t_n)\|Pu_n\|^2 + t_n(T(u_n), u_n)$$

$$\geq -\frac{1}{3}\|Lu_n\|^2 + (1 - t_n)\|Pu_n\|^2 + \frac{t_n}{\gamma'}\|T(u_n)\|^2 - t_nC$$

$$\geq \left(\frac{1}{\gamma'} - \frac{1}{3}\right)t_n^2\|T(u_n)\|^2 + (1 - t_n)\|Pu_n\|^2 - C.$$

If $\|Lu_n\| \to \infty$, then $t_n \|T(u_n)\| \to \infty$ and we have a contradiction. If $\{Lu_n\}$ is bounded, then $\{Qu_n\}$ is also bounded. Thus $\|Pu_n\| \to \infty$ and since $\|T(u_n)\| \to \infty$ we again have a contradiction. Hence we can conclude by Theorem 1 (ii) and Lemma 1 the existence of $u \in D(L)$ with $Lu + T(u) = 0$.

References

[1] J. Berkovits and V. Mustonen, Nonlinear mappings of monotone type, Part I. Classification and degree theory., Report No. 2/88, Mathematics, University of Oulu (1988).

[2] J. Berkovits and V. Mustonen, An extension of Leray-Schauder degree and applications to nonlinear wave equations, in preparation.

[3] H. Brézis and L. Nirenberg, Forced vibrations for a nonlinear wave equation, Comm. Pure Appl. Math., 31 (1978), 1-30.

[4] H. Brézis and L. Nirenberg, Characterizations of the ranges of some nonlinear operators and applications to boundary value problems, Ann. Scuola Norm. Sup. Pisa (4), 5 (1978), 225-326.

[5] F.E. Browder, Degree theory for nonlinear mappings, Proc. Symp. Pure Math., vol. 45, Part 1, AMS, (1986), 203-226.

[6] H. Hofer, A multiplicity result for a class of nonlinear problems with applications to a nonlinear wave equation, J. Nonlinear Anal., 5 (1981), 1-11.

[7] N. Krylova and O. Vejvoda, A linear and weakly nonlinear equation of a beam. The boundary value problem for free extremities and its periodic solution, Czech. Math. J., 21 (1971), 535-566.

[8] J. Mawhin, Nonlinear functional analysis and periodic solutions of semilinear wave equations, Nonlinear Phenomena in Mathematical Sciences (Lakshmikantham, ed.), Academic Press (1982), 671-681.

[9] J. Mawhin and M. Willem, Perturbations non-linéaires d'opérateurs linéaires à noyau de dimension infinie, C.R. Acad. Sci. Paris, 287A (1978), 319-322.

[10] J. Mawhin and M. Willem, Operators of monotone type and alternative problems with infinite dimensional kernel, Recent Advances in Differential Equations (Trieste 1978), Academic Press (1981), 295-307.

[11] P.H. Rabinowitz, Some global results for nonlinear eigenvalue problems, J. Funct. Anal., 7 (1971), 487-513.

[12] P.H. Rabinowitz, Free vibrations for a semilinear wave equation, Comm. Pure Appl. Math., 31 (1978), 31-68.

Juha Berkovits
Department of Mathematics
Faculty of Science
University of Oulu
SF-90570 Oulu
Finland

B. BOJARSKI

Geometric properties of the Sobolev mapping

This conference is mainly concentrated on the discussion of various function spaces and their properties either from the point of view of general functional analysis or as a convenient and important tool in handling problems related to partial differential equations or mathematical analysis in general. Thus the operational properties of function spaces, their behaviour under the action of differential, pseudodifferential and other natural classes of operators or other analytical processes, arising in various branches of analysis, are the main object of interest.

I want to attract your attention to a variety of problems related with the geometric information conveyed by the transformations between manifolds locally described by functions in Sobolev spaces.

Geometry was intimately related with basic facts of the theory of Sobolev spaces from the very beginning. Important cases of Sobolev inequalities are essentially equivalent to isoperimetric inequalities. Also the general Sobolev imbedding and trace theorems are geometric in nature.

The general setting of our considerations will be the Sobolev space $W^{\ell,p}(M,N)$, $\ell \geqq 0$, $p \geqq 1$ of mappings

$$f : M \to N \tag{1}$$

between smooth (sufficiently smooth) manifolds such that, roughly speaking, the local coordinate description of the mapping f is realised by functions belonging to the Sobolev spaces $W^{\ell,p}$. In what follows $m = \dim M$, $n = \dim N$.

If M is an open bounded domain in the Euclidean space R^m and $N = R$ or $N = R^n$ we use the usual local definitions for $W^{\ell,p}(M,N)$ [1], [26], [30]. The spaces $W^{\ell,p}(M,R^n)$ are defined then by the global requirement that the "derivatives" $D^\alpha f^i$, $|\alpha| \leqq \ell$, $i = 1,\dots,n$ of coordinate functions f^i, of the map f with respect to the admissible coordinates are in the Lebesgue space $L^p(M)$. Also clear is the case of an arbitrary compact Riemannian manifold M, with boundary or closed, and $N = R^n$, when we get the linear Banach space $W^{\ell,p}(M,R^n)$.

In the general case of M and N arbitrary, we may use the Nash embedding of N in some Euclidean space R^k and define $W^{\ell,p}(M,N)$ as the subset in the Banach space $W^{\ell,p}(M,R^k)$ of those maps $f : M \to R^k$ which satisfy the condition

$$f(m) \subset N \tag{2}$$

for almost all $m \subset M$. This also holds if the boundary of the closure $\bar{M}$ of M, ∂M, is not a regular manifold, e.g. for ∂M Lipschitz or piecewise smooth.

With rather obvious though may be troublesome modifications the Sobolev spaces $W^{\ell,p}(M,N)$ may be defined for the case when M and N have only the structure of a regular (finite) polyhedra, stratified into a finite number of simple building blocks (pieces of various dimensions $\leq m$, or $\leq n$ respectively) of the elementary type mentioned above.

Note that for $\ell p \leq m$ the space $W^{\ell,p}(M,N)$ cannot be defined in terms of local coordinates on M and N only. Indeed in this case the functions in $W^{\ell,p}(M,R)$ may be even locally unbounded and therefore it is not possible to reduce the description of the map f to local pieces $f : U \to V$, where $U \subset M$ and $V \subset N$ are domains of local charts on M and N respectively. The a priori requirement that the maps considered admit a localisation of this type would be too restrictive and troublesome. However this is possible for $\ell p > m$, since then, by the local Sobolev imbedding inequality, the general function in $W^{\ell,p}(M,R)$ is continuous.

Having defined the admissible classes of mappings $W^{\ell,p}(M,N)$ i.e. morphisms between the admissible classes of objects - Riemannian manifolds, submanifolds or regular polyhedra in Euclidean spaces - we have a framework which should be compared with the similar frameworks of various branches of topology, when one takes as objects compact or locally compact topological spaces or finite dimensional topological spaces and as morphisms continuous transformations. We can think also about related theories with some richer structure - like differential topology - with the class of all C^∞-maps as admissible morphisms or piecewise linear topology with the class of piecewise linear or, more generally, Lipschitzian morphism. What comes to mind in our situation is a theory which could be named Sobolev topology, with a natural class of objects, closely related with the family of rather regular, not pathological finite polyhedra and manifolds, and the class of morphisms: maps in the Sobolev spaces $W^{\ell,p}$.

In what follows I shall try to present a number of examples with the hope to show that there is a place in mathematics for such a Sobolev topology. Meaningful examples will be presented which show the similarities as well as peculiarities of such a theory when compared with the above mentioned classical geometric topology of regular finite dimensional manifolds or polyhedra and the differential topology of smooth manifolds and smooth mappings.

It is natural to expect that in this approach the proper modification of the basic notions of differential topology like the genericity, the transversality, applications of the Sard's theorem or the notions of good Morse functions and many others will find their analogues. Like in those theories to reveal the geometrical or topological information conveyed by a given Sobolev map it may be necessary to put the map by a "small" deformation in a form of some kind of "general position". These phenomena are known in the geometric topology as well as in the differential topology.

What follows should be considered only as a preliminary introduction to the topic. Because of lack of time and space and a variety of other reasons we shall be able to supply only some ideas and sketches of proofs, often in conjectural form. Also the references can be supplied now only in a manner far from satisfactory.[1]

Hopefully a more detailed and systematic discussion will be prepared in cooperation with my colleagues.

1. CONTINUITY AND DIFFERENTIABILITY

As is well known the parameter $\gamma = pl - m$ measures the modulus of continuity of a Sobolev function $f \in W^{\ell,p}(R^m,R)$ or $f \in W^{\ell,p}(M,R)$, $m = \dim M$. For $\gamma = 0$ and $\ell = 1$ we have $f \in C^{0,\alpha}_{loc}(M,R)$ with $\alpha = \frac{\gamma}{p} = \ell - \frac{m}{p}$. For $\gamma < 0$ the function f may admit discontinuities. However the requirement $f \in W^{\ell,p}(R^m,R)$ imposes rather serious restrictions on the geometric character of these discontinuities. The class $W^{\ell,1}(R,R)$ $(m = 1)$ is identical with the class of absolutely continuous real valued functions of a real variable and the class $W^{\ell,1}(R^m,R)$,

---

1 Some of the topics discussed below seem to arouse increasing research interest recently. In particular, the references [4] and [36] have been available to me only after the Sodänkylä meeting.

for $m > 1$, may be described in terms of the absolutely continuous behaviour on almost each line parallel to the coordinate axes or as the class ACL [19], [20]. A theorem by S.M. Nikolski [28], [20] generalizes this to the effect that functions in $W^{\ell,p}(R^m,R)$ or in $W^{\ell,p}(\Omega,R)$ for an open set $\Omega \subset R^m$, are in the Sobolev class $W^{\ell,p}(R^k,R)$ on almost each k-dimensional coordinate subspace or, more generally, on almost each k-dimensional subspace Π_k (or $\Pi_k \cap \Omega$), from a (m-k)-dimensional family P_{m-k} of parallel subspaces of R^m. In particular, for $k < \ell p$, the restrictions $f|_{\Pi_k}$, $\Pi_k \subset P_{m-k}$ are Hölder continuous (locally) on almost each Π_k. We can say that on a "generic" k-hyperplane of a (m-k)-dimensional family of parallel hyperplanes a function in the Sobolev class $W^{\ell,p}(R^m)$, $k < \ell p$, is continuous. It is necessary to stress the difference between the Nikolski theorem and the Sobolev trace theorem. Tr: $W^{\ell,p}(R^m) \to W^{\ell-(m-n/p),p}(R^n)$, $n < m$, when the value of the continuity parameter $\gamma = pl - m$ is conserved [1], [26] [31] and the trace operator Tr is meaningful for every n-dimensional hyperplane (hypersurface) imbedded in R^m.

Simple examples show that Nikolski's theorem does not hold for an arbitrary (m-k)-dimensional family P_{m-k}. Some extra geometric condition is necessary. The notion of the module of a family of k-dimensional surfaces in R^m [19] and the related notion of "exceptional systems of hypersurface" are useful tools to describe the families of k-dimensional subspaces for which Nikolski's theorem holds.

Since under bi-Lipschitzian transformations of the ambient space R^m, (or Ω) exceptional systems are invariant, Nikolski's theorem immediately generalizes to rather general classes of curvilinear families of hypersurfaces, or even general classes of finite curvilinear polyhedra. In particular holds

PROPOSITION 1: Let $f \in W^{\ell,p}(M,N)$, where M is a compact manifold, dim M = m. Let $k < \ell p$ be an integer ≥ 1. Then on the k-skeleton $T^{(k)}$ of a generic triangulation T of M the mapping $f|_{T^{(k)}}$ is in the Sobolev class $W^{\ell,p}(T^{(k)},N)$. In particular it is Hölder continuous, as a map from $T^{(k)}$ to N.

In other words the above states that if we consider an arbitrary triangulation T of M and jiggle it a little in some general way, roughly "transversally", then we get a triangulation T' such that f is continuous on the k-skeleton of T'.

While Sobolev trace theorem is a basic tool in the theory of boundary value problems for p.d.e. the Nikolski type theorems up till now seem to have rather few applications. In the geometric analysis of Sobolev mappings it has been first used probably in the papers by Goldstein and Vodopianov, [32], [33], [34]. Another immediate corollary of the Nikolski theorem and of the differentiability theorem of Calderon [12], is the fact that the graph of an arbitrary f in the Sobolev class $W^{\ell,p}(\Omega,R)$, $\omega \subset R^m$, has at least a k-dimensional "tangent hyperplane" at almost every point of Ω. We recall that the Calderon theorem [12] asserts that the functions in $W^{\ell,p}(\Omega)$, $p > m$, have a total differential for almost each $x \in \Omega$.

Another type of continuity properties of functions in the Sobolev classes was discovered by Calderon and Zygmund in 1961 [13] as a generalization of Luzin's property of measurable functions. We recall this result in the form of the theorem of F.C. Liu [24].

PROPOSITION 2: Let Ω be a strongly Lipschitz domain in R^m and let $f \in W^{\ell,p}(\Omega,R)$ with $1 \leqq p < +\infty$. Then for any $\varepsilon > 0$ there exist a compact set F of Ω and a function $g \in C^{\ell}(\Omega)$, such that

(1) $|\Omega \setminus F| < \varepsilon$

(2) $f = g$ on F (3)

(3) $\|f-g\|_p^{\ell} < \varepsilon.$

Note that this result can also be viewed as an approximation theorem of functions in the Sobolev class $W^{\ell,p}(\Omega,R)$ by functions in the class $C^{\ell}(\Omega)$ (ℓ-times continuously differentiable functions in Ω), with an extra rather deep condition that the approximating function g arises from f by a modification on a set of arbitrary small measure. Though in general the open set $\Omega \setminus F$, being a neighbourhood of the "proper" singularity carrier Σ_f of the function f, is rather complicated, many natural questions on the structure of the singularity set Σ_f, its geometric and metric properties are meaningful, undoubtedly interesting and seem to be rather poorly understood or unknown in the literature. The notions of capacities of various nature and Hausdorff dimension are useful and important tools in this area. It is natural to ask questions about the Hausdorff dimension of Σ_f, the asymptotics of the C^ℓ norms of the functions g_ε for $\varepsilon \to 0$, the more precise understanding of the statement (1) and (3) in Proposition 2 and a lot of others.

2. HOMOTOPY AND GLOBAL EXTENSION PROPERTIES. APPROXIMATION BY SMOOTH MAPS

The property (3) of the Proposition 2 is a special case of the general approximating properties of functions in the Sobolev classes $W^{\ell,p}(M,N)$ by smooth, i.e. C^∞ functions or sufficiently differentible functions.

The simplest and best understood is the case of Sobolev classes $W^{\ell,p}(\Omega,R)$ with Ω an open subset of the Euclidean space R^m. In this respect the Meyers-Serrin theorem [1], [27], [10] is the most general result.

PROPOSITION 3: For an arbitrary open subset $\Omega \subset R^m$ the space $C^\infty(\Omega,R)$ is dense in $W^{\ell,p}(\Omega,R)$ i.e. for each $\varepsilon > 0$ and each $f \in W^{\ ,p}(\Omega,R)$ there exists a function $g_\varepsilon \in C^\infty(\Omega,R)$, such that $\|f-g\|_{W^{\ell,p}(\Omega)} < \varepsilon$.

If $\Omega = R^m$ the choice of g_ε can be presented in the form of a linear operator $g_\varepsilon = J_\varepsilon * f$, [1] (mollifiers). The linear operator-type choice of g_ε is also possible in some other cases, if Ω satisfies some extra conditions, e.g. has sufficiently regular boundary [1], [31]. However, in the case of an arbitrary Ω the smooth approximation process cannot be represented in a linear operator form.

The Proposition 3 may be also interpreted as the statement that the "singularities" of an arbitrary mapping $f : \Omega \to R$ in the Sobolev class $W^{\ell,p}(\Omega,R)$ are unstable, or removable, by a small perturbation that is by an arbitrarily small deformation $f \to g_\varepsilon$ the mapping f may be globally "mollified" to a smooth mapping. While this smooth approximating property immediately generalizes to the classes $W^{\ell,p}(M,R^n)$ or to the spaces $W^{\ell,p}(M,N)$ if N is e.g. a convex open subset of R^n or a curvilinear, say, Lipschitz image of a convex submanifold, it does not hold in the general case of Sobolev mappings in $W^{\ell,p}(M,N)$ ($\ell p < m$) between arbitrary compact manifolds if the topological structure of the image domain N is more complicated, i.e. the homotopy type of N is not that of the point

This fact was probably first noticed by Vodopyanov and Goldstein [32] who in their papers study the homotopic invariants associated with some discontinuous mappings into spheres or in the classes $W^{\ell,n}(M,N)$, $\dim M = n = \dim N$ or $\dim N = n - 1$ ($N = S^{n-1}$) [33]. Also Schoen and Uhlenbeck in [29] discuss the case of $W^{\ell,2}(B^3,S^2)$ and show that the mapping $\chi : x \to \frac{x}{|x|}$ cannot be approximated by smooth, or even C^1 mappings of B^3 into S^2. Their argument is rather obvious, since χ restricted to $\partial B^3 = S^2 \to S^2$ is the identity and has the topological degree 1 [22] while the smooth map in $C(\bar{B}^3,S^2)$, considered at the boundary would necessarily have the degree 0. Though the precise proof is a little more subtle, essentially the same argument is used in the discussion by Vodopyanov and Goldstein [32], [33].

The mapping

$$p_n : B^n \to S^{n-1}, \qquad p_n(x) = \frac{x}{\|x\|} \tag{4}$$

is a model mapping interesting in many respects.

Let us formulate some of its analytical and geometrical properties $W^{1,p}(B^n,S^{n-1})$ for $p < n$. Indeed we see that the gradient ∇p_n admits the estimate $|\nabla p_n| \leq \frac{C}{\|x\|}$ for $x \in B^n$. Actually $p_n \in W^{r,p}(B^n,S^{n-1})$ for $rp < n$.

The mapping p_n realizes a retraction of the closed unit ball B^n onto its boundary S^{n-1}. Here we see the essential distinction between the Sobolev mappings topology and the continuous or smooth topology: it is a classical and fundamental fact that the sphere S^{n-1} is not a continuous retract of B^n. This follows from Sperner's lemma and it immediately implies the Brouwer fixed point theorem for the ball B^n [2], [16].

The mapping $p_n(x)$ is a harmonic mapping of the ball B^n onto S^{n-1}, with the given boundary map

$$\partial B^n = S^{n-1} \to S^{n-1}, \quad p_n|_{\partial B^n} = \text{id}.$$

This means that $p_n(x)$ is a stationary map (in fact a minimizer) for the Dirichlet integral

$$E(f) = \int_{B^n} |\nabla f|^2 \, dx$$

in the class of all maps $f : B^n \to S^{n-1}$ of the class $W^{1,2}(B^n,S^{n-1})$ satisfying the boundary condition $f|_{\partial B^n} = \text{id}$, [14]. Thus the discontinuous mapping $p_n(x)$ appears in analysis and geometry as a solution of a natural variational problem. It should be noted that the general theory of harmonic maps between

Riemannian manifolds, initiated by J. Eells in 1964 occupies an important role in analysis and has broad applications in geometry and mathematical physics [14].

We shall also consider a deformation of $p_n(x)$ given by the formula

$$y = p_{n,\varepsilon}(x,s) = (1 - s)x + \frac{s\varepsilon x}{\|x\|} = \frac{x}{\|x\|}((1 - s)\|x\| + s\varepsilon) \tag{5}$$

for $0 < \varepsilon < 1$ and $0 \leqq s \leqq 1$ which is a mapping of the ball $B^n(\varepsilon)$ onto the ring $s\varepsilon < |y| \leqq \varepsilon$, fixing the boundary $|x| = \varepsilon$. The family of mappings $p_{n,\varepsilon}(x,s)$ for $0 \leqq s \leqq 1$ realizes the boundary $\partial B^n(\varepsilon)$ as the (strong) Sobolev deformation retract of $B^n(\varepsilon)$. We also have $\chi_\varepsilon(x) = p_{n,\varepsilon}(x,s) - x = \frac{sx}{|x|}(\varepsilon - |x|)$. The p-energy of the difference $p_{n,\varepsilon}$-id is estimated as follows

$$E_p(\nabla\chi_\varepsilon) = \int_{|x|<\varepsilon} |\nabla\chi_\varepsilon|^p \, dx \leqq Cs^p\varepsilon^n \quad (p < n) \tag{6}$$

with the constant C depending on n and p only. This estimation implies that by a small perturbation in the sense of the Sobolev norm in $W^{\ell,p}$ it is possible to perturb the identity mapping of the closed ball $B^n(\varepsilon)$ into a Sobolev mapping of the ball $B^n(\varepsilon)$ onto a semiclosed ring $s\varepsilon < |y| \leqq$ and thus change its homotopy type.

For $0 < s < 1$ we can reverse this tearing process of the closed ball $B^n(\varepsilon)$ into the ring, by "sewing" the ring $s\varepsilon < |y| \leqq \varepsilon$ into the closed ball $B^n(\varepsilon)$ back. Note that the "sewing" mapping $x = p_n^{-1}(y) = \frac{y}{1-s}\left(1 - \frac{\varepsilon s}{\|y\|}\right)$ is Lipschitz for $s\varepsilon < |y| \leqq \varepsilon$, $p_n^{-1}(y) = 0$ for $|y| = \varepsilon s$. We remark also that the retraction mapping $p_{n,\varepsilon}(x)$ is in the Sobolev class $W^{r,p}(B^n(\varepsilon))$ for $rp < n$.

Borsuk's theory of separation of points by subcontinua of Euclidean space R^n is based on the study of the mapping

$$p_{n,x_0}(x) = \frac{x - x_0}{\|x - x_0\|}, \; x, x_0 \in R^n, \quad x \neq x_0$$

into the sphere S^{n-1}. If X is an arbitrary continuum in R^n and the point x_0 lies outside X then, by Borsuk's result, the continuous mapping $p_{x_0}(x) : X \to S^{n-1}$ is non-essential or null-homotopic (i.e. homotopic to a constant map) iff the point x_0 lies in the unbounded component of $R^n \setminus X$.

Otherwise the continuum X "separates" the point x_0 from in R^n and the mapping $p_{x_0}(x)$ is essential [2], [7], [16].

The mapping $p_{n,\varepsilon}(x,s)$ or its analogues for $p_{n,x_0}(x)$ may be extended outside the ball $\|x\| < \varepsilon$ as $y \equiv x$ producing a $W^{\ell,p}_{loc}$ mapping of the full space R^n onto the open exterior $\|y\| > s\varepsilon$. For $\varepsilon = 1$ $y \equiv x$ outside the ball $\|x\| \leq 1$. This property makes possible the transfer of the retracting mappings of the type $p_n(x,s)$ on arbitrary manifolds M onto M/B(m,s) (here $B(m_0,s)$ is, say, the geodesic ball on **M** with centre at the point m_0 and radius s) which differ from the identity only in the ball $B(m_0,1)$ i.e. Sobolev retractions of M onto M with the open ball $B(m_0,1)$ deleted.

It is important to note that analogous constructions of retracting mappings and deformations hold for the case when instead of balls we consider linear or curvilinear simplices Δ_m or cubes Q_m as subsets of a manifold M or, more generally, some classes of "regular" polyhedra, modelled on polyhedra obtained as triangulations of sufficiently smooth manifolds (or "piecewise" smooth manifolds). This is so since simplices or cubes considered are bi-Lipschitz equivalent to balls with the (bi)Lipschitz constants invariant under dilations. Also the constants in the local p-energy estimates (6) depend on the bi-Lipschitz constants only. All this holds with the complete control of constants involved in the estimates of deformations if we consider polyhedra build up from simplices, cubes, etc. under the condition that all triangulation (or cubulation) blocks admitted in the construction do not change their metric shape in an uncontrolled way i.e. they should not "degenerate" into (for n = 2), say, triangles with very small angles, etc.

Let **M** now be a compact manifold possibly with boundary - or more generally, let **M** be a m-dimensional polyhedron - triangulated into m-dimensional regular (curvilinear) simplices (cubes, bi-Lipschitz balls). If we apply the above considerations to each simplex of the triangulation T we obtain a family p_Δ, $\Delta \in T$, of local deformations of **M** which may be composed in the Sobolev class $W^{1,p}(\mathbf{M},\mathbf{M})$, since the singularities of the individual retractions p_Δ are "separated". Notice also that because of the local estimates we may use arbitrarily fine triangulations i.e. triangulations T with max diam $\Delta < \varepsilon$, $\Delta \in T$, for ε arbitrarily small.

In this way we obtain a retraction p_s of **M**, first, for $0 < s < 1$ onto a tubular neighbourhood of the (m-1)-skeleton T^{m-1} of the triangulation T and, for $s = 1$, onto T^{m-1} itself. The singularity set Σ_{p_s} does not depend on s and consists of isolated points, one in the interior of each simplex Δ of T.

Since $p_1(m) \equiv m$ for each (m-1)-simplex $\Delta' \in T^{m-1}$ ($p_s(m) \equiv m$ for all $0 \leqq s \leqq 1$) we can repeat the retraction process of $\Delta' \setminus \{m_{\Delta'}\}$, onto a neighbourhood of the boundary $\partial\Delta'$, with $m_{\Delta'}$ - an arbitrarily chosen interior point of Δ', as long as $m - 1 > p$, obtaining the retraction $p^1(1,m)$, of T^{m-1} onto T^{m-2} (for $s = 1$) and the retraction $p^1(s,m)$, for $s < 1$, of T^{m-1} onto a tubular neighbourhood of T^{m-2}. The composition $p^1 \circ p$ will give then a retraction of $|T| = M$ onto T^{m-2}, or, for $0 < s < 1$ a retraction of M onto a tubular neighbourhood of T^{m-2}. The singularity set $\Sigma_{p^1 \circ p}$ will be then a 1-dimensional subpolyhedron in **M** disjoint to T^{m-2}, actually "transverse" to T^{m-2} in the sense of the definition introduced by K. Borsuk in [9]. This process may be repeated to construct $p^2(s,m),\ldots,p^k(s,m)$ as long as $m-k > p$.

What was said above can be considered a sketch of the proof of the following general

PROPOSITION 4: Let M be a compact Riemannian manifold $m = \dim \mathbf{M} > p$. For any s, $1 \geqq s > 0$, any $\varepsilon > 0$ and any (regular) sufficiently fine triangulation T, with $\operatorname{diam} \Delta < \eta$, $\Delta \in T$ for η small enough, there exists a $f_{s,\varepsilon} \in W^{\ell,p}(M,M)$ such that

(a) $|f_{s,\varepsilon}(m) - m| < \varepsilon$ for each $m \in M$

(b) $\|f_{s,\varepsilon} - Id\|_{W^{1,p}(M,M)} \leqq Cs$ for some constant C

(c) $f_{s,\varepsilon}$ is a Sobolev retraction of **M** onto a tubular neighbourhood $\mathcal{U}_{p,s,\varepsilon}$ of the [p]-skeleton $T^{[p]}$ of T. The mapping $f_{1,\varepsilon}$ can be chosen as a Sobolev strong deformation retraction of **M** onto $T^{[p]}$

(d) The singularity set Σ_f of the constructed retraction $f_{1,\varepsilon}$ lies on a subpolyhedron of M of dimension $\leqq m - [p] - 1$, which may be situated outside a tubular neighbourhood of the skeleton $T^{[p]}$ with measure arbitrarily close to the full measure of **M**.

(e) The singularity set Σ_f and the skeleton $T^{[p]}$ are "transverse" in M in the sense of Borsuk, i.e. $T^{[p]}$ is a (strong) deformation retract of $M \setminus \Sigma_f$ and Σ_f is a deformation retract of $M \setminus T^{[p]}$.

Thus we see that, for $p < m$, in an arbitrarily small neighbourhood of the identity map in $W^{\ell,p}(M,M)$ there exist Sobolev maps of M into M Lipschitz homotopic to the Sobolev retraction χ_p of M onto $T^{[p]}$. This gives

PROPOSITION 5: If f is a Lipschitz map, $f \in Lip(M,N) \subset W^{\ell,p}(M,N)$ then, after an arbitrarily small perturbation in $W^{\ell,p}(M,N)$ norm, the perturbed map $\tilde{f}$ may be deformed in $W^{\ell,p}(M,N)$ into the composition $f \circ \chi_p$, i.e. into a map in $Lip(T^{[p]},N)$.

In other words the homotopy classes in $W^{\ell,p}(M,N)$ of maps in Lip(M,N) are described by the homotopy classes of their restrictions to the [p]-skeletons $T^{[p]}$ of triangulations of M. Proposition 5 is a special case of the theory of [p]-homotopy types of B. White [35], [36]. Proposition 4 is also complementing some constructions of [35].

The main construction in the proof of Proposition 4 may have some consequences for the global approximation problems of Sobolev maps by smooth, i.e. C^∞ (or just continuous) maps. As mentioned above in the arbitrarily small neighbourhood in $W^{\ell,p}(S^n,S^n)$ of the identity $I : S^n \to S^n$, for $p < n$, there exists a retraction r_ε of S^n onto S^n with a small n-ball (or spherical cap) $B^n(x_0,\varepsilon)$ deleted. $S^n \setminus B^n(x_0,\varepsilon)$ is bi-Lipschitz (C^∞) equivalent in R^{n-1} to a convex open set Ω in R^n : $L : S^n \setminus B^n(x_0,\varepsilon) \longleftrightarrow \Omega$. Now if $f \in W^{\ell,p}(M,S^n)$ is a Sobolev map such that

(a) $r_\varepsilon \circ f$ is in $W^{\ell,p}(M,S^n)$

and (7)

(b) $\|r_\varepsilon \circ f - f\|_{W^{\ell,p}}$ is small for $\varepsilon \to 0$

we can find, by the generalized Meyers-Serrin theorem, the smooth map which approximates the map $L \circ r_\varepsilon \circ f$ and then $L^{-1} \circ \phi$ will be a smooth approximation in $W^{\ell,p}$ of $r_\varepsilon \circ f$ and, hence, of f. However for an arbitrary $f \in W^{1,p}(M,S^n)$ (7) (a) and (b) do not hold. Nevertheless Betheuel and Zheng [3], [4] prove

<u>PROPOSITION 6</u>: For $1 \leq p < n$ every map in $W^{\ell,p}(M,S^n)$ can be approximated by smooth maps in the $W^{\ell,p}$ norm.

The proof in [3] relies on the observation that there exists, for $p < n$, in an arbitrarily small $W^{\ell,p}$ neighbourhood of the identity, a retraction $\hat{r}_\varepsilon$ of S^n onto $S^n \setminus B^n(x_0,\varepsilon)$ which is Lipschitz. The Lipschitz retraction $\hat{r}_\varepsilon$ is obtained by elementary dilation type transformations from the Lipschitz retraction $r_0(x) = (|x_1|,x_2,\ldots,x_n)$ of the sphere S^n onto the hemisphere $x_1 \geq 0$. A judicious choice of the point x_0 on S^n makes possible to satisfy conditions (7(a)) and (7(b)). The disadvantage of the use of $\hat{r}_\varepsilon$ instead of r_ε lies in the fact that $\hat{r}_\varepsilon$ - "changes orientation" and cannot satisfy the condition $|r_\varepsilon(x) - x| < \varepsilon$ for ε small. A possible way to get around the difficulty related with conditions (7) would be, first, to find, in an arbitrarily small neighbourhood of f in $W^{\ell,p}(M,S^n)$, a mapping $\tilde{f}$ satisfying (7). Thus a kind of Sard's theorem or transversality theorem [21] for Sobolev mappings would be needed. Though such a theorem does not seem to be known so far, some results of Borsuk in his study of geometric properties of continuous mappings of polyhedra suggest that a certain weak version of Sard's theorem holds.

Approximation problems are known to be related with smooth extension problems. In our basic example of the maps $p_{n,x_0}(x)$ (5) the singularity at the point x_0 creates an obstacle to smooth extensions and smooth approximation as well. The full description of the closure $H^{\ell,p}(M,N)$ of $C^\infty(M,N)$ in the Sobolev spaces $W^{\ell,p}(M,N)$ seems to be a rather difficult problem. The notion of the local degree and the linking numbers considered in [32], [33] may be used to describe necessary conditions for maps in $H^{\ell,p}(M,N)$. However the condensation of singularities of local retractions may produce rather complicated examples of functions in $H^{\ell,p}(M,N)$. In [3], [4] is proved

<u>PROPOSITION 7</u>: For $n \leq p < n + 1$ every map $f \in W^{\ell,p}(B^{n+1},S^n)$ can be approximated in the $W^{\ell,p}$ norm by maps f_k, $f_k \to f$, smooth except at most at a finite number of points.

In this connection, in the effort to understand more clearly the phenomena of global extension and approximation of Sobolev mappings, it may be helpful to recall two theorems of Borsuk and Eilenberg belonging to the geometric theory of continuous mappings [8], [9], [15].

<u>PROPOSITION 8</u>: Let A be a closed subset of a polyhedron X $\dim(X \setminus A) \leq m$.

Let Y be a polyhedron locally connected in dimension < m and connected in dimensions < n. Then for each continuous mapping $f : A \to Y$, there exists a polyhedron $E \subset X \setminus A$, of dimension $\leq m - n - 1$, closed in X, such that f can be continuously extended to a map $\tilde{f} : X \setminus E \to Y$.

PROPOSITION 9: With assumptions of Proposition 8, E may be chosen such that the set A is a (continuous) retract of $X \setminus E$ ($\dim E \leq m - n - 1$).

Propositions 8 and 9 are closely related. It is immediately seen that the underlying geometry of the Propositions 4, 5, 6, 7, is intimately related with the geometry of Propositions 8 and 9. Thus in the Proposition 4 A is the skeleton $T^{[p]}$, $Y = T^{[p]}$ and E - the singularity set $\Sigma_{p_{\varepsilon,s}}$, $\dim \Sigma_{p_{\varepsilon,s}} = m - [p] - 1$. In Proposition 7 we have $Y = S^n$, A = the n-skeleton of a triangulation of B^{n+1} and the point singularities may be chosen in the interior of the simplices of the triangulation.

The retraction properties of the Proposition 9 have been mentioned also in the Proposition 4. Through the Propositions 8 and 9 and the Borsuk's notion of transversality [9] the Proposition 4 and Sobolev imbeddings theorems somehow relate with the Alexander's duality theory (see [15]).

Proposition 7 can be interpreted as describing the "generic" singularities of the maps of B^{n+1} into S^n of class $W^{\ell,p}(B^{n+1},S^n)$ for $n \leq p < n + 1$. These pointwise singularities are "stable" in the sense that they cannot disappear by "small perturbation" in $W^{\ell,p}$ of the given map $f : B^{n+1} \to S^n$. Propositions 4, 7, 8 and 9 suggest the geometric form of "stable" and "generic" singularities of Sobolev maps of higher dimensional manifolds into lower dimensional, say spheres S^n, for proper values of p.

We shall not try to state many other natural and interesting questions arising in this area. One may expect that the interaction between the geometric polyhedral topology and the theory of Sobolev mappings can go both ways and may be an interesting area of current research.

3. COMPOSITION OF MAPPINGS

There are rather few cases when unrestricted composition of Sobolev-mappings is possible; in the case of discontinuous maps, $f \in W^{\ell,p}(M,N)$, $\ell p < m$, practically we can compose f only with sufficiently smooth, invertible maps. We could see also above that the lack of manageable composition rules may cause serious difficulties (e.g. in the proof of the Proposition 6 and its generalizations). We restrict our considerations to the case $W^{\ell,p}(M,N)$, $p < m$. Then it is known that we can compose f, "from the left", by arbitrary Lipschitzian maps: if $h \in Lip(N,Z)$ then $h \circ f \in W^{\ell,p}(M,Z)$. The following theorem of Goldstein and Vodopyanov [32], [33], [20] describes the composition "from the right".

PROPOSITION 10: Let ϕ be a map $\phi: M \to M$. Then the induced map

$$\phi^* : W^{\ell,p}(M,R) \to W^{\ell,p}(M,R), \quad \phi^*(f) = f \circ \phi$$

is a bounded operator $\|\phi^*(f)\|_{W^{\ell,p}(M,R)} \leq C \|f\|_{W^{\ell,p}(M,R)}$ iff ϕ is a homeomorphic quasiisometry for $p \neq \dim M$ and ϕ is quasiconformal homeomorphism of M, for $p = \dim M$.

M is a quasiisometry if

$$\frac{1}{L} \leq \lim_{x_1 \to x_2} \frac{\|\phi(x_1) - \phi(x_2)\|}{\|x_1 - x_2\|} \leq L$$

for some constant L.

The definition of quasiconformality - see [6] or [20].

For $p \neq m$ the Sobolev mappings compose well essentially only with locally bi-Lipschitzian mappings. The class of quasiconformal mappings is however much broader.

Thus we see that in general we cannot compose two Sobolev mappings $f \in W^{\ell,p}(M,M)$ and $g \in W^{\ell,p}(M,M)$ to obtain $f \circ g \in W^{\ell,p}(M,M)$. However there is one obvious, but important case when the composition is possible: f - is Lipschitzian on the g-image of the singularity set of g and g is locally bi-Lipschitzian on the universe g^{-1}-image of the singularity set of f. We will say in this case that the singularities of f and g are separated. This condition has been satisfied for the composition of local retractions p_Δ for different simplices Δ in a triangulation of M.

A natural and important question which arises in this respect is the following: given two maps f and g in $W^{\ell,p}(M,M)$ do there exist two maps $\tilde{f}$ and $\tilde{g}$, in arbitrary small neighbourhoods of f and g respectively, such that the $\Sigma_{\tilde{f}}$ and $\Sigma_{\tilde{g}}$ are separated? Is the [p]-homotopy class of $\tilde{f} \circ \tilde{g}$ dependent on the [p]-homotopy classes of f and g only?

The lack of a reasonable composition rules for Sobolev mappings is a serious obstacle in developing the theory. A remedy could, may be, come if the notion of the generic map and the generic singularities of mappings in $W^{\ell,p}(M,M)$ or $W^{\ell,p}(M,N)$ would be further elaborated.

The list of topics of interest in the Sobolev topology could be enlarged. Not touched upon here, but very promising, are, in my opinion, the topics related with the study of applications of such notions as the Brouwer's degree or the intersection theory and the linking coefficients, ideas of H. Federer related with the general area and coarea formulae or the curvature and others, which may be expressed by classical integral formulae [2], [17], [8].

I hope that the examples described above give some evidence that the merging of classical geometric topology, à la Hopf - Brouwer-Borsuk, of the differential topology and the geometric integration theory with the analytical theory of Sobolev may be fruitful and produce valuable new results.

References

[1] R.A. Adams, Sobolev Spaces, Acad. Press, 1975.

[2] P.S. Aleksandroff and H. Hopf, Topologie, Springer, 1935.

[3] F. Betheuel and X. Zheng, Sur la densité de fonctions régulières entre deux variétés dans des espaces de Sobolev, C.R. Acad. Sci. Paris 303 (1986) pp. 447-449.

[4] F. Betheuel and X. Zheng, Density of smooth functions between two manifolds in Sobolev spaces, Journal of Functional Analysis, vol.80 1988.

[5] B. Bojarski, Remarks on local function spaces, Proceedings of the Function Space Conference, Lund, 1986, Springer Lecture Notes 1302, 1988.

[6] B. Bojarski and T. Iwaniec, Analytical Foundations of the theory of quasiconformal mappings in R^n, Ann. Acad. Sci. Feen. Ser AI, Math 8, 1983.

[7] K. Borsuk, Theory of retracts, Warszawa 1967.

[8] K. Borsuk, Un théorème sur les prolongements des transformations, Fund. Math. 29 (1937).

[9] K. Borsuk, Quelques relations entre la situation des ensembles et la rétraction dans les espaces Euclidiens, Fund. Math. 29 (1937).

[10] V.I. Burenkov, O plotnosti beskonečno differenciruyemyh funkcji... in "Teoriya Kubaturnyh Formul", Novosibirsk, Inst. of Math., 1975.

[11] A.P. Calderon, On the differentiability of absolutely continuous functions, Ren. Mat. Univ. Parma, 1951.

[12] A.P. Calderon, Lebesgue space of differentiable functions. Proc. Symp. Pure Math., v. 4, 1961.

[13] A.P. Calderon and A. Zygmund, Local properties of solutions of elliptic partial differential equations, Studia Math., vol. 20, 1961.

[14] J. Eells and L. Lemaire, Report on harmonic maps, Bull. London Math. Soc., v. 10 (1978).

[15] S. Eilenberg, Un théorème de dualite, Fund. Math. 26 (1936).

[16] R. Engelking and K. Sieklucki, Wstęp do topologl Bibl. Matem., PWN, Warszawa 1986.

[17] H. Federer, Geometric measure theory, Springer-Verlag, Berlin-New York 1969.

[18] H. Federer, Curvature measures, Trans. Amer. Math. Soc., vol. 93. 1959.

[19] B. Fuglede, Extremal length and functional completion, Acta Math., vol. 90, 1957.

[20] V.M. Goldstein and Yu.G. Resetnyak, Vvedenye v teoriyu funkcji s obobščennymi proizvodnymi i kvazikonformnyje otobrazenya. Izdat. "Nauka" Moskva 1983.

[21] M. Hirsch, Differentiable Topology, Springer-Verlag, New York 1976.

[22] S.T. Hu, Homotopy Theory, Acad. Press, 1959.

[23] W. Hurewicz, Über Abbildungen topologischer Räume auf die n-dimensionale Sphäre, Fund. Math. 24 (1985).

[24] F.C. Liu, A Lusin type property of Sobolev functions, Indiana Univ. Math. Journ., vol. 26, No. 4 (1977).

[25] W.G. Mazja, Einbettungssatz für Sobolewsche Räume, Teubner Texte zur Math. t. 1,2, Leipzig 1979.

[26] W.G. Mazja, Prostranstva S.L. Soboleva, Leningrad, Izdat. Lening. Univ., 1985 - Springer-Verlag 1986.

[27] N. Meyers and J. Serrin, H = W, Proc. Nat. Acad. Sci. USA, vol. 51, 1964.

[28] S.M. Nikolski, Svoistva nekotoryh klasov funkcji mnogih peremennyh na differencirujemyh mnogo-obrazijah. Mat. sbor., v. 33, 1953.

[29] R. Schoen and K. Uhlenbeck, Boundary regularity and the Dirichlet problem for harmonic maps, J. Diff. Geom. 18 (1983).

[30] S.L. Sobolev, Applications of Functional Anaysis in Mathematical Physics, Leningrad 1950.

[31] E.M. Stein, Singular Integrals and Differentiability properties of functions, Princeton 1970.

[32] S.K. Vodopyanov and V.M. Goldŝtein, Kvazikonformyje otobrazeniya i prostranstva funkcii s pervymi obobŝĉennymi proizvodnymi, Sybir. Mat. Zhurnal, vol. 17, No. 3, 1976.

[33] S.K. Vodopyanov and V.M. Goldŝtein, O geometriceskih svoistvah funkcii s pervymi obobŝĉennymi proizvodnymi. Uspekhi Mat. Nauk, v. 34, No. 1, 1979.

[34] S.K. Vodopyanov and V.M. Goldŝtein, Prostranstva Soboleva i specyalnye klassy otobrazhenii, Novosib. Gos. Univ., Novosibirsk, 1981.

[35] B. White, Infima of energy functionals in homotopy classes of mappings, J. Diff. Geometry, vol. 23, 1986.

[36] B. White, Homotopy classes in Sobolev spaces and the existence of energy minimizing maps, Acta Math. 160, 1988.

B. Bojarski
Department of Mathematics
Polish Academy of Sciences
00950 Warsaw
Poland

H. GAJEWSKI

Nonlinear equations describing carrier transport in semiconductors

A system of nonlinear partial differential equations describing carrier transport in semiconductors was proposed by van Roosbroeck [11] in 1950 and is now generally accepted. The first significant report on numerical techniques for solving this system was given by Gummel [5] in 1964. Since then numerical modelling of semiconductors proved to be an efficient tool for device designers (see [10] and the literature quoted therein).

In spite of their physical and technical relevance the device equations received little attention from the viewpoint of mathematical analysis during a long period of time. Mock [8] was the first to prove an existence and uniqueness result in 1974. His book [9] covers the state of the art up to 1983. A more recent survey has been given in [7]; however it is restricted to the stationary equations. In this case a maximum principle works and the existence of solutions can be proved via Schauder's fixed point theorem under quite natural conditions. Uniqueness in general cannot be expected by physical reasons. The situation is different in the nonstationary case. Here the usual technique of proof rests on hypotheses excluding physically interesting cases. We list some of them (see [9]).

(i) The Laplace operator works as an isomorphism between the Sobolev space $H^{2,p}(G)$ and the Lebesgue space $L^p(G)$ for a $p > N$, where N is the dimension of the domain G occupied by the semiconductor.

(ii) The carrier mobilities are constant.

(iii) The energy distribution of the carriers is governed by Boltzmann statistics.

Using Moser's iteration technique to get L^∞-estimates Groeger and the author [3] could cancel (i). By means of a new technique of time discretization we could cancel also (ii) and (iii) recently [4].

In our analysis of semiconductor equations some physically motivated ideas play an important role. In spite of this fact the detailed proofs of existence, uniqueness, asymptotic behavior and numerical approximation of solutions are rather involved. It is the main aim of this lecture to make these ideas transparent. Accordingly, in the first section we study a reduced version of the equations in order to exclude technicalities as far as possible. The complete system of equations is presented together with the assumptions in section 2. In section 3 the system is transformed into an evolution equation in a suitable function space and some results are formulated. Finally, section 4 contains a sketch of the proof.

1. SOME QUALITATIVE CONSIDERATIONS

In this section we are concerned with the following initial boundary value problem

$$
\begin{aligned}
&-\Delta u = f-n, \quad n = e^{v}, \\
&n' = \nabla\cdot J, \quad J = -nd(\nabla(u-v)) \text{ in } \mathbb{R}_+ \times G, \\
&u = u_\Gamma, \; v = v_\Gamma \text{ on } \mathbb{R}_+ \times \Gamma_0, \\
&\nu\cdot(\nabla u'-J) + b(u-u_\Gamma) = 0, \; v = v_\Gamma \text{ on } \mathbb{R}_+ \times \Gamma_1, \\
&\nu\cdot\nabla u+a(u-a_0) = \nu\cdot J = 0 \text{ on } \mathbb{R}_+ \times \Gamma_2, \\
&v = v_0 \text{ on } \{0\} \times G.
\end{aligned}
\tag{1}
$$

Here $\mathbb{R}_+ = [0,\infty[$, G is the N-dimensional domain occupied by the semiconductor, $\partial G = \Gamma_0 \cup \Gamma_1 \cup \Gamma_2$, $n' = \partial n/\partial t$, ∇ is the gradient operator, ν is the outer unit normal on ∂G; a, b and a_0 are constants, a, b $\geqq 0$.

The unknown functions u and v represent electrostatic potential and the chemical potential of electrons, respectively, and f is a given density of impurities. The densities of electrons and the electron current are denoted by n and J, respectively. The quantities are scaled so that the dielectric permittivity becomes unity. The vector-valued function d models the mobility of electrons and is assumed to be monotone with respect to the gradient of the electrochemical (quasi-Fermi) potential u-v (see [10]), d(0) = 0. Finally, the functions u_Γ, v_Γ and v_0 represent boundary and initial values, respectively. All functions arising in this section are assumed to be sufficiently smooth.

In what follows we essentially use properties of the functional

$$F(u,v) = \int [n(\log(n/n_\Gamma)-1)+n_\Gamma + (1/2)|\nabla(u-u_\Gamma)|^2]dG + (a/2)\int (u-u_\Gamma)^2 d\Gamma_2,$$

where $n = e^v$, and $n_\Gamma = e^{v_\Gamma}$.

PROPOSITION 1: The functional F is nonnegative and convex. More precisely

(i) $F(u,v) \geqq F(u_\Gamma,v_\Gamma) = 0$,

(ii) $F(u_1,v_1) + F(u_2,v_2)-2F((u_1+u_2)/2,(v_1+v_2)/2)$

$$\geqq (1/4)[\int (n^2/(n_1+n_2) + |\nabla u|^2)dG + a\int u^2 d\Gamma_2].$$

PROOF: (i) follows from the elementary inequality $\log s \geqq 1-1/s$, $s > 0$. On the other hand for smooth real functions r it holds

$$r(2x)+r(2y)-2r(x+y) = \int_y^x (r'(s+x)-r'(s+y))ds = \int_y^x\int_y^x (r''(s+t)dtds$$

$$\geqq (x-y)^2 \min_{2y\leqq s\leqq 2x} r''(s).$$

Applying this with respect to $r(s) = s(\log(s/s_0-1) + s_0$ and $r(s) = a/2s^2$ we get (ii). □

REMARK: The functional F may be interpreted as the free energy of the system.

A function h with h = 0 on Γ_0 is called a test function. For test functions h_1, h_2 we define the following scalar product

$$\langle h_1,h_2\rangle = \int \nabla h_1\cdot\nabla h_2 dG + a\int h_1h_2\, d\Gamma_2.$$

PROPOSITION 2: Let (u,v) be a solution of (1). Then for each test function

$$\langle u',h\rangle = \int J\cdot\nabla h\, dG - b\int (u-u_\Gamma)h\cdot d\Gamma_1.$$

Moreover, the time derivative of the function

$$L(t) = F(u(t),v(t))$$

satisfies the estimate

$$L'(t) \leqq \int J\cdot\nabla(v_\Gamma - u_\Gamma)\; dG.$$

PROOF: Firstly, integrating by parts and using (1), we get

$$\langle u',h\rangle - \int J\cdot\nabla h\; dG + b\int (u-u_\Gamma)h\; d\Gamma_1 = -\int \nabla\cdot(\nabla u'-J)h\; dG$$

$$= \int (\Delta u'-n')h\; dG = 0.$$

From this, setting $h = u-u_\Gamma$ and using the monotonicity of d, it follows that

$$L'(t) = \int n'(v-v_\Gamma)\; dG + \langle u',h\rangle$$

$$= \int (\nabla\cdot J(v-v_\Gamma)+J\cdot\nabla h)\; dG - b\int h^2 d\Gamma_1 \leqq \int J\cdot\nabla(h-(v-v_\Gamma))dG$$

$$= -\int (nd(\nabla(u-v))\cdot\nabla(u-v)+J\cdot\nabla(u_\Gamma - v_\Gamma))dG \leqq \int J\cdot\nabla((v_\Gamma - u_\Gamma))dG. \qquad \square$$

REMARK: If the boundary values satisfy the thermal equilibrium condition $u_\Gamma - v_\Gamma$ = const, then we have $L'(t) \leqq 0$. Hence the function L is monotonically decreasing. Therefore L is called a Lyapunov function. L' may be looked at as the dissipation rate of the system. If additionally u_Γ is the solution of the boundary value problem

$$-\Delta u = f-e^{u-c}, \qquad c = \text{const},$$

$$u = u_\Gamma \text{ on } \Gamma_0 \cup \Gamma_1,\; \nu\cdot\nabla u+a(u-a_0) = 0 \text{ on } \Gamma_2,$$

then one can show that there exist constants c_1, $\lambda > 0$ such that

$$L(t) \leqq c_1 e^{-\lambda t}.$$

REMARK: The function L turns out to be the key for getting *a priori* estimates.

To give an easy example, let $\nabla(u_\Gamma - v_\Gamma)$ and d be bounded. Then we have

$$L(t) = L(0) + \int_0^t L'(s)ds \leqq L(0) + \int_0^t \int n\ d(\nabla(u-v))\cdot\nabla(u_\Gamma - v_\Gamma)dG\ ds$$

$$\leqq c(1 + \int_0^t \int n\ dG\ ds) \leqq c(1 + t\int_0^t L(s)\ ds.$$

Hence by means of Gronwall's lemma, we conclude the *a priori* estimate $L(t) \leqq (1 + t)e^{ct}$.

Now we turn to the problem of uniqueness of solutions.

PROPOSITION 3: Let (u_i, v_i), be solutions of problem (1). Set

$$q(t) = 2[F(u_1,v_1) + F(u_2,v_2) - 2F((u_1+u_2)/2,(v_1+v_2)/2)].$$

Then the following inequality holds

$$\int (nM + |\nabla u|^2)(t)dG \leqq 2[q(0) + \int_0^t \int (M\ (J_1+J_2)\cdot\nabla v)(s)dG\ ds],$$

where

$$u = u_1 - u_2,\ v = v_1 - v_2,\ n_i = e^{v_i},\ n = n_1 - n_2,\ M = n/(n_1+n_2),$$

$$J_i = -n_i d(\nabla(u_i - v_i)).$$

PROOF: Set $N_i = n_i/(n_1+n_2)$. We note that

$$\nabla \log(2N_1) = N_2\nabla v,\ \nabla \log(2N_2) = -N_1\nabla v.$$

Now, using the monotonicity of d, we obtain

$$q'(t) = 2\int [n_1' \log n_1 + n_2' \log n_2 - (n_1'+n_2')\log((n_1+n_2)/2)]dG + \langle u',u\rangle$$

$$= 2\int [n_1' \log(2N_1) + n_2'\log(2N_2)]dG + \langle u',u\rangle$$

$$= 2\int [\nabla\cdot J_1\ \log(2N_1) + \nabla\cdot J_2\ \log(2N_2)]dG + \langle u',u\rangle$$

$$= \int [-2(N_2J_1-N_1J_2)\cdot\nabla v + (J_1-J_2)\cdot\nabla u]dG - b\int u^2\, d\Gamma_1$$

$$= \int [2(N_2J_1-N_1J_2)\cdot\nabla(u-v)+((1-2N_2)J_1 + (2N_1-1)J_2)\cdot\nabla u]dG - b\int u^2 d\Gamma_1$$

$$= \int [1/(n_1+n_2)(-n_1n_2(d_1-d_2))\cdot\nabla(u-v)+n(J_1+J_2)\cdot\nabla u]dG - b\int u^2\, d\Gamma_1$$

$$\leq \int n/(n_1+n_2)(J_1+J_2)\cdot\nabla u\, dG.$$

Hence the assertion follows by Proposition 1(ii). □

REMARK: In the case that v_i and d are bounded, it follows from Proposition 3 that

$$q'(t) \leq c_1(1 + \int_0^t q(s)\, ds).$$

In view of Gronwall's lemma and the convexity of q (see Proposition 1), this estimate implies uniqueness if $v_i(0) = v_0$ and continuous dependence of the initial values $v_i(0)$ if these are different.

Let $T > 0$ be a compact interval of time. The proof of existence given in [4] is based on the following discretization of time

$$-\Delta u_{i+1} = f-n_{i+1}, \quad 0 \leq i \leq I,$$

$$(n_{i+1}-n_i)/\tau = \nabla J_i, \ J_i = -n_i d(\nabla(u_{i+1}-v_{i+1})), \ n_i = e^{v_i},$$

where $\tau = T/I$ is the size of the time step. This scheme of time discretization differs from the usual Euler backwards rule [9]. However, in view of the following proposition it appears to be quite natural.

PROPOSITION 4: The discrete Lyapunov function $L_i = F(u_i,v_i)$ satisfies the estimation

$$(L_{i+1}-L_i)/\tau \leq \int J_i\cdot\nabla(v_\Gamma-u_\Gamma)\, dG.$$

PROOF: Setting $h = (u_{i+1}-u_\Gamma)$, we get (see the proof of Proposition 2)

$$(L_{i+1}-L_i)/ = \{\int [\int_{n_i}^{n_{i+1}} \log(s/n_\Gamma)\, ds + (1/2)(|\nabla h|^2 - |\nabla(u_i-u_\Gamma)|^2)]dG$$

$$+ (a/2)\int (h^2 - (u_i-u_\Gamma)^2)d\Gamma_2\}/\tau$$

$$= [\int (n_{i+1}-n_i)(v_{i+1}-v_\Gamma)dG + \langle u_{i+1}-u_i, h\rangle - (a/2)\int (u_{i+1}-u_i)^2 d\Gamma_2$$

$$- \int (\int_{n_i}^{n_{i+1}} \log(n_{i+1/s})ds + (1/2)\ |\nabla(u_{i+1}-u_i)|^2)dG]/\tau$$

$$\leq \int (\nabla\cdot J_i(v_{i+1}-v_\Gamma) + J_i\cdot\nabla h)\ dG - b\int h^2\ d\Gamma_1$$

$$\leq \int J_i\cdot\nabla(h-(v_{i+1}-v_\Gamma))dG \leq \int J_i\cdot\nabla(v_\Gamma-u_\Gamma)\ dG.$$

REMARK: Similarly as in the continuous case, Proposition 5 can be applied to get *a priori* estimates. Moreover, it was proved in [4] that the discrete solution (u_I, v_I) converges to a solution of (1) provided $\tau \to 0$. In the noncompact case $T = \tau I \to \infty$, the discretization scheme can be understood as an iteration procedure for calculating steady states. This procedure converges under thermal equilibrium conditions. This can be proved by using the fact that the L_i's decrease monotonically. However, it is one of the most relevant open questions whether there are more general conditions ensuring convergence.

2. THE SYSTEM OF SEMICONDUCTOR EQUATIONS. ASSUMPTIONS

The complete system of semiconductor equations reads:

$$-\Delta u = f + p - n,\quad n = e_n(v),\quad p = e_p(v),$$

$$n' - \nabla\cdot J_p = R,\qquad J_n = -nd_n(\nabla(u-v))\ \text{in}\ \mathbf{R}_+ \times G,$$

$$p' + \nabla\cdot J_p = R,\qquad J_p = -pd_p(\nabla(u+w))\ \text{in}\ \mathbf{R}_+ \times G,$$

$$u = u_\Gamma,\quad v = v_\Gamma,\quad w = w_\Gamma\ \text{on}\ \mathbf{R}_+ \times \Gamma_0,$$

$$\nu\cdot(\nabla u' - J_n - J_p) + \beta(u) = 0,\ v = v_\Gamma,\ w = w\quad \text{on}\ \mathbf{R}_+ \times \Gamma_1,$$

$$\nu\cdot\nabla u+\alpha(u) = \nu\cdot J_n = \nu\cdot J_p = 0 \text{ on } R_+ \times \Gamma_2,$$

$$v = v_o,\ w = w_o \text{ on } \{0\} \times G.$$

Here in addition to section 1 the functions w and p represent chemical potential and density of holes, respectively. R is the rate of recombination.

We make the following assumptions:

Domain. $G \subseteq R^N$, $N \leq 3$, is a bounded Lipschitzian domain, $\partial G = \Gamma_o \cup \Gamma_1 \cup \Gamma_2$, Γ_j are pairwise disjoint, Γ_o, Γ_1 are closed, the surface measure of Γ_o is positive.

Data. $f, v_o, w_o \in L^\infty(G)$, $u_\Gamma, v_\Gamma, w_\Gamma \in H^{1,\infty}(G)$, $\alpha, \beta \in (R \to R)$ monotone and Lipschitzian.

Carrier densities. $e_i \in C^1(R)$, $i = n,p$, $e_i(s) \geq e_o e_i'(s) > 0$, $e_o = \text{const} > 0$,

$$\lim_{s\to-\infty} e_i(s) = 0, \qquad \lim_{s\to-\infty} e_i(s) = \infty.$$

Mobility. $d_i \in (R^N \to R^N)$, $i = n,p$, strongly monotone and Lipschitzian, $d_i(0) = 0$.

Recombination. $R = r(n,p)(g-\exp(v+w))$, $0 \leq g \in L^\infty$, $r \in (R_+ \times R_+ \to R_L)$ Lipschitzian.

REMARKS:

(i) Boltzmann statistics, where $e_i(s) = \exp(s)$, as well as Fermi-Dirac statistics, where

$$e_i(s) = C_i F_{1/2}(s-c_i),\ F_{1/2}(t) = (2/\sqrt{\pi})\int_{R_+} (\sqrt{s})(1+\exp(s-t))^{-1}ds,\ i=n,p,$$

satisfy our assumptions.

(ii) The standard Shockley-Real-Hall recombination rate

$$g \equiv 1,\ r = (c_1+c_2n+c_3p)^{-1},\ c_i = \text{const},\ i = 1,2,3,$$

as well as the Auger recombination rate

$$g \equiv 1,\ r = c_1 n + c_2 p, \quad c_i = \text{const},\ i = 1,2,$$

satisfy our assumptions.

(iii) Our results remain valid if the functions α, β, d_i, e_i and r depend on the space variable reasonably.

(iv) Monotonicity properties are to be understood in the sense of the theory of monotone operators [2].

3. FORMULATION AS EVOLUTION EQUATION

We want to transform the initial boundary value problem into an evolution equation. For this end we need some function spaces.

$$V_0 = \{h \in H^1(G),\ h = 0 \text{ on } \Gamma_0\},$$

$$V_1 = \{h \in V_0, \quad h = 0 \text{ on } \Gamma_1\}, \quad V = V_0 \times V_1 \times V_1,$$

$$X = (x_\Gamma + V) \cap (L^\infty(G) \times L^\infty(G) \times L^\infty(G)), \qquad x_\Gamma = (u_\Gamma, v_\Gamma, w_\Gamma),$$

$$H = V_0^* \times L^2(G) \times L^2(G), \qquad Y = V_0^* \times L^\infty(G) \times L^\infty(G),$$

where V_0^*, V^* are the duals of V_0, V, respectively. Since V is densely and continuously imbedded into H the space H is imbedded into V^*. We denote by $(\cdot,\cdot)$, $(\cdot,\cdot)_i$ the scalar product on $L^2(G)$, $L^2(\Gamma_i)$, respectively. Further, $\langle\cdot,\cdot\rangle$ is the dual pairing between V^* and V.

For brevity we set $x = (u,v,w) \in X$, and $\bar{x} = (\bar{u},\bar{v},\bar{w}) \in V$. Now we are able to define the needed operators. Firstly, we define an operator $E \in (X \to Y)$ by

$$\langle Ex,\bar{x}\rangle = (\nabla u,\nabla\bar{u}) + (\alpha(u),\bar{u})_2 + (e_n(v),\bar{v}) + (e_p(w),\bar{w}),\ \forall\ \bar{x} \in V.$$

Secondly, for each $y = (y_1,n,p) \in Y$ we define an operator $A(y) \in (X \to V^*)$ by

$$\langle A(y)x,\bar{x}\rangle = (nd_n(\nabla(u-v)),\nabla(\bar{u}-\bar{v})) + (pd_p(\nabla(u+w)),\nabla(\bar{u}+\bar{w}))$$

$$+ (\beta(u),\bar{u})_1 + (r(n,p)(\exp(v+w)-g),\bar{v}+\bar{w}), \ \forall\ \bar{x} \in V.$$

Then the initial boundary value problem can be rewritten as an evolution equation

$$y' + A(y)x = 0, \quad y = Ex, \quad y(0) = y_o,$$
$$x \in L^2_{loc}(\mathbb{R}_+;X), \quad y \in C(\mathbb{R}_+;H), \quad y' \in L^2_{loc}(\mathbb{R}_+;V^*), \tag{2}$$

where $y_o = (f+p_o-n_o,n_o,p_o)$, $n_o = e_n(v_o)$, $p_o = e_p(w_o)$. Besides (2) we consider the following time discretization

$$y_{i+1}-y_i + \tau A(y_i)x_{i+1} = 0, \quad y_i = Ex_i,$$
$$0 \leqq i \leqq I, \ \tau = T/I, \quad 0 < T < \infty. \tag{3}$$

We estract some results from [4]:

<u>THEOREM</u>:

(i) For each I there exists a unique solution $x_I = \{x_i\}$ of (3).

(ii) Let $I \to \infty$. Then x_I converges to a solution x of the continuous problem (2).

(iii) In the case of Boltzmann statistics x is unique.

(iv) Let the following conditions of thermal equilibrium be satisfied

$$u_\Gamma-v_\Gamma = \text{const}, \quad u_\Gamma+w_\Gamma = \text{const}, \quad g = \exp(v_\Gamma+w_\Gamma), \quad b(u_\Gamma) = 0.$$

Then, there exists a unique $x^* \in X$ such that

$$A(y^*)x^* = 0, \ y^* = Ex^*.$$

Moreover, it holds that

$$\|y(t)-y^*\|_H \leqq c \exp(-\lambda t),\ c,\ \lambda > 0.$$

OPEN QUESTIONS:

(i) We do not know if the solution x is unique in the case of Fermi-Dirac statistics. Only under some *ad hoc* assumptions concerning the regularity of x we can prove uniqueness in this case.

(ii) Apart from thermal equilibrium conditions we know little about the global behavior of solutions. However, in some situations we can prove the existence of a compact global attractor in the sense of Babin and Visik [1].

(iii) From the numerical point of view it would be extremely desirable to find conditions ensuring that x_I converges to a steady state in the non-compact case $T = \tau I \to \infty$.

4. SKETCH OF PROOF

For the sketch of the proof of the announced results some properties of the operators E and A are needed. First of all, the operator E is the gradient of the convex functional

$$e(x) = \int [\int_{v_\Gamma}^{v} e_n(s)ds + \int_{w_\Gamma}^{w} e_p(s)ds + (1/2)|\nabla(u-u_\Gamma|^2]dG + \int\int_{u_\Gamma}^{u} \alpha(s)ds d\Gamma_2.$$

The corresponding conjugate functional e* satisfies the relation [2]

$$e^*(Ex) = \langle Ex, x-x_\Gamma\rangle - e(x).$$

Now, introducing the Lyapunov function

$$L(t) = e^*(Ex(t)),$$

where x is a solution of (2), we get

$$L'(t) = \langle Ex, x-x_\Gamma\rangle = - \langle A(Ex)x, x-x_\Gamma\rangle.$$

The proof of our existence result consists of the following main steps:

(i) We cut off the functions e_i such that even for each $y \in H \subseteq Y$ is defined.

(ii) Taking into account the properties of e_i and β it becomes clear that $A(y) \in (x_\Gamma + V \to V^*)$ is strongly monotone and continuous. Hence the existence of a unique time discrete solution x_I follows from the theory of monotone operators [2].

(iii) Using the properties of L, *a priori* estimates can be proved for x_I.

(iv) The map $(y,x) \to A(y)x$ can be understood as a pseudo-monotone operator [6]. Applying standard techniques of the theory of such operators, we can take the limit $\tau \to 0$ in order to get a continuous solution x.

(x) Finally, by means of Moser's iteration technique we show that x is actually bounded. Hence we can cancel the cut-off of the e_i's.

References

[1] A.V. Babin and M.I. Visik, Attractors of evolution equations with partial derivatives and estimates of their dimension, Usp. Mat. Nauk, 38 (1983), 133-137.

[2] H. Brezis, Operators maximaux monotones et semi-groupes de contractions dans les spaces de Hilbert, Math. Stud. 5, Amsterdam (1973).

[3] H. Gajewski and K. Groeger, On the basic equations for carrier transport in semiconductors, J. Math. Anal. Appl., 113 (1986), 12-35.

[4] H. Gajewski and K. Groeger, Semiconductor equations for variable mobilities based on Boltzmann statistics and Fermi-Dirac statistics, Math. Nachr. (in press).

[5] H.K. Gummel, A selfconsistent iterative scheme for one-dimensional steady state transistor claculations, IEEE Trans. Electron Devices, ED-11 (1964), 456-465.

[6] J.L. Lions, Quelque Méthodes de Resolution des Problemes aux Limites non Linéaires, Dunod, Gauthier-Villars (1969).

[7] P.A. Markovich, The Stationary Semiconductor Device Equations, Springer (1986).

[8] M.S. Mock, An initial value problem from semiconductor device theory, SIAM J. Math. Anal., 5 (1974), 597-612.

[9] M.S. Mock, Analysis of Mathematical Models of Semiconductor Devices, Boole Press (1984).

[10] S. Selberherr, Analysis and Simulation of Semiconductor Devices, Springer (1984).

[11] W. Van Roosbroeck, Theory of the flow of electrons and holes in germanium and other semiconductors, Bell Syst. Tech. J., 29 (1950), 560-607.

Herbert Gajewski
Karl-Weierstrass-Institut für Mathematik
Mohrenstrasse 39
Berlin 1086
DDR

A. KUFNER

Weighted Sobolev spaces and nonlinear boundary value problems

1. THE WEIGHTED SOBOLEV SPACES: Let $1 < p < \infty$ and let Ω be a domain in R^N. For $k \in \mathbb{N}$ and for N-dimensional multi-indices α such that $|\alpha| \leq k$ let $w_\alpha = w_\alpha(x)$ be *weight functions*, i.e. measurable and almost everywhere in Ω positive functions, and let us denote $S = \{w_\alpha, |\alpha| \leq k\}$. The *weighted Sobolev space*

$$W^{k,p}(\Omega;S) \tag{1}$$

is defined as the set of all functions $u = u(x)$, $x \in \Omega$, such that

$$\|u\|^p_{k,p,S} = \sum_{|\alpha|\leq k} \int_\Omega |D^\alpha u(x)|^p \, w_\alpha(x) \, dx < \infty, \tag{2}$$

the derivatives $D^\alpha u$ being considered in the sense of distributions. Further, let

$$W_0^{k,p}(\Omega;S) \tag{3}$$

be the closure (if it is meaningful) of the set $C_0^\infty(\Omega)$ with respect to the norm from (2).

2. REMARK: Let us recall that under the condition

$$w_\alpha^{-1/(p-1)} \in L^1_{loc}(\Omega) \text{ for } |\alpha| \leq k \tag{4}$$

the linear set $W^{k,p}(\Omega;S)$ is a Banach space with respect to the norm $\|\cdot\|_{k,p,S}$ defined by (2), and that under the additional condition

$$w_\alpha \in L^1_{loc}(\Omega) \text{ for } |\alpha| \leq k \tag{5}$$

the linear set $W_0^{k,p}(\Omega;S)$ is a Banach space with respect to the same norm.

This result can be found in [1], where it is also shown how to modify the definition of the weighted spaces if (4) or (5) are not fulfilled.

3. The aim of this lecture is to show *how* (and *why*) the weighted Sobolev spaces introduced in section 1 can be used for solving boundary value problems for *nonlinear* partial differential operators. The case of *linear* operators of elliptic, degenerate-elliptic and singular-elliptic types was dealt with in the author's communication at the conference Equadiff 6 (Brno 1985) (see [2]; see also [3]). Here, we try to concentrate on nonlinear analogues of the above-mentioned linear problems and to give a survey of results obtained by the author and his colleagues, to extend these results and to describe a modified approach to nonlinear boundary value problems.

4. THE DIFFERENTIAL OPERATOR AND THE WEAK FORMULATION OF THE BOUNDARY VALUE PROBLEM: Let us consider a nonlinear differential operator of order 2k of the form

$$(Au)(x) = \sum_{|\alpha|\leq k} (-1)^{|\alpha|} D^{\alpha} a_{\alpha}(x;\delta_k u(x)) \tag{6}$$

where $\delta_k u(x) = \{D^{\beta}u(x);\ |\beta| \leq k\}$. We suppose that the 'coefficients' $a_{\alpha} = a_{\alpha}(x;\xi)$ of the operator A are defined on $\Omega \times \mathbb{R}^m$ ($x \in \Omega \subset \mathbb{R}^N$, $\xi = \{\xi_{\beta};\ |\beta| \leq k\} \in \mathbb{R}^m$, $m = (N + k)!/(N!k!)$) and fulfil the *Carathéodory conditions* (see, e.g. [4], section 12.2). Together with the operator A, let us consider the form

$$a(u,v) = \sum_{|\alpha|\leq k} \int_{\Omega} a_{\alpha}(x;\delta_k u(x))\ D^{\alpha}v(x)\ dx. \tag{7}$$

Let us suppose that the form a(u,v) is defined on the Cartesian product $V_1 \times V_2$ of two Banach spaces V_1, V_2. In a little rougher form, the *weak formulation* of a boundary value problem for the operator A can be expressed as follows: Given a function $u_0 \in V_1$ (which 'covers' the boundary conditions in our boundary value problem of orders $k_i \leq k - 1$) and a functional $F \in V_1^*$ (which 'covers' the remaining boundary conditions as well as the right-hand sides in the equation and in the boundary conditions), we call the function $u \in V_1$ a *weak solution* (of the boundary value problem considered) if

$$
\begin{aligned}
&u - u_0 \in V_1, \\
&a(u,v) = \langle F,v\rangle \text{ for every } v \in V_2,
\end{aligned} \tag{8}
$$

(where $\langle F,v\rangle$ is the duality relation in V_2).

5. REMARK: Here we will deal with the case when V_1, V_2 are suitable *weighted* spaces. These spaces are a useful tool for deriving assertions about the existence of a weak solution especially in the following cases:

(i) if the 'geometry' of Ω is unsatisfactory (the boundary $\partial\Omega$ of Ω contains corner points, edges, etc.);

(ii) if the ellipticity of the operator A is violated (i.e. if degenerating operators or operators with singular coefficients appear);

(iii) if the right-hand sides of the equation and of the boundary value problem are 'badly behaved'.

We will not deal with the case (i); the influence of the nonsmoothness of the domain Ω is described in detail in [5] and in [3], part I. As concerns the cases (ii) and (iii), let us illustrate the situation by two examples.

6. EXAMPLE (DEGENERATE OR SINGULAR ELLIPTIC OPERATOR): Let us consider the second-order differential operator

$$
(A_S u)(x) = - \sum_{i=1}^{N} \frac{\partial}{\partial x_i} [w_i(x) \left|\frac{\partial u}{\partial x_i}\right|^{p-1} \operatorname{sgn} \frac{\partial u}{\partial x_i}] + w_0(x)|u|^{p-1} \operatorname{sgn} u. \tag{9}
$$

This is in fact the operator A from (6) with the coefficients $a_i(x;\xi) = a_i(x;\xi_0,\xi_1,\ldots,\xi_N)$ given by the formulae

$$
a_i(x;\xi) = w_i(x)|\xi_i|^{p-1} \operatorname{sgn} \xi_i, \; i = 0,1,\ldots,N. \tag{10}
$$

The corresponding form a(u,v) from (7) is now given by

$$
a_S(u,v) = \sum_{i=0}^{N} \int_\Omega w_i(x) \; |D^i u|^{p-1} \operatorname{sgn} D^i u \; D^i v \, dx \tag{11}
$$

(where $D^i u = \partial u/\partial x_i$ for $i = 1,\ldots,N$, $D^0 u = u$).

A straightforward application of the Hölder inequality shows that the form $a_S(u,v)$ from (11) is defined on $W^{1,p}(\Omega;S) \times W^{1,p}(\Omega;S)$ with $S = \{w_0,w_1,\ldots,w_N\}$. Indeed, the typical summand in (11) can be estimated (for $w_i \geqq 0$) as follows:

$$|J_i| = \left|\int_\Omega w_i \; |D^i u|^{p-1} \operatorname{sgn} D^i u \; D^i v \; dx\right|$$

$$\leqq \int_\Omega |D^i u|^{p-1} \; w_i^{(p-1)/p} \; |D^i v| \; w_i^{1/p} \; dx$$

$$\leqq \left(\int_\Omega |D^i u|^p \; w_i \; dx\right)^{(p-1)/p} \left(\int_\Omega |D^i v|^p \; w_i \; dx\right)^{1/p}$$

$$\leqq \quad \|u\|^{p-1}_{1,p,S} \cdot \|v\|_{1,p,S}.$$

Here, the situation from section 4 arises with $V_1 = V_2 = W^{1,p}(\Omega;S)$. Therefore, the concept of a weak solution in $W^{1,p}(\Omega;S)$ can be introduced; it is meaningful and becomes a *natural* tool if one of the following two cases occurs:

(i) $w_i(x) \geqq 0$ (*degenerating* coefficients),

(ii) $w_i(x)$ unbounded (*singular* coefficients).

Then the weighted space $W^{1,p}(\Omega;S)$ is the *necessary* and *best possible* space.

The approach just mentioned, i.e. the investigation of degenerate or singular elliptic operators, has three characteristic features:

(a) the weight $S = \{w_0,w_1,\ldots,w_N\}$ is determined directly by the operator A_S (i.e. by its coefficients);

(b) the weight functions w_i are rather general (no requirements concerning their special form are given);

(c) the degeneration ($w_i(x) \to 0+$ for $x \to x_0$) as well as the singularity ($w_i(x) \to \infty$ for $x \to x_0$) can appear at any point $x_0 \in \bar{\Omega}$ (i.e. inside Ω as well as on the boundary $\partial\Omega$); moreover, *both* situations can appear at the *same* point x_0.

These features are characteristic not only for the example mentioned, but even for general operators; see sections 9 and 10.

7. EXAMPLE (STRONGLY ELLIPTIC OPERATOR): Let us consider the second-order differential operator

$$A_1 u = - \sum_{i=1}^{N} \frac{\partial}{\partial x_i} \left(\left| \frac{\partial u}{\partial x_i} \right|^{p-1} \operatorname{sgn} \frac{\partial u}{\partial x_i} \right) + |u|^{p-1} \operatorname{sgn} u \tag{12}$$

which is in fact the operator A_S from (9) with $S = \{1,1,\dots,1\}$. According to the ideas developed in Example 6, the necessary and best possible space should be the *nonweighted* (= classical) Sobolev space $W^{1,p}(\Omega)$. This, of course, requires some additional assumptions concerning the right-hand sides: If we deal, for simplicity, with the *Dirichlet problem*

$$A_1 u = f \text{ in } \Omega, \; u = g \text{ on } \partial\Omega , \tag{13}$$

then we need that $g \in W^{1-1/p,p}(\partial\Omega)$ and $f \in [W_0^{1,p}(\Omega)]^*$. If one of these two conditions is violated, the space $W^{1,p}(\Omega)$ cannot be used.

Therefore, it is natural to ask whether some weighted space could be used. Investigating from this point of view the form corresponding to the operator A_1 from (12),

$$a_1(u,v) = \sum_{i=0}^{N} \int_\Omega |D^i u|^{p-1} \operatorname{sgn} D^i u \, D^i v \, dx \; (= \sum_{i=0}^{N} J_i), \tag{14}$$

we easily obtain that the form $a_1(u,v)$ from (14) is defined on $V_1 \times V_2$ where $V_1 = W^{1,p}(\Omega;S)$ and $V_2 = W^{1,p}(\Omega;S^{1-p})$. Here, $S = \{w_0, w_1, \dots, w_N\}$ is an (arbitrarily chosen) set of weight functions and $S^{1-p} = \{w_0^{1-p}, w_1^{1-p}, \dots, w_N^{1-p}\}$. Indeed, the Hölder inequality implies that

$$\begin{aligned} |J_i| &\leq \int_\Omega |D^i u|^{p-1} w_i^{(p-1)/p} \, |D^i v| \, w_i^{(1-p)/p} \, dx \\ &\leq \left(\int_\Omega |D^i u|^p w_i \, dx \right)^{(p-1)/p} \left(\int_\Omega |D^i v|^p w_i^{1-p} \, dx \right)^{1/p} \\ &\leq \|u\|_{1,p,S}^{p-1} \cdot \|v\|_{1,p,S^{1-p}} . \end{aligned}$$

Consequently, we can again introduce the concept of a weak solution in $V_1 = W^{1,p}(\Omega;S)$, taking now the 'test' functions from $V_2 = W^{1,p}(\Omega;S^{1-p})$, V_2 being in general different from V_1.

The characteristic features of the approach just mentioned, i.e. the investigation of strongly elliptic operators, are (not only for our special operator A, but for general operators as well) as follows:

(a) *two* (in general different) Banach spaces V_1, V_2 are used, which are mutually connected by means of the corresponding weight functions: w_i appear in V_1, w_i^{1-p} in V_2;

(b) the weight $S = \{w_0, w_1, \ldots, w_N\}$ is *not* determined by the operator A but by the right-hand sides of the boundary problem (in the case of the Dirichlet problem (13), the weight S is determined by the function f and g).

8. REMARK: In the foregoing examples, we have been concerned with (weak) *formulation* of the boundary value problem. From this point of view, in Example 7 it was possible to choose rather general weight functions. In order to obtain *existence results*, we have to restrict the class of admissible weight functions substantially, at least for the case of strongly elliptic operators as in Example 7; see section 13.

Now, let us consider the general operator A from (6); for simplicity, we will deal with the Dirichlet problem.

9. DEGENERATE OR SINGULAR ELLIPTIC OPERATORS: Let us consider the differential operator A from (6) and suppose that its 'coefficients' $a_\alpha = a_\alpha(x;\xi)$ fulfil the following *weighted growth conditions*, where $w_\alpha = w_\alpha(x)$, $|\alpha| \leq k$, are given weight functions:

$$|a_\alpha(x;\xi)| \leq w_\alpha^{1/p}(x)[g_\alpha(x) + c_\alpha \sum_{|\beta|\leq k} |\xi_\beta|^{p-1} w_\beta^{1/p'}(x)], \quad |\alpha| \leq k, \tag{15}$$

where $g_\alpha \in L^{p'}(\Omega)$, $c_\alpha \geq 0$ are given functions and constants, $p' = p/(p-1)$.

Conditions (15) guarantee that the form $a(u,v)$ from (7) is defined on

$$W^{k,p}(\Omega;S) \times W^{k,p}(\Omega;S), \quad S = \{w_\alpha; |\alpha| \leq k\}$$

and induces a continuous operator $\tilde{T} : W^{k,p}(\Omega;S) \to [W^{k,p}(\Omega;S)]^*$ via the formula

$$a(u,v) = \langle \tilde{T}u, v\rangle. \tag{16}$$

Indeed, it follows from (15) (similarly as in Example 6) that

$$|a(u,v)| \leqq c(1 + \|u\|_{k,p,S})^{p-1} \|v\|_{k,p,S}.$$

The function $u \in W^{k,p}(\Omega;S)$ is called a *weak solution of the Dirichlet problem*

$$Au = f \text{ in } \Omega,\ D^\beta u = f_\beta \text{ on } \partial\Omega,\ |\beta| \leqq k - 1 \tag{17}$$

(where $f \in [W_0^{k,p}(\Omega;S)]^*$ and f_β are such that there is a function $u_0 \in W^{k,p}(\Omega;S)$ with $f_\beta = D^\beta u_0$ in the sense of traces on $\partial\Omega$) if

$$\begin{aligned} &u - u_0 \in W_0^{k,p}(\Omega;S), \\ &a(u,v) = \langle f,v\rangle \text{ for every } v \in W_0^{k,p}(\Omega;S). \end{aligned} \tag{18}$$

In other words, to solve the Dirichlet problem (17) (in the weak sense) means to solve the operator equation

$$Tu = f \tag{19}$$

where $T : V \to V^*$ with $V = W_0^{k,p}(\Omega;S)$ is given by the formula $Tu = \tilde{T}(u + u_0)$; see (16).

10. <u>EXISTENCE THEOREM</u>: Let us suppose that $f \in [W_0^{k,p}(\Omega;S)]^*$ and $u_0 \in W^{k,p}(\Omega;S)$ are given and that the operator A from (6) fulfils the growth conditions (15) and, moreover, the monotonicity condition

$$\sum_{|\alpha|\leqq k} [a_\alpha(x;\xi) - a_\alpha(x;\eta)](\xi_\alpha - \eta_\alpha) \geqq 0,\ \xi, \eta \in \mathbf{R}^m, \tag{20}$$

and the coercivity condition

$$\sum_{|\alpha|\leqq k} a_\alpha(x;\xi)\xi_\alpha \geqq c_0 \sum_{|\alpha|\leqq k} |\xi_\alpha|^p w_\alpha(x). \tag{21}$$

Then there exists at least one weak solution $u \in W^{k,p}(\Omega;S)$ of the Dirichlet problem (17) (in the sense of (18)). If the inequality in (20) is strict for $\xi \neq \eta$, then the weak solution u is uniquely determined.

11. REMARK: The proof of the foregoing theorem is based on the usual procedure of the theory of monotone operators: under the assumptions (15), (20) and (21), the operator T from (19) has all necessary properties. Theorem 10 can be easily extended to other boundary value problems and the assumptions can be substantially weakened. For details, see [6] or [3].

12. STRONGLY ELLIPTIC OPERATORS: Let us again consider the operator A from (6), but instead of (15) let us assume that the following (nonweighted) growth conditions are fulfilled:

$$|a_\alpha(x,\xi)| \leq g_\alpha(x) + c_\alpha \sum_{|\beta|\leq k} |\xi_\beta|^{p-1}, \quad |\alpha| \leq k. \tag{22}$$

Analogously as in Example 7 it can be shown that - given a set $S = \{w_\alpha; |\alpha| \leq k\}$ of weight functions - the form a(u,v) is defined on

$$W^{k,p}(\Omega;S) \times W^{k,p}(\Omega;S^{1-p})$$

with $S^{1-p} = \{w_\alpha^{1-p} ; \ |\alpha| \leq k\}$ and induces a continuous operator $\tilde{T}:W^{k,p}(\Omega;S) \to [W^{k,p}(\Omega;S^{1-p})]^*$ via the formula (16) (where, of course, the brackets $\langle\cdot,\cdot\rangle$ now express the duality in $W^{k,p}(\Omega;S^{1-p})$).

The function $u \in W^{k,p}(\Omega;S)$ is called a *weak solution of the Dirichlet problem* (17) (where now $f \in [W_0^{k,p}(\Omega;S^{1-p})]^*$ and f_β are such that there is a function $u_0 \in W^{k,p}(\Omega;S)$ with $f_\beta = D^\beta u_0$ in the sense of traces on $\partial\Omega$) if

$$\begin{aligned} &u - u_0 \in W_0^{k,p}(\Omega;S), \\ &a(u,v) = \langle f,v\rangle \text{ for every } v \in W_0^{k,p}(\Omega;S^{1-p}). \end{aligned} \tag{23}$$

In other words, we again have to solve the operator equation (19) with $Tu = \tilde{T}(u + u_0)$, but now $T : V_1 \to V_2^*$ with $V_1 = W_0^{k,p}(\Omega;S)$ and $V_2 = W_0^{k,p}(\Omega;S^{1-p})$.

13. <u>SPECIAL WEIGHT FUNCTIONS</u>: Up to now, satisfactory results about the existence of a weak solution of the just defined Dirichlet problem are known only for weighted spaces with special weight functions

$$w_\alpha(x) = [\text{dist}\,(x,M)]^\varepsilon, \quad M \subset \partial\Omega, \ \varepsilon \in \mathbb{R} \tag{24}$$

for all α, $|\alpha| \leq k$. In this special case, we denote the corresponding weighted Sobolev space $W^{k,p}(\Omega;S)$ by

$$W^{k,p}(\Omega;M,\varepsilon), \tag{25}$$

so that $W^{k,p}(\Omega;S^{1-p}) = W^{k,p}(\Omega;\,M,\ \varepsilon(1-p))$.

For these special weighted spaces we have

14. <u>EXISTENCE THEOREM</u>: Let us suppose that $f \in [W_0^{k,p}(\Omega;M,\ \varepsilon(1-p))]^*$ and $u_0 \in W^{k,p}(\Omega;M,\varepsilon)$ are given and that the operator A from (6) fulfils the growth condition (22), the monotonicity condition (20) and the (nonweighted) coercivity condition

$$\sum_{|\alpha|\leq k} a_\alpha(x;\xi)\xi_\alpha \geq c_0 \sum_{|\alpha|\leq k} |\xi_\alpha|^p. \tag{26}$$

Then there is a nonempty interval $I = (-c,\tilde{c})$, c, $\tilde{c} > 0$, such that for $\varepsilon \in I$ there exists at least one weak solution $u \in W^{k,p}(\Omega;M,\varepsilon)$ of the Dirichlet problem (17) (in the sense of (23)).

15. <u>REMARK</u>: (i) Let us point out that Theorem 14 states among other things the *existence of an interval* I *of admissible powers* ε in (24). The size of this interval depends on the data of our boundary value problem and also on the set $M \subset \partial\Omega$, from which the distance in (24) is measured. If we take $\varepsilon = 0$ (zero is also an admissible value), we are working with classical (nonweighted) Sobolev spaces $W^{k,p}(\Omega)$. Consequently, our result contains the classical existence results for weak solutions of nonlinear equations via the theory of monotone operators (see, e.g. [7]).

(ii) Theorem 14 was proved by Voldřich [8] for the case $M = \partial\Omega$, $k = 1$, via a rather sophisticated modification of the method of pseudomonotone operators. The generalization to general sets M and to higher-order equations will appear shortly in a paper by Voldřich and Kufner.

(iii) We concentrated on the Dirichlet problem, since in this case we are working with the space $W_0^{k,p}(\Omega;M,\varepsilon)$. An extension of the results to other boundary value problems, e.g. to the Neumann problem, where the space $W^{k,p}(\Omega;M,\varepsilon)$ is needed, is possible, but similar problems arise as in the case of linear operators (see [9], or [3], section 12).

16. A MODIFIED APPROACH: In the case of degenerate or singular elliptic operators A the weight functions w_α appear in the corresponding form a(u,v) in a *natural way* and the form a(u,v) is considered on the product $V \times V$. In the case of strongly elliptic operators, the weight functions w are introduced into a(u,v) artificially (see Example 7) and the form a(u,v) is considered on a pair of Banach spaces $V_1 \times V_2$ with $V_1 \neq V_2$. This fact also causes difficulties in deriving existence results: we do not know a suitable analogue of the theory of monotone operators for equations of the type (19) with $T : V_1 \to V_2^*$.

Now, on a simple example, we will shortly describe another approach that can be used for strongly elliptic operators. It is characterized by an artificial introduction of the weight function into the boundary problem but allows one to consider the corresponding form a(u,v) on the product $V \times V$, i.e. to work with *one space only*. By this approach, some of the problems arising when we consider strongly elliptic operators can be avoided.

Its substance is a modified concept of the weak solution similar to that introduced for *linear* operators in [10]. We define the so-called w-*weak solution* of our boundary value problem, where w is a (for the moment arbitrary) weight function.

Let A be a (strongly elliptic) differential operator of the form (6) and let a(u,v) be the corresponding form given by (7). Let us introduce a new form $a_w(u,v)$ by the formula

$$a_w(u,v) = a(u,wv), \tag{27}$$

i.e.

$$a_w(u,v) = \sum_{|\alpha|\leq k} \int_\Omega a_\alpha(x;\delta_k u(x))\, D^\alpha(wv)\, dx. \tag{28}$$

It can be shown that under certain additional assumptions concerning the weight function w the form $a_w(u,v)$ is defined on $V \times V$, where V is an appropriate subspace of the weighted Sobolev space $W^{k,p}(\Omega;S)$ with the special weight $S = \{w,w,\dots,w\}$; we will denote this space by

$$W^{k,p}(\Omega;w). \tag{29}$$

A function $u \in V$ is called a w-*weak solution of a certain boundary value problem for the* (nonlinear) *operator* A, if (similarly as in (8))

$$u - u_0 \in V,$$

$$a_w(u,v) = \langle F,v\rangle \text{ for every } v \in V$$

with given $u_0 \in V$ and $F \in V^*$.

Let us illustrate the situation with an example.

17. EXAMPLE (STRONGLY ELLIPTIC OPERATOR): Let us again consider the simple strongly elliptic second-order operator A_1 from (12). Assuming that the weight w is differentiable, in view of formula (14) we have

$$a_w(u,v) = a_1(u,wv) = \sum_{i=0}^{N} \int_\Omega |D^i u|^{p-1} \operatorname{sgn} D^i u \, D^i(wv)\, dx = K_1 + K_2 \tag{30}$$

where

$$\begin{aligned} K_1 &= \sum_{i=0}^{N} \int_\Omega |D^i u|^{p-1} \operatorname{sgn} D^i u \, D^i v \, w \, dx \ \left(= \sum_{i=0}^{N} J_{1i}\right), \\ K_2 &= \sum_{i=1}^{N} \int_\Omega |D^i u|^{p-1} \operatorname{sgn} D^i u \, v \, D^i w \, dx \ \left(= \sum_{i=1}^{N} J_{2i}\right). \end{aligned} \tag{31}$$

Similarly as in Example 6 we can show that

$$|K_1| \le (N + 1)\, \|u\|^{p-1}_{1,p,w} \|v\|_{1,p,w} \tag{32}$$

where $\|\cdot\|_{1,p,w}$ is the norm in the weighted Sobolev space $W^{1,p}(\Omega;w)$, i.e. in $W^{1,p}(\Omega;S)$ with $S = \{w,w,\dots,w\}$. Indeed, it follows from the Hölder inequality that

$$|J_{1i}| \leq \int_\Omega |D^i u|^{p-1} |D^i v| \; w \, dx = \int_\Omega |D^i u|^{p-1} \, w^{(p-1)/p} \, |D^i v| \; w^{1/p} \, dx$$

$$\leq \left(\int_\Omega |D^i u|^p \, w \, dx \right)^{(p-1)/p} \left(\int_\Omega |D^i v|^p \, w \, dx \right)^{1/p} \leq \|u\|^{p-1}_{1,p,w} \cdot \|v\|_{1,p,w} .$$

As concerns the term K_2, we have (again by the Hölder inequality)

$$|J_{2i}| \leq \int_\Omega |D^i u|^{p-1} \, |v| \, |D^i w| \, dx$$

$$= \int_\Omega |D^i u|^{p-1} \, w^{(p-1)/p} \, |v| \, |D^i w| \, w^{(1-p)/p} \, dx$$

$$\leq \left(\int_\Omega |D^i u|^p \, w \, dx \right)^{(p-1)/p} \left(\int_\Omega |v|^p \, \tilde{w}_i \, dx \right)^{1/p} \tag{33}$$

where

$$\tilde{w}_i = w^{1-p} \, |D^i w|^p. \tag{34}$$

Denoting by $\|\cdot\|_{p,\tilde{w}}$ the norm in the weighted Lebesgue space $L^p(\Omega;\tilde{w})$, i.e.

$$\|u\|_{p,\tilde{w}} = \left(\int_\Omega |u(x)|^p \, \tilde{w}(x) \, dx \right)^{1/p},$$

we have

$$|K_2| \leq \|u\|^{p-1}_{1,p,w} \sum_{i=1}^N \|v\|_{p,\tilde{w}_i} . \tag{35}$$

(i) Let us now assume that the weight function w has the following property, which will be called *property* (P): There exists a constant $c > 0$ such that for every $v \in W^{1,p}(\Omega;w)$, the inequality

$$\|v\|_{p,\tilde{w}_i} \leq c \, \|v\|_{1,p,w} \tag{36}$$

holds.

Then it follows from (35) that

$$|K_2| \leq cN \, \|u\|^{p-1}_{1,p,w} \, \|v\|_{1,p,w} \tag{37}$$

and from formulae (30), (32) and (37) we immediately have the following assertion:

(ii) If the weight function w has property (P), then the form $a_w(u,v)$ from (30) is defined on

$$W^{1,p}(\Omega;w) \times W^{1,p}(\Omega;w)$$

and induces a continuous operator $\tilde{T} : W^{1,p}(\Omega;w) \to [W^{1,p}(\Omega;w)]^*$ via the formula

$$a_w(u,v) = \langle\tilde{T}u,v\rangle.$$

18. CONCERNING WEIGHTS WITH PROPERTY (P): An assertion analogous to that from section 17(ii) holds not only for our special form $a_w(u,v)$ from (30), but for every form corresponding to a *second-order* operator of the type (6) whose 'coefficients' a_α fulfil the growth conditions (22). On the other hand, property (P) looks a little strange and without apparent purpose. Therefore, let us show that weights w with property (P) really exist.

(i) Let us suppose that there exists a constant $c > 0$ such that

$$|D^i w(x)| \leqq c w(x) \text{ for } x \in \Omega. \tag{38}$$

Then the weight w has property (P). Indeed, in view of (38) it follows from (34) that

$$|\tilde{w}_i| \leqq c^p w$$

and consequently

$$\|v\|_{p,\tilde{w}_i} \leqq c\,\|v\|_{p,w} \leqq c\,\|v\|_{1,p,w},$$

which is inequality (36).

(ii) Weight functions of the special form

$$w(x) = \exp(\varepsilon \operatorname{dist}(x,M)), \quad M \subset \partial\Omega \quad , \quad \varepsilon \in R \tag{39}$$

obviously fulfil condition (38) with $c = |\varepsilon|$ and, consequently, have property (P).

(iii) The special weight functions

$$w(x) = [\text{dist}\,(x,M)]^{\varepsilon}, \; M \subset \partial\Omega \;, \quad \varepsilon \in R \tag{40}$$

introduced in section 13 also have (at least for certain values of ε) property (P). Indeed, here we have

$$\tilde{w}_i = |D^i w|^p \, w^{1-p} \leqq |\varepsilon|^p [\text{dist}\,(x,M)]^{p(\varepsilon-1)+\varepsilon(1-p)}$$

$$= |\varepsilon| [\text{dist}\,(x,M)]^{\varepsilon-p}$$

and inequality (36) is a consequence of the imbedding

$$W^{1,p}(\Omega;M,\varepsilon) \hookrightarrow L^p(\Omega;M,\varepsilon - p)$$

which is described in detail in [11]. (We used the notation (25) for the spaces considered.)

19. <u>REMARK</u>: A weight w with property (P) enables us to give a *w-weak formulation* of the boundary value problem considered (see section 17(ii)). If we want to obtain *existence results*, we have to make further restrictive assumptions about the weight w (e.g. that the constants c appearing in (36) or (38) are 'small enough', or that the number $|\varepsilon|$ in the weights (39) or (40) is 'small'). We will not go into details here. We formulate only one special result, namely the existence of a w-weak solution of the *Dirichlet problem* for a *second-order* operator for the case when w is of the type (40) so that we are working with the spaces $W^{1,p}(\Omega;M,\varepsilon)$ and $W_0^{1,p}(\Omega;M,\varepsilon)$.

20. <u>THEOREM</u>: Let us consider the Dirichlet problem

$$Au = f \text{ in } \Omega \;, \quad u = g \text{ on } \partial\Omega \tag{41}$$

where

$$(Au)(x) = -\sum_{i=1}^{N} \frac{\partial}{\partial x_i} a_i(x;\, u(x),\, \nabla u(x)) + a_0(x;\, u(x),\, \nabla u(x)).$$

Assume that the functions $a_i(x;\xi) = a_i(x;\xi_0,\xi_1,\ldots,\xi_N)$ fulfil the Carathéodory condition, the growth conditions

$$|a_i(x;\xi)| \leqq g_i(x) + c_i \sum_{j=0}^{N} |\xi_j|^{p-1}, \quad i = 0,1,\ldots,N$$

with $g_i \in L^{p'}(\Omega)$ and $c_i \geqq 0$, the monotonicity condition

$$\sum_{i=0}^{N} [a_i(x;\xi) - a_i(x;\eta)](\xi_i - \eta_i) \geqq 0$$

and the coercivity condition

$$\sum_{i=0}^{N} a_i(x;\xi)\xi_i \geqq \tilde{c} \sum_{i=0}^{N} |\xi_i|^p.$$

Then there is a nonempty interval $I = (-c^*, c^{**})$, $c^*, c^{**} > 0$ such that for $\varepsilon \in I$, $f \in [W_0^{1,p}(\Omega;M,\varepsilon)]^*$, and for $u_0 \in W^{1,p}(\Omega;M,\varepsilon)$ such that $u_0 = g$ in the sense of traces on $\partial\Omega$ there exists at least one w-weak solution (with $w(x) = [\mathrm{dist}\,(x,M)]^\varepsilon$) $u \in W^{1,p}(\Omega;M,\varepsilon)$ of the Dirichlet problem (41) (i.e. at least one function $u \in W^{1,p}(\Omega;M,\varepsilon)$ such that $u - u_0 \in W_0^{1,p}(\Omega;M,\varepsilon)$ and

$$a_w(u,v) \equiv \sum_{i=0}^{N} \int_\Omega a_i(x;\, u(x),\, \nabla u(x))\, D^i(w(x)v(x))\, dx = \langle f,v\rangle$$

for every $v \in W_0^{1,p}(\Omega;M,\varepsilon)$).

21. <u>REMARK</u>: The proof of this theorem uses the same ideas as Voldřich's proof of the result mentioned in Remark 15(ii), i.e. a modification of the method of pseudomonotone operators. A paper of Voldřich and Kufner containing this proof as well as an extension to the case of other weight functions (not only of power type) will appear shortly.

Let us note that an extension to other boundary value problems and to higher-order equations is also possible. Of course, the very *formulation* of a w-weak solution for equations of order 2k makes it necessary to extend the conditions imposed on the weight function w. Among other things we have to assume that not only the function itself but also its derivatives up to order k-1 have property (P).

References

[1] A. Kufner and B. Opic, How to define reasonably weighted Sobolev spaces, Comment. Math. Univ. Carolin., 25:3 (1984), 537-554.

[2] A. Kufner, Boundary value problems in weighted spaces, Equadiff 6 (Proceedings of a conference), Lecture Notes in Math. no. 1192, Springer-Verlag (1986), pp. 35-48.

[3] A. Kufner and A.-M. Sändig, Some applications of weighted Sobolev spaces, Teubner-Texte zur Mathematik Bd. 100, Teubner (1987).

[4] S. Fučík and A. Kufner, Nonlinear Differential Equations, Elsevier (1980).

[5] P. Grisvard, Elliptic Problems in Nonsmooth Domains, Pitman (1985).

[6] A. Kufner and B. Opic, The Dirichlet problem and weighted spaces, I, II, Časopis Pěst. Mat., 108:4 (1983), 381-408; 111:3 (1986), 242-253.

[7] J.-L. Lions, Quelques Méthodes de Résolution des Problèmes aux Limites Non Linéaires, Dunod (1969).

[8] J. Voldřich, On the Dirichlet boundary value problem for nonlinear elliptic partial differential equations in Sobolev power weight spaces, Časopis Pěst. Mat., 110:3, (1985), 250-269.

[9] A. Kufner and J. Voldřich, The Neumann problem in weighted Sobolev spaces, C.R. Math. Rep. Acad. Sci. Canada, 7:4 (1985), 239-243.

[10] A. Kufner and J. Rákosník, Linear elliptic boundary value problems and weighted Sobolev spaces: A modified approach, Math. Slovaca, 34:2 (1984), 185-197.

[11] A. Kufner, Weighted Sobolev Spaces, Teubner (1980) (first edition), Wiley (1985) (second edition).

Alois Kufner
Mathematical Institute Academy of Sciences
Zitna 25
11567 Prague 1
Czechoslovakia

O. MARTIO

Harmonic measures for second order non-linear partial differential equations

1. INTRODUCTION

For the Laplace equation $\Delta u = 0$ the harmonic measure provides a useful tool to estimate other solutions. For a wide class of elliptic partial differential equations $\nabla\cdot A(x, \nabla u(x)) = 0$ there exists a similar solution with 0 - 1 boundary values. This solution, the A-harmonic measure, can be used as well to estimate other solutions. However, for nonlinear operators A the A-harmonic measure does not define a measure on the Borel subsets of the boundary.

Let G be a bounded domain in R^n, $\nabla\cdot A(x,\nabla u(x)) = 0$ an elliptic equation in G with $|A(x,h)| \approx |h|^{p-1}$, $p > 1$, and $\omega(E) = \omega(E,G;A)$ the A-harmonic measure of $E \subset \partial G$ in G. For the assumptions on A and the construction of $\omega(E)$ see section 2. The most interesting problem associated with $\omega(E)$ is to determine the class of subsets E of ∂G such that $\omega(E) = 0$. This class depends on A. However, in this paper we study the following problem:

PROBLEM: Characterize the class of subsets E in ∂G such that $\omega(E) = 0$ for all operators A.

Observe that we have not fixed the number $p > 1$.

A set $E \subset \partial G$ with $\omega(E) = 0$ for all A can be said to be of *total harmonic measure zero*, i.e. the set E does not carry non-zero solutions for any reasonable elliptic partial differential operators.

We give, in the case G = B, the unit ball, several sufficient conditions for a set $E \subset \partial B$ to be of total harmonic measure zero. In particular porous subsets of ∂B and subsets with self-similar structure have this property. A typical example is a Cantor 1/3-set constructed on the boundary of the unit disk.

2. OPERATORS A AND HARMONIC MEASURES

Suppose that G is a bounded domain in R^n and that $p > 1$. We shall study partial differential operators $A : G \times R^n \to R^n$ which satisfy the following assumptions:

(a) For each $\varepsilon > 0$ there exists a compact subset F of G such that $A|F \times \mathbb{R}^n$ is continuous and $m(G \setminus F) < \varepsilon$.

(b) There exist positive constants γ_1 and γ_2 such that for a.e. $x \in G$

$$|A(x,h)| \leq \gamma_1 |h|^{p-1} \tag{2.1}$$

$$A(x,h)\cdot h \geq \gamma_2 |h|^p \tag{2.2}$$

for all $h \in \mathbb{R}^n$.

(c) For a.e. $x \in G$

$$(A(x,h_1) - A(x,h_2))\cdot(h_1 - h_2) > 0,$$

$h_1 \neq h_2$.

(d) For a.e. $x \in G$

$$A(x,\lambda h) = |\lambda|^{p-2}\lambda A(x,h)$$

for $\lambda \in \mathbb{R}\setminus\{0\}$ and $h \in \mathbb{R}^n$.

A continuous function $u:G \to \mathbb{R}$ is a solution of the equation

$$\nabla\cdot A(x,\nabla u(x)) = 0 \tag{2.3}$$

if u belongs to the Sobolev space loc $W_p^1(G)$, i.e. u is ACL^p, and if

$$\int_G A(x,\nabla u(x))\cdot\nabla\phi(x)\,dm(x) = 0 \tag{2.4}$$

for all $\phi \in C_0^\infty(G)$. We call solutions of (2.3) A-*harmonic*. A lower semi-continuous function $u:G \to \mathbb{R} \cup \{\infty\}$ is A-*superharmonic* if it satisfies the A-comparison principle, i.e. if for every domain $D \subset\subset G$ and every A-harmonic function $h \in C(\bar{D})$ in D, $h \leq u$ in ∂D implies $h \leq u$ in D. These functions form a similar, but in general nonlinear, potential theory as ordinary harmonic and superharmonic functions; see [2] and [5].

Finally, let E be a subset of ∂G. The upper class $\mathcal{U}$ consists of all A-superharmonic functions $u:G \to \mathbb{R} \cup \{\infty\}$ such that

$$\varliminf_{x \to y} u(x) \geq \chi_E(y)$$

for each $y \in \partial G$. Here χ_E is the characteristic function of E. It can be shown that

$$\omega(E,G;A)(x) = \inf_{u \in \mathcal{U}} u(x), \quad x \in G,$$

defines an A-harmonic function $\omega = \omega(E,G;A)$, called the A-harmonic measure of E with respect to G. For this construction see [5] and [4]. The set E has zero A-harmonic measure, if $\omega(x) = 0$ for some $x \in G$, or equivalently $\omega(x) = 0$ for all $x \in G$. The last assertion follows from Harnack's inequality. In this case we simply write $\omega = 0$.

3. SETS OF TOTAL HARMONIC MEASURE ZERO ON B

Let $B = B(0,1)$ denote the unit ball in $\mathbb{R}^n$ and suppose that an operator A satisfying (a)-(d) is given in B. If E is a subset of ∂B, then for $x_1, x_2 \in \mathbb{R}^n \setminus E$ we let

$$k_E(x_1,x_2) = \inf_\gamma \int_\gamma d(x,E)^{-1}\, ds$$

denote the quasihyperbolic distance of x_1 and x_2 in $\mathbb{R}^n \setminus E$. Here the infimum is taken over all rectifiable curves γ joining x_1 and x_2 in $\mathbb{R}^n \setminus E$ and $d(x,E)$ is the distance from x to E. If no such curves exist, i.e. $E = \partial B$ and x_1 and x_2 belong to different components of ∂B, then we set $k_E(x_1,x_2) = \infty$; this case is of no interest to us. For the properties of the quasihyperbolic distance see [1].

In [7] the following sufficient condition for $\omega(E,B;A) = 0$ was given; note that the unit ball B satisfies a p-capacity density condition for all $p > 1$; see [7, section 3].

3.1. THEOREM: Suppose that E is a closed subset of ∂B such that there exist a sequence of neighborhoods U_i, $i = 1,2,\ldots$, of E and $M < \infty$ with

(a) $\cap \mathcal{U}_i \cap B = \emptyset$ and

(b) for each $i = 1,2,\ldots$ and $z \in \partial\mathcal{U}_i \cap B$ there is $y \in \partial B$ with $k_E(z,y) \leq M$.

Then $\omega(E,B;A) = 0$.

3.2. REMARKS: The proof of Theorem 3.1 is based on three facts; see [7]:

(i) The Harnack inequality for solutions of $\nabla\cdot A(x,\nabla u(x)) = 0$.

(ii) $\omega = 0$ if and only if $\sup_{x\in B}\omega(x) < 1$.

(iii) The boundary estimate for bounded solutions which are zero on an open set in ∂B.

Next we use Theorem 3.1 to derive sufficient conditions for total harmonic measure zero.

Let E be a subset of ∂B. For $r > 0$ write $E(r) = E + B(0,r)$. Then $E(r)$ is an open set, the r-inflation of E. Define

$$\delta(r) = \sup_{x\in E} d(x,\partial B \setminus E(r)). \tag{3.1}$$

3.3 THEOREM: Suppose that $E \subset \partial B$. If

$$\liminf_{r\to 0} \delta(r)/r < \infty, \tag{3.2}$$

then E is of total harmonic measure zero.

PROOF: Note that $E(r) = \bar{E}(r)$; hence we may assume that E is closed. By (3.2) there is $C < \infty$ and a sequence $r_i \to 0$ such that $\delta(r_i) \leq Cr_i$. We may assume that $r_i \leq 1/2$, $i = 1,2,\ldots$. Set $\mathcal{U}_i = E + B(0,r_i)$. Then $\mathcal{U}_i$ is a neighborhood of E and $\cap\mathcal{U}_i \cap B = \emptyset$.

Next let $z \in \partial\mathcal{U}_i \cap B$. Then $z \in \partial B(w,r_i)$ for some $w \in E$. Now there is $y \in \partial B\setminus E(r_i)$ such that

$$|w - y| \leq \delta(r_i) \leq Cr_i.$$

Let $\gamma = \gamma_1 \cup \gamma_2 \cup \gamma_3$ where

$$\gamma_1 = \{ty : 1 - r_i \leqq t \leqq 1\},$$

$$\gamma_3 = \{tz : (1-r_i)/|z| \leqq t \leqq 1\}$$

are radial segments and γ_2 is a circular arc in $\partial B(0,1 - r_i)$ joining $(1 - r_i)y$ to $(1 - r_i)z/|z|$. Then γ joins to z in $\bar{B} \setminus E$. Since $d(y,E) \geqq r_i$ and $d(z,E) \geqq r_i$, we have

$$d(x,E) \geqq r_i, \; x \in \gamma_1 \cup \gamma_3;$$

also $d(x,E) \geqq r_i$ for each $x \in \gamma_2$, hence we obtain

$$\begin{aligned} k_E(y,z) &\leqq \int_\gamma \frac{ds}{d(x,E)} = \int_{\gamma_1} + \int_{\gamma_2} + \int_{\gamma_3} \\ &\leqq (\ell(\gamma_1) + \ell(\gamma_2) + \ell(\gamma_3))/r_i \\ &\leqq (r_i + \tfrac{\pi}{2}(1 - r_i)|y - z/|z|| + r_i)/r_i \\ &\leqq 2 + \pi|y - z/|z||/(2r_i). \end{aligned}$$

On the other hand

$$\begin{aligned} |y - z/|z|| &\leqq |y - z| + |z - z/|z|| \\ &\leqq |y - w| + |w - z| + r_i \\ &\leqq Cr_i + r_i + r_i = (C + 2)r_i \end{aligned}$$

and this yields

$$k_E(y,z) \leqq 2 + \pi(C + 2)/2 = M.$$

The result now follows from Theorem 3.1. □

Let $E \subset A$ be subsets of R^n. The set E is *porous* in A if there are positive constants c and r_0 such that for each $x \in E$ and $0 < r \leqq r_0$ there is $y \in A$ with $|x - y| \leqq r$ and $B(y,cr) \cap E = \emptyset$.

3.4. THEOREM: If E is porous in ∂B, then E is of total harmonic measure zero.

PROOF: Let δ be as in (3.1). Fix $r > 0$ such that $r/c \leqq r_0$ where c and r_0 are the porosity constants of E. Let $x \in E$. Since E is porous in ∂B, there is $y \in \partial B$ with $|x - y| \leqq r/c$ and $B(y,r) \cap E = \emptyset$. Now $y \in \partial B \setminus E(r)$, hence

$$d(x, \partial B \setminus E(r)) \leqq |x - y| \leqq r/c.$$

Thus $\delta(r) \leqq r/c$ and we obtain

$$\liminf_{r \to 0} \frac{\delta(r)}{r} \leqq 1/c.$$

The result follows from Theorem 3.3. □

3.5. REMARKS: (a) Theorem 3.4 was proved in [3] for variational integrals in the case $p = n$.

(b) If E is porous in ∂B, then $\dim_H E < n - 1$ where $\dim_H$ is the Hausdorff dimension, see e.g. [3]. It is interesting to note that sets E of total harmonic measure zero cannot be characterized using the Hausdorff dimension. A recent example of Tukia [8] can be used to construct for each $\varepsilon > 0$ a linear operator A satisfying for $p = 2$ the assumptions (a)-(d) in the unit disk and a closed set E with $\dim_H E < \varepsilon$ and $\omega(E,B;A) > 0$. Observe that, by Theorem 3.4, a Cantor 1/3-set in ∂B has the A-harmonic measure zero for this operator A because it has total harmonic measure zero.

4. SELF-SIMILAR FRACTALS AND A-HARMONIC MEASURE

Let $F = \{S_1, \ldots, S_N\}$ be a finite set of similarities of R^n with

$$|S_i(x) - S_i(y)| = r_i |x - y|,$$

$0 < r_i < 1$. A compact set K in R^n is called a *self-similar fractal* associated with F if

$$K = \bigcup_{i=1}^{N} S_i(K),$$

see [10]. The self-similar fractal K is said to satisfy the *open set condition* [10, p. 735], if there exists a non-empty open set 0 such that

$$\cup S_i(0) \subset 0, \quad S_j(0) \cap S_i(0) = \emptyset, \quad i \neq j.$$

We shall consider self-similar fractals in $\mathbb{R}^{n-1}$, which we identify with $\{x \in \mathbb{R}^n : x_n = 0\}$.

Suppose that $A \subset \mathbb{R}^n$ and $f: A \to \mathbb{R}^n$ is a non-constant mapping. The mapping f is called *quasisymmetric* if there is a homeomorphism $\eta:[0,\infty) \to [0,\infty)$ such that $|a - x| \leq t|b - x|$ implies $|f(a) - f(x)| \leq \eta(t)|f(b) - f(x)|$ for all a, b, x in A and for all $t > 0$. For example, if f is L-bi-Lipschitz, that is,

$$|x - y|/L \leq |f(x) - f(y)| \leq L|x - y|$$

for all $x,y \in A$, then f is quasisymmetric with $\eta(t) = L^2 t$.

4.1. <u>THEOREM</u>: Suppose that K is a self-similar fractal in $\mathbb{R}^{n-1}$ satisfying the open set condition. If $\dim_H K < n - 1$, then for every quasisymmetric mapping f of K into ∂B, $E = f(K)$ is of total harmonic measure zero with respect to the unit ball B.

<u>PROOF</u>: We first show that K is porous in $\mathbb{R}^{n-1}$. Let 0 be an open set in $\mathbb{R}^{n-1}$ given by the open set condition for K. Since $\dim_H K < n-1$, $0 \setminus K$ is an open, non-empty set and we choose an open ball $B(z,\varepsilon) \subset 0 \setminus K$. By [10, (3)(iii), p. 736] for each $i_1,\ldots,i_p \in \{1,\ldots,N\}$

$$K \cap S_{i_1 \ldots i_p}(B(z,\varepsilon)) = \emptyset. \tag{4.1}$$

Here $S_{i_1 \ldots i_p} = S_{i_1} \circ \ldots \circ S_{i_p}$ and $S_1,\ldots,S_N$ are the similarities associated with K; for $A \subset \mathbb{R}^{n-1}$ we also write $A_{i_1 \ldots i_p} = S_{i_1 \ldots i_p}(A)$.

Let $M = \mathrm{diam}(K \cup \{z\})$, $\theta = 1 + 2M/\mathrm{diam}(K)$ and $r_0 = \mathrm{diam}(K)/2$. Fix $x \in K$

and $0 < r \leqq r_0$. Now $\{x\} = \bigcap_{p=1}^{\infty} K_{i_1 \ldots i_p}$, see [10, 3.1(iii), p. 724], and choose the least p such that $K_{i_1,\ldots,i_p} \subset B(x,r/\theta)$. Write $s = r_{i_1} \ldots r_{i_p}$ where r_i is the similarity constant of S_i (we assume $r_1 \leqq r_2 \leqq \ldots \leqq r_N$) and let $S = S_{i_1 \ldots i_p}$. Then

$$s \operatorname{diam} K \geqq r_1 r/\theta \tag{4.2}$$

and, on the other hand,

$$s \operatorname{diam} K \leqq 2r/\theta. \tag{4.3}$$

Next, since s is the similarity constant of S, we obtain

$$|S(x) - S(z)| \leqq s|x - z| \leqq sM$$

and (4.3) implies

$$\begin{aligned} |x - S(z)| &\leqq |x - S(x)| + |S(x) - S(z)| \\ &\leqq r/\theta + sM \\ &\leqq r/\theta + 2r\, M/(\theta \operatorname{diam} K) \\ &= r\theta^{-1}(1 + 2M/\operatorname{diam} K) = r. \end{aligned}$$

This inequality means that the centre $S(z)$ of the ball $S(B(z,\varepsilon)) = B(S(z),s\varepsilon)$ lies in $\bar{B}(x,r)$. By (4.2)

$$s\varepsilon \geqq r(\varepsilon r_1/\theta \operatorname{diam} K)$$

and, by (4.1), $B(S(z),s\varepsilon)$ does not meet K. Thus K is porous in $\mathbb{R}^{n-1}$ with constants

$$c = \varepsilon r_1/\theta \operatorname{diam} K, \quad r_0 = \operatorname{diam} K/2.$$

To prove that E = f(K) is of total harmonic measure zero we use a result of Väisälä [9]: If f is a quasisymmetric mapping of a porous set K in R^{n-1} into ∂B, then E = f(K) is porous in ∂B. Actually, Väisälä's result was formulated for quasisymmetric mappings into R^{n-1} but a simple application of the stereographic projection takes care of our case. Theorem 3.4 now completes the proof. □

References

[1] F.W. Gehring and B. Palka, Quasiconformally homogeneous domains, J. Analyse Math., 30 (1976), 172-199.

[2] S. Granlund, P. Lindqvist and O. Martio, Conformally invariant variational integrals, Trans. Amer. Math. Soc., 277 (1983); 43-73.

[3] S. Granlund, P. Lindqvist and O. Martio, F-harmonic measure in space, Ann. Acad. Sci. Fenn. A I Math., 7 (1982), 233-247.

[4] S. Granlund, P. Lindqvist and O. Martio, Note on the PWB-method in the non-linear case, Pacific J. Math., 125 (1986), 381-395.

[5] J. Heinonen and T. Kilpeläinen, A-superharmonic functions and supersolutions of degenerate elliptic equations, Ark. Mat., 26 (1988), 87-105.

[6] O. Martio, F-harmonic measures, quasihyperbolic distance and Milloux's problem, Ann. Acad. Sci. Fenn. A I Math., 12 (1987), 151-162.

[7] O. Martio, Sets of zero elliptic harmonic measures, Ann. Acad. Sci. Fenn. A I Math. (to appear)

[8] P. Tukia, Hausdorff dimension and quasisymmetric maps, Math. Scand. (to appear)

[9] J. Väisälä, Porous sets and quasisymmetric maps. Trans. Amer. Math. Soc., 229 (1987), 525-533.

[10] J. Hutchinson, Fractals and self similarity. Indiana Univ. Math. J. 30 (1981), 713-747.

O. Martio
Department of Mathematics
University of Jyväskylä
Seminaarinkatu 15
SF-40100 Jyväskylä
Finland

V. MUSTONEN

Examples and applications of mappings of monotone type

1. INTRODUCTION

Since the pioneering work of Minty in 1962 the theory of monotone nonlinear mappings from a real reflexive Banach space X into its dual space X* has been extensively generalized by Brezis, Browder, Hess and many other mathematicians. In its present form this theory makes possible an efficient treatment of existence problems for solutions of nonlinear elliptic and parabolic equations and variational inequalities. Also some hyperbolic equations can be tackled (see [3]). As prototypes of such problems we mention

$$(E)\quad \begin{cases} \sum_{|\alpha|\leq m} (-1)^{|\alpha|} D^{\alpha} A_{\alpha}(x,u(x),\ldots,D^m u(x)) = f(x),\ x \in \Omega \subset \mathbb{R}^n \\ \text{with boundary conditions of Dirichlet or Neumann type} \end{cases}$$

$$(P)\quad \begin{cases} \frac{\partial u(t,x)}{\partial t} + \sum_{|\alpha|\leq m} (-1)^{|\alpha|} D_x^{\alpha} A_{\alpha}(t,x,u(t,x),\ldots,D_x^m u(t,x)) = f(t,x), \\ (t,x) \in (0,T) \times \Omega, \text{ with initial boundary conditions} \end{cases}$$

$$(H)\quad \begin{cases} \frac{\partial^2 u(t,x)}{\partial t^2} - \frac{\partial^2 (t,x)}{\partial x^2} + g(t,x,u(t,x)) = f(t,x),\ (t,x) \in \mathbb{R} \times (0,\pi) \\ u(t,0) = u(t,\pi) = 0\ \forall t \in \mathbb{R} \text{ and} \\ u \text{ is } 2\pi\text{-periodic in } t \end{cases}$$

where the coefficient functions A_{α} in (E) and (P) and the function g in (H) satisfy, in addition to certain growth restrictions, some monotonicity conditions.

Previously the Galerkin method was the basic tool in producing existence results for the abstract equation $T(u) = h$, involving a mapping $T : X \to X^*$ of monotone type and the method has given the motivation for various generalizations of the class of monotone mappings. Most important of these

extensions are the concepts of pseudomonotone and quasimonotone mappings and mappings of class (S_+) and type (M).

More recently, the topological degree theory has been extended for some classes of mappings of monotone type. For two different constructions we refer to the papers [1] and [2,4], respectively. Topological degree is a good tool for obtaining information about solutions of the equation $T(u) = h$.

In this lecture we shall survey briefly the basic results on the theory of mappings of monotone type and illuminate the surjectivity results by some examples.

2. MAPPINGS OF MONOTONE TYPE AND A SURJECTIVITY THEOREM

We start with recalling the definitions of some classes of mappings of monotone type. Let X be a real reflexive Banach space, $\langle\cdot,\cdot\rangle$ the continuous pairing between X and X*. A mapping $T : X \to X^*$ is:

(i) monotone (denoted $T \in$ (MON)), if $\langle T(u)-T(v),u-v\rangle \geq 0$ for all $u, v \in X$;

(ii) quasimonotone ($T \in$ (QM)), if for any sequence $\{u_n\}$ in X with $u_n \rightharpoonup u$, $\limsup \langle T(u_n) - T(u), u_n - u\rangle \geq 0$:

(iii) pseudomonotone ($T \in$ (PM)), if for any sequence $\{u_n\}$ in X with $u_n \rightharpoonup u$ and $\limsup \langle T(u_n),u_n - u\rangle \leq 0$, it follows that $T(u_n) \rightharpoonup T(u)$ in X* and $\langle T(u_n),u_n\rangle \to \langle T(u),u\rangle$;

(iv) of class (S_+) ($T \in (S_+)$), if for any sequence $\{u_n\}$ in X with $u_n \rightharpoonup u$ and $\limsup \langle T(u_n),u_n - u\rangle \leq 0$, it follows that $u_n \to u$ in X;

(v) of type (M) ($T \in$ (M)), if for any sequence $\{u_n\}$ in X with $u_n \rightharpoonup u$, $T(u_n) \rightharpoonup f$ and $\limsup \langle T(u_n),u_n\rangle \leq \langle f,u\rangle$, it follows that $T(u) = f$;

(vi) bounded, if it takes bounded sets of X into bounded sets of X*:

(vii) demicontinuous, if $u_n \to u$ in X implies $T(u_n) \rightharpoonup T(u)$ in X*.

It is easy to see that the classes (MON), (S_+) and (QM) fit well together in the sense that $T_1 \in$ (QM) if and only if $T_1 + T_2 \in (S_+)$ for any $T_2 \in (S_+)$.

Moreover, the class of compact mappings (COMP) is contained in (QM) and, if we denote by (SM) the class of all demicontinuous mappings T with the property

$$\lim_{u_n \to u} \langle T(u_n), u_n - u \rangle = 0,$$

then (COMP) $\subset$ (SM) $\subset$ (QM). Assuming all mappings demicontinuous we have the following containments

$$\begin{array}{ccccc}
 & & (M) & & \\
 & & \uparrow & & \\
(S_+) & \rightarrow & (PM) & \rightarrow & (QM) \\
 & & \uparrow & & \uparrow \\
 & & (MON) & & (SM) \\
 & & & & \uparrow \\
 & & & & (COMP)
\end{array}$$

For more details we refer to [5].

A good classical unique topological degree theory can be constructed for the class (S_+) ([1], [4]). In a slightly weakened form it can be extended up to the class (QM) ([2]). Existence and surjectivity theorems can be established via a homotopy argument. It is enough to consider affine homotopies between the duality map J and the given mapping T from the class (S_+), (PM) or (QM), respectively. We recall here the following (see [5]).

<u>THEOREM 1</u>: Let $F : X \to X^*$ be a bounded demicontinuous quasimonotone mapping satisfying the conditions

(a) $\|F(u_n)\| \to \infty$ for any sequence $\{u_n\}$ in X with $\|u_n\| \to \infty$,

(b) there exists $R > 0$ such that

$$\frac{\langle F(u), u \rangle}{\|u\|} + \|F(u)\| > 0 \text{ for all } u \in X \text{ with } \|u\| \geq R.$$

Then $F(X)$ is dense in X^*, i.e. the equation $F(u) = f$ is almost solvable for any $f \in X^*$. In particular, if F is pseudomonotone, then $F(X) = X^*$, i.e. the equation $F(u) = f$ admits a solution u for any $f \in X^*$.

3. EXAMPLES AND APPLICATIONS

<u>3.1</u>. To indicate the meaning of the conditions (a) and (b) in Theorem 1 we

consider first mappings from ℓ_p to $(\ell_p)^* = \ell_{p'}$, with $1 < p < \infty$ and $p' = p/(p-1)$. Indeed, the duality mapping J is given by

$$J(u) = \begin{cases} 0, & \text{if } u = 0 \\ \|u\|^{2-p}(|\xi_j|^{p-1} \operatorname{sgn} \xi_j), & \text{if } u \neq 0 \end{cases}$$

where $u = (\xi_j)_{j=1}^{\infty} \in \ell_p$. Then J is strictly monotone, continuous, $J \in (S_+)$, $\|J(u)\| = \|u\|$ and $\langle J(u),u\rangle = \|u\|^2$. We denote by A the mapping

$$A(u) = \|u\|^{2-p}(0,0,|\xi_3|^{p-1} \operatorname{sgn} \xi_3, |\xi_4|^{p-1} \operatorname{sgn} \xi_4,\ldots)$$

and by B_α the mapping

$$B_\alpha(u) = \|u\|^{2-p}(\cos \alpha|\xi_1|^{p-1} \operatorname{sgn} \xi_1 - \xi_2 \sin \alpha, \xi_1 \sin \alpha + \\ + \cos \alpha|\xi_2|^{p-1} \operatorname{sgn} \xi_2,0,0,\ldots)$$

where $0 \leqq \alpha \leqq \pi$ for all $u = (\xi_j) \in \ell_p$. Then $F_\alpha = A + B_\alpha$ is a compactly perturbed duality map, hence $F \in (S_+)$. In fact, $F_0 = J$.

For $\alpha = \pi/2$ we have $\langle F_{\pi/2}(u),u\rangle \geqq 0$ for all $u \in \ell_p$ and $\|F_{\pi/2}(u)\| = \|u\|$. Hence the conditions (a) and (b) of Theorem 1 are satisfied.

For $\alpha = 3\pi/4$ it is not hard to see that $\|F_{3/4\pi}(u_n)\| \to \infty$ for any sequence $\{u_n\}$ with $\|u_n\| \to \infty$. Moreover, for $u_n = (\xi_1^{(n)}, \xi_2^{(n)},0,0,\ldots)$ we have

$$\frac{\langle F_{3\pi/4}(u_n),u_n\rangle}{\|u_n\|} \to -\infty \text{ as } \|u_n\| \to \infty .$$

On the other hand, since

$$F_{3\pi/4}(u) \neq -\lambda J(u) \text{ for all } \lambda > 0 \text{ and } u \neq 0$$

we can conclude that

$$\langle F_{3\pi/4}(u),u\rangle > -\|F_{3\pi/4}(u)\| \, \|u\| \text{ for all } u \neq 0.$$

Therefore $F_{3\pi/4}$ satisfies the conditions (a) and (b) of Theorem 1. In fact the conclusion holds for all α with $\pi/2 < \alpha < \pi$.

3.2. Secondly we consider partial differential equation of the form

$$- \sum_{i=1}^{N} D_i(g_i(D_i u(x))) + g_0(x,u(x),Du(x)) = f(x) \text{ in } \Omega \tag{1}$$

with Dirichlet or Neumann type boundary conditions, where Ω is a bounded open set in R^N ($N \geq 2$). We assume that the functions g_i ($i = 1,2,\ldots,N$) are continuous and nondecreasing and satisfy the growth condition

$$|g_i(x)| \leq C_1(|t|^{p-1} + 1) \quad \forall t \in R \tag{2}$$

for some constants $C_1 > 0$ and $1 < p < \infty$. The function $g_0 : \Omega \times R \times R^N \to R$ is a Caratheodory function satisfying

$$\begin{aligned}
&|g_0(x,\eta,\zeta)| \leq C_0(|\zeta|^{p/q'} + |\eta|^{q-1} + k_0(x)),\\
&\quad \text{where } C_0 > 0 \text{ and } 1 < q < Np/(N-p),\ k_0 \in L^{q'}(\Omega), \text{ if } p < N,\\
&\quad 1 < q < \infty \text{ arbitrary, if } p = N\\
&\quad \text{or } |g_0(x,\eta,\zeta)| \leq C_0(|\eta|)(|\zeta|^p + k_0(x)),\\
&\quad \text{where } C_0 \text{ is some continuous function: } R_+ \to R_+\\
&\quad \text{and } k_0 \in L^1(\Omega), \text{ if } p > N.
\end{aligned} \tag{3}$$

A *weak solution* of (1) is any u in V for which

$$\int_\Omega \sum_{i=1}^{N} g_i(D_i u)D_i v + \int_\Omega g_0(x,u,Du)v = \int_\Omega fv \quad \forall v \in V, \tag{4}$$

where $V = W_0^{1,p}(\Omega)$ for Dirichlet boundary conditions and $V = W^{1,p}(\Omega)$ for Neumann type boundary conditions. In view of the Sobolev embedding theorem the integrals in (4) are well-defined and bounded, if (2) and (3) are met, and if $f \in L^{p'}(\Omega)$.

Defining the mappings F_1 and $F_2 : V \to V^*$ by

$$\langle F_1(u),v\rangle = \int_\Omega \sum_{i=1}^{N} g_i(D_iu)D_iv \tag{5}$$

and

$$\langle F_2(u),v\rangle = \int_\Omega g_0(x,u,Du)v \tag{6}$$

for all u, v $\in$ V, we are led to deal with the equation

$$F_1(u) + F_2(u) = f, \tag{7}$$

where F_1 and F_2 are continuous. Moreover, since g_i is nondecreasing for all i = 1,2,...,N, F_1 is monotone and hence also pseudomonotone. In the case where g_i is strictly increasing for all i = 1,2,...,N, i.e.

$$(g_i(t) - g_i(s))(t - s) > 0 \text{ for all } t \neq s \tag{8}$$

and the condition

$$g_i(t)t \geq C_2|t|^p - C_3 \text{ for all } i = 1,2,\ldots,N \tag{9}$$

is satisfied for some constants $C_2,C_3 > 0$, we can conclude that $F_1 \in (S_+)$. Indeed, let $u_n \rightharpoonup u$ in V and lim sup $\langle F_1(u_n), u_n - u\rangle \leq 0$. Then $u_n \to u$ in $L^p(\Omega)$ and $u_n(x) \to u(x)$ a.e. in Ω for a subsequence. Moreover,

$$\limsup \int_\Omega \sum_{i=1}^{N} \{g_i(D_iu_n(x) - g_i(D_iu(x))\}(D_iu_n(x) - D_iu(x)) \leq 0$$

and (8) implies that $Du_n(x) \to Du(x)$ a.e. in Ω for a further subsequence. Since $F_1 \in$ (PM), $\langle F_1(u_n),u_n\rangle \to \langle F_1(u),u\rangle$. Denoting

$$h_n = \sum_{i=1}^{N} g_i(D_iu_n)D_iu_n + C_3$$

we conclude by (9) that $h_n \geq 0$, $\|h_n\|_1 \to \|h\|_1$ and $h_n(x) \to h(x)$ a.e. in Ω, where

$$h = \sum_{i=1}^{N} g_i(D_iu)D_iu + C_3.$$

Invoking a result in [6, p. 208] we get $h_n \to h$ in $L^1(\Omega)$. Hence there exists a function $k \in L^1(\Omega)$ such that

$$\sum_{i=1}^{N} g_i(D_i u_n(x)) D_i u_n(x) \leq k(x) \text{ a.e. in } \Omega$$

for some further subsequence. Employing (9) again we obtain by the dominated convergence theorem that $D_i u_n \to D_i u$ in $L^p(\Omega)$ for all $i = 1,2,\ldots,N$ and $u_n \to u$ in V follows.

Finally we consider the mapping F_2. Indeed, if $u_n \rightharpoonup u$ in V, the Sobolev embedding theorem implies that

$$\lim \langle F_2(u_n), u_n - u\rangle = 0.$$

Hence $F_2 \in (SM) \subset (QM)$. Assuming $F_1 \in (S_+)$ we therefore have $F_1 + F_2 \in (S_+)$ and Theorem 1 can be applied provided the conditions (a) and (b) are satisfied.

If $V = W_0^{1,p}(\Omega)$, the Poincare inequality holds and then (9) implies that

$$\frac{\langle F_1(u_n), u_n\rangle}{\|u_n\|} \to \infty \text{ , as } \|u_n\| \to \infty.$$

Therefore, for instance a condition

$$g_0(x,\eta,\zeta)\eta \geqq -k_1(x) \quad \forall x \in \Omega,\ \eta \in \mathbb{R},\ \zeta \in \mathbb{R}^N \qquad (10)$$

for some $k_1 \in L^1(\Omega)$ is enough to make also $F_1 + F_2$ coercive and Theorem 1 can be applied. If $V = W^{1,p}(\Omega)$, the condition (10) is not sufficient for coercivity of $F_1 + F_2$. However, for instance in the particular case where F_1 is given by (5) and F_2 is defined by

$$\langle F_2(u), v\rangle = \int_\Omega g_0(x,u)v, \quad u,v \in W^{1,p}(\Omega),$$

with g_i satisfying (2), (8), (9) and $g_i(t)t \geqq 0$ for all $t \in \mathbb{R}$, and g_0 satisfying (3) we have the following existence result. The equation (7) admits a solution u in $W^{1,p}(\Omega)$ for all those $f \in L^{p'}(\Omega)$ satisfying the condition

$$\int_\Omega g_0^- < \int_\Omega f < \int_\Omega g_0^+,$$

where

$$g_0^+(x) = \liminf_{\eta \to +\infty} g_0(x,\eta) \quad \text{and} \quad g_0^- = \limsup_{\eta \to -\infty} g_0(x,\eta).$$

For the details we refer to [5].

References

[1] F.E. Browder, Fixed point theory and nonlinear problems, Bull Amer. Math. Soc., 9 (1983), 1-39.

[2] J. Berkovits, On the degree theory for nonlinear mappings of monotone type, Ann. Acad. Sci. Fenn. Ser. A1, Dissertationes, 58 (1986).

[3] J. Berkovits, A degree theoretic approach to the semilinear wave equation (elsewhere in this publication).

[4] J. Berkovits and V. Mustonen, On the topological degree for mappings of monotone type, Nonlinear Anal. TMA, 10 (1986), 1373-1383.

[5] J. Berkovits and V. Mustonen, Nonlinear mappings of monotone type, I. Classification and degree theory, Report No. 1/88, Mathematics, University of Oulu (1988).

[6] E. Hewitt and K. Stromberg, Real and Abstract Analysis, Springer-Verlag, (1965).

Vesa Mustonen
University of Oulu
Department of Mathematics
SF-90570 Oulu
Finland

T. RUNST

Multiple solutions for some semilinear elliptic equations in function spaces

1. INTRODUCTION

The aim of this paper is the study of the numbers of solutions of the Dirichlet problem for semilinear elliptic equations. Namely we consider the following problem:

$$(P)\quad \begin{cases} \Delta u + f(u) = g & \text{in } \Omega \\ u = 0 & \text{in } \partial\Omega, \end{cases}$$

where Ω is a bounded connected smooth domain in R^n and g belongs to a function space of Triebel-Lizorkin and Besov type, respectively. Denote by $0 < \lambda_1 < \lambda_2 \leqq \dots \leqq \lambda_k \leqq \dots$ the sequence of eigenvalues of $-\Delta$ in Ω under the Dirichlet condition. Here f is a sufficiently smooth function with linear growth at infinity, and more precisely:

$$a \leqq f'(t) \leqq b \text{ for all } t \in [-\infty,\infty], \text{ where } f'(\pm\infty) = \lim_{t\to\pm\infty} f'(t).$$

It is known that our problem (P) admits multiple solutions depending on the interaction between the value of f' and the spectrum of $-\Delta$ if g belongs to the Hölder space $C^\alpha(\bar{\Omega})$, $0 < \alpha < 1$.

Subsequent works of Dolph [5] showed that if the interval [a,b] contains no eigenvalue λ_k, then (P) is uniquely solvable in $C^{2+\alpha}(\bar{\Omega})$. In [2], Ambrosetti and Prodi considered the case in which the range of f' contains only the first eigenvalue λ_1. They showed that the conditions $0 < f'(-\infty) < \lambda_1 < f'(\infty) < \lambda_2$ and $f'' > 0$ on $(-\infty,\infty)$ imply the existence of a closed connected C^1-manifold M of codimension 1 in the Banach space $C^\alpha(\bar{\Omega})$ such that $C^\alpha(\bar{\Omega})\setminus M$ has exactly two components A and B with the property that (P) has no solution if $g \in A$; exactly two solutions if $g \in B$; and exactly one solution if $g \in M$. Subsequently, Manes and Micheletti [8] were able to replace the condition $0 < f'(-\infty) < \lambda_1$ by $-\infty < f'(-\infty) < \lambda_1$. In [7], Kazdan and Warner considered the more general case where λ_1 is not necessarily the only eigenvalue in [a,b]. They showed that if we decompose $g = g_1 + t\phi_1$ (ϕ_1 = normalized eigenfunction corresponding to λ_1), then there exists a $t_0 = t_0(g_1)$ such that if $t > t_0$ the problem (P) has at least one solution, and if $t < t_0$ it has no solutions. This result was obtained by using the method of upper and lower solutions. Simultaneously, Dancer [4] and Amann and Hess [1] proved that if f satisfies a certain growth condition at infinity, then (P) has at least two solutions if $t > t_0(g_1)$ and is solvable for $t = t_0(g_1)$. The more general problem of estimating the number of solutions of (P) with respect to the number of eigenvalues contained in [a,b] has been investigated too. We mention also the paper of Lazer and McKenna [9], Ruf [10] and Cafagna and Tarantello [3].

We consider the problem (P) in Triebel-Lizorkin spaces, $F^s_{p,q}$, and Besov spaces, $B^s_{p,q}$, with methods going back to [2]. For $0 < q < 1$ and/or $0 < p < 1$ $B^s_{p,q}$ and $F^s_{p,q}$ are quasi-Banach spaces, i.e. the topology is generated by a quasi-norm $\|\cdot\|$ satisfying the usual conditions of a norm except the triangle inequality which is replaced by $\|x+y\| \leqq K\,(\|x\| + \|y\|)$ with some $K \geqq 1$. The Ambrosetti-Prodi theory was elaborated for Banach spaces. For general quasi-Banach spaces some striking tools fail: not locally convex, the quasi-norm is not necessarily continuous, the dual space may be empty,... . However, if there are 'enough' continuous linear functionals over the quasi-Banach space the theory of [2] can be carried over.

2. DUAL RICH QUASI-BANACH SPACES

In the following, we describe the concept of dual rich quasi-Banach spaces. We explain 'how many' continuous linear functionals a space should have to become a 'good' space in our considerations.

DEFINITION 1: A quasi-Banach space X is said to be dual rich if for all $x \in X$, $x \neq 0$, there exists a continuous linear functional x^* with $\langle x^* | x\rangle = 1$.

REMARK 1: Clearly every Banach space is dual rich; on the other hand, $L_p(R)$ with $0 < p < 1$ is not dual rich because its dual is trivial.

REMARK 2: In dual rich quasi-Banach spaces we can define the factorization with respect to spaces of finite dimension or finite codimension, i.e. these factor spaces are isomorphic to a closed subspace of the dual rich quasi-Banach space itself.

DEFINITION 2: Let X be a quasi-Banach space. A set $M \subset X$ is called a C^k-manifold of codimension 1 if for all $m \in M$ one can find a neighbourhood U and a C^k map $F : U \to R$ such that (i) $F'(m) \neq 0$ and (ii) $M \cap U = \{x \in U | F(x) = 0\}$.

REMARK 3: With respect to dual rich quasi-Banach spaces we are able to prove a corresponding result to [2] about the inversion of some differentiable mapping with singularities, for instance:

If M is a closed connected C^1-manifold of codimension 1 in a dual rich quasi-Banach space X then $X \setminus M$ has at most two components.

For our purposes Besov spaces $B^s_{p,q}$ and Triebel-Lizorkin spaces $F^s_{p,q}$ are of special interest; for definitions and basic properties see [12], which covers Hölder-Zygmund spaces, Sobolev-Slobodeckij spaces, Besov-Lipschitz spaces, Bessel-potential spaces and spaces of Hardy type.

We get: For $-\infty < s < \infty$, $0 < p, q \leqq \infty$ $F^s_{p,q}(\Omega)(p < \infty)$ and $B^s_{p,q}(\Omega)$ are dual rich quasi-Banach spaces, where Ω is a bounded open subset of R^n with smooth boundary.

It is well known that for $0 < p, q \leqq \infty$, $s > 1/p + \max(0,(n-1)(1/p-1))$ the spaces $B^s_{p,q}(\Omega)$ and $F^s_{p,q}(\Omega)$ (here $p < \infty$) admit traces on the boundary $\partial\Omega$. Then (for admissible couples (s,p)) the Laplacian yields an isomorphic mapping

$$\text{from } B^{s+2}_{p,q,0}(\Omega) = \{f \in B^{s+2}_{p,q}(\Omega),\ \ f|\partial\Omega = 0\} \text{ onto } B^s_{p,q}(\Omega)$$

and

$$\text{from } F^{s+2}_{p,q,0}(\Omega) = \{f \in F^{s+2}_{p,q}(\Omega),\ \ f|\partial\Omega = 0\} \text{ onto } F^s_{p,q}(\Omega).$$

3. STATEMENT OF THE RESULTS

THEOREM: Let $\Omega \in R^n$ be a bounded connected domain with smooth boundary. Assume that the real function $f \in C^{\rho+2}$ fulfils the following conditions: (i) $f(0) = 0$, (ii) $f''(t) > 0$ for all $t \in R$ and $0<f'(-\infty)<\lambda_1<f'(\infty)<\lambda_2, \lambda_1, \lambda_2$

as above. Then for the solutions of the boundary value problem (P) the following assertions are valid:

(i) If $0 < p, q \leqq \infty$, $\rho > s > \max(n/p, 1)$, then there exists a closed connected C^1-manifold M of codimension 1 in $B^{s-2}_{p,q}(\Omega)$ such that $B^{s-2}_{p,q}(\Omega) \setminus M$ consists of two connected components A and B with the following properties.

(a) if $g \in A$ then (P) has no solution in $B^{s}_{p,q,0}(\Omega)$;

(b) if $g \in B$ then (P) has exactly two solutions in $B^{s}_{p,q,0}(\Omega)$ and

(c) if $g \in M$ then (P) has exactly one solution.

(ii) If $p < \infty$ then a corresponding result holds also in the case of Triebel-Lizorkin spaces $F^{s}_{p,q}(\Omega)$.

4. SOME REMARKS ABOUT THE PROOF

The proof will be carried out in four steps, see [6].

STEP 1: The map $F : B^{s}_{p,q,0}(\Omega) \to B^{s-2}_{p,q}(\Omega)$, $F(u) = \Delta u + f(u)$ is proper; i.e. the preimage of every compact set $K \subset B^{s-2}_{p,q}(\Omega)$ is compact in $B^{s}_{p,q,0}(\Omega)$. Therefore we use the fact that F is a continuous mapping from $B^{s}_{p,q,0}(\Omega)$ in $B^{s-2}_{p,q}(\Omega)$; see [11]. The rest of the proof is more or less standard.

STEP 2: The singular set S_F is non-empty, closed and connected. S_F consists of ordinary singular points only (for definitions see [2]).

STEP 3: If $g \in F(S_F)$, then (P) has exactly one solution in $B^{s}_{p,q,0}(\Omega)$. $F(S_F)$ is also closed, connected and a C^1-manifold of codimension 1.

STEP 4: Applying the results of the above steps and the idea of [2], see Remark 3, then we get our theorem.

References

[1] H. Amann and P. Hess, A multiplicity result for a class of elliptic boundary value problems, Proc. R. Soc. Edin. Sect. A, 84 (1979), 145-151.

[2] A. Ambrosetti and G. Prodi, On the inversion of some differentiable mappings with singularities between Banach spaces, Ann. Mat. Pura Appl., 93 (1972), 231-246.

[3] V. Cafagna and G. Tarantello, Multiple solutions for some semilinear elliptic equations, Math. Ann., 276 (1987), 643-656.

[4] E.N. Dancer, On the range of certain weakly nonlinear elliptic partial differential equations, J. Math. Pure Appl., 57 (1978), 351-366.

[5] C.L. Dolph, Nonlinear integral equations of the Hammerstein type, Trans. Am. Math. Soc., 60 (1949), 289-307.

[6] M. Geisler and T. Runst, On a superlinear Ambrosetti-Prodi problem in Besov and Triebel-Lizorkin spaces (to appear).

[7] J.L. Kazdan and F.W. Warner, Remarks on some quasilinear elliptic equations, Comm. Pure Appl. Math., 28 (1975), 837-846.

[8] A. Manes and A.M. Micheletti, Un estensione della teori variazionale classica degli auto-valori per operatori ellittici del sucundo ordine, Boll. Un. Mat. Ital., 7 (1973), 285-301.

[9] A.C. Lazer and P.J. McKenna, On the number of solutions of a nonlinear Dirichlet problem, J. Math. Anal. Appl., 84 (1981), 282-294.

[10] B. Ruf, On nonlinear elliptic boundary value problems with jumping nonlinearities, Ann. Mat. Pura Appl. VI, 128 (1980), 133-151.

[11] T. Runst, Mapping properties of nonlinear operators in spaces of Triebel-Lizorkin and Besov type, Analysis Math., 12 (1986), 313-346.

[12] H. Triebel, Theory of Function Spaces, Birkhäuser (1983).

T. Runst
Sektion Mathematik
Universität Jena
Universitätshochhaus
DDR-6900 Jena
DDR